# Giant Vesicles

# Editorial Board

# Giant Vesicles

*Perspectives in Supramolecular Chemistry*
*Volume 6*

*Edited by* PIER LUIGI LUISI *and* PETER WALDE
*ETH, Zürich, Switzerland*

JOHN WILEY & SONS, LTD
Chichester • New York • Weinheim • Brisbane • Singapore • Toronto

Baffins Lane, Chichester,
West Sussex PO19 1UD, England

National 01243 779777
International (+44) 1243 779777

e-mail (for orders and customer service enquiries): cs-books@wiley.co.uk
Visit our Home Page on http://www.wiley.co.uk
or http://www.wiley.com



*Other Wiley Editorial Offices*

John Wiley & Sons, Inc., 605 Third Avenue,
New York, NY 10158-0012, USA

WILEY-VCH Verlag GmbH, Pappelallee 3,
D-69469 Weinheim, Germany

Jacaranda Wiley Ltd, 33 Park Road Milton,
Queensland 4064, Australia

John Wiley  Sons (Asia) Pte Ltd, Clementi Loop #02-01,
Jim Xing Distripark, Singapore 129809

John Wiley & Sons (Canada) Ltd, 22 Worcester Road,
Rexdale, Ontario M9W 1L1, Canada

***Library of Congress Cataloging-in-Publication Data***
Giant vesicles / edited by Pier Luigi Luisi and Peter Walde.
p. cm. – (Perspectives in supramolecular chemistry ; v. 6)
Includes bibliographical references and index.
ISBN 0471-97986-4 (alk. paper)
1. Liposomes Congresses. I. Luisi, Pier Luigi. II. Walde, Peter. III. Series.
QH602.G53 1999
571.6′.55–dc21 98-24961
CIP

***British Library Cataloguing in Publication Data***

A catalogue record for this book is available from the British Library

ISBN 0 471 97986 4

Typeset in Times by Techset Composition Ltd, Salisbury, Wiltshire
Printed and bound in Great Britain by Biddles Ltd, Guildford and King's Lynn
This book is printed on acid-free paper responsibly manufactured from sustainable forestry, in which at least two trees are planted for each one used for paper production.

# Contents

# List of Contributors

**Ken-ichirou Akashi,** Department of Physics, Faculty of Science and Technology, Keio University, Hiyoshi 3-14-1, Kohoku-ku, Yokohama, 223-8522, Japan; Core Research for Evolutional Science and Technology, Genetic Programming Team 13, Nogawa 907, Miyamae-ku, Kawasaki 216-0001, Japan.

**Rossitsa G. Alargova,** Laboratory of Thermodynamics and Physicochemical Hydrodynamics, University of Sofia, Faculty of Chemistry, 1126 Sofia, Bulgaria.

**Anne-Marie Albrecht,** Ecole Européenne de Chimie, Polymères et Matériaux, 1 rue Blaise Pascal, F-67008 Strasbourg, France.

**Miglena I. Angelova,** Institute of Biophysics, Bulgarian Academy of Sciences, Acad. G. Bonchev Str., Bl. 21, 1113 Sofia, Bulgaria.

**Kenji Arakawa,** Department of Chemistry, Tokyo Institute of Technology, O-okayama, Meguro-ku, Tokyo 152, Japan.

**Jean-Claude Bacri,** Laboratoire des Milieux Désordonnés et Hétérogènes, Universités Paris 6 et Paris 7, Case courrier 78, 4 place Jussieu, F-75252 Paris CEDEX 05, France.

**Patricia Bassereau,** Physico-Chimie Curie UMR 168, Section Recherche, Institut Curie, 11 Rue Pierre et Marie Curie, F-75231 Paris CEDEX 05, France.

**Anne-Laure Bernard,** Collège de France, Laboratoire de Physique de la Matière Condensée, URA 792, 11 Place Marcelin Berthelot, F-75231 Paris CEDEX 05, France.

**Isak Bivas,** Institute of Solid State Physics, Bulgarian Academy of Sciences, Laboratory of Liquid Crystals, 72 Tzarigradsko Chaussee blvd., 1784 Sofia, Bulgaria.

**Sylvie Blanc,** Ecole Européenne de Chimie, Polymères et Matériaux, 1 rue Blaise Pascal, F-67008 Strasbourg, France.

**Pierre Bothorel,** Centre de Recherche Paul Pascal – CNRS, Av. Schweitzer, F-33600 Pessac, France.

**Jean-Claude Bradley,** Drexel University, Department of Chemistry, 32nd and Chestnut Street, Philadelphia, PA 19104, USA.

**Valérie Cabuil,** Laboratoire LI2C, Université Pierre et Marie Curie (Paris 6), Case courrier 63, 4 place Jussieu, F-75252 Paris CEDEX 05, France.

**Parthapratim Chandaroy,** Membrane Biophysics Laboratory, Roswell Park Cancer Institute, Buffalo, NY 14263, USA.

**Andrejs Cebers,** Laboratoire des Milieux Désordonnés et Hétérogènes, Universités Paris 6 et Paris 7, Case courrier 78, 4 place Jussieu, F-75252 Paris CEDEX 05, France.

**Olivier Dannenmuller,** Laboratoire de Chimie Organique des Substances Naturelles, associé au CNRS, centre de Neurochimie, Université Louis Pasteur, 5 rue Blaise Pascal, F-67084 Strasbourg, France.

**Rumiana Dimova,** Laboratory of Thermodynamics and Physicochemical Hydrodynamics, University of Sofia, Faculty of Chemistry, 1126 Sofia, Bulgaria.

**Christian Dietrich,** Centre de Recherche Paul Pascal – CNRS, Av. Schweitzer, F-33600 Pessac, France.

**Hans-Günther Döbereiner,** Max-Planck-Institut für Kolloid- und Grenzflächenforschung, Am Mühlenberg 2, D-14476 Golm, Germany.

**Toshifusa Doi,** Department of Materials Science and Engineering, Kochi National College of Technology, 200-1 Monobe, Nankoku, Kochi 783-8508, Japan.

**Tadashi Eguchi,** Department of Chemistry, Tokyo Institute of Technology, O-okayama, Meguro-ku, Tokyo 152, Japan.

**Aline Fischer,** Institut für Polymere, ETH-Zürich, Universitätstrasse 6, CH-8092 Zürich, Switzerland.

**Susanne Gangl,** Institute for Theoretical Chemistry and Radiation Chemistry, University of Vienna, UZAII, Althanstrasse 14, A-1090 Vienna, Austria.

**Claire Gerbeaud,** Centre de Recherche Paul Pascal – CNRS, Av. Schweitzer, F-33600 Pessac, France.

**Ayako Goto,** School of Informatics, University of Shizuoka, Yada 52-1, Shizuoka-shi 422-8526, Japan.

**Rensuke Goto,** Graduate School of Nutritional and Environmental Sciences, University of Shizuoka, Yada 52-1, Shizuoka-shi 422-8526, Japan.

**Marie-Alice Guedeau-Boudeville,** Collège de France, Laboratoire de Physique de la Matière Condensée, URA 792, 11 Place Marcelin Berthelot, F-75231 Paris CEDEX 05, France.

**Wolfgang Helfrich,** Freie Universität Berlin, Fachbereich Physik, Arnimallee 14, D-14195 Berlin, Germany.

**Juha M. Holopainen,** Helsinki Biophysics & Biomembrane Group, Department of Medical Chemistry, Institute of Biomedicine, University of Helsinki, POB 8, FIN-00014 University of Helsinki, Finland.

**Sek Wen Hui,** Membrane Biophysics Laboratory, Roswell Park Cancer Institute, Buffalo, NY 14263, USA.

**Toyoko Imae,** Department of Chemistry, Faculty of Science, Nagoya University, Furo-cho, Chikusa-ku, Nagoya-shi 464-0814, Japan.

**Masanao Imai,** Department of Food Science and Technology, College of Bioresource Sciences, Nihon University, 34-1 1-Chome, Simo-uma, Setagaya-ku, Tokyo 154-8513, Japan.

**Hiroyasu Itoh,** Tsukuba Research Laboratory, Hamamatsu Photonics K. K., Tokodai 5-9-2, Tsukuba, 300-2635, Japan; Core Research for Evolutional Science and Technology, Genetic Programming Team 13, Nogawa-907, Miyamae-ku, Kawasaki 216-0001, Japan.

**Ludovic Jullien,** Département de Chimie de l'Ecole Normale Supérieure, URA 1679, 24 rue Lhomond, F-75231 Paris CEDEX 05, France.

**Katsumi Kakinuma,** Department of Chemistry, Tokyo Institute of Technology, O-okayama, Meguro-ku, Tokyo 152, Japan.

**Paavo K. J. Kinnunen,** Helsinki Biophysics & Biomembrane Group, Department of Medical Chemistry, Institute of Biomedicine, University of Helsinki, POB 8, FIN-00014 University of Helsinki, Finland.

**Kazuhiko Kinosita, Jr.,** Department of Physics, Faculty of Science and Technology, Keio University, Hiyoshi 3-14-1, Kohoku-ku, Yokohama, 223-8522, Japan; Core Research for Evolutional Science and Technology, Genetic Programming Team 13, Nogawa 907, Miyamae-ku, Kawasaki 216-0001, Japan.

**Beate Klösgen,** Freie Universität Berlin, Fachbereich Physik, Institut für Experimentalphysik, Arnimallee 14, D-14195 Berlin, Germany.

**Gottfried Köhler,** Institute for Theoretical Chemistry and Radiation Chemistry, University of Vienna, UZAII, Althanstrasse 14, A-1090 Vienna, Austria.

**Ryoichi Kuboi,** Department of Chemical Science and Engineering, Graduate School of Engineering Science, Osaka University, 1-3 Machikaneyama-cho, Toyonaka, Osaka 560-8531, Japan.

**Danilo D. Lasic,** Liposome Consultations, 7512 Birkdale Dr, Newark, CA 94560, USA.

**Daniel Lévy,** Physico-Chimie Curie UMR 168, Section Recherche, Institut Curie, 11 Rue Pierre et Marie Curie, F-75231 Paris CEDEX 05, France.

**Pier Luigi Luisi,** Institut für Polymere, ETH-Zürich, Universitätstrasse 6, CH-8092 Zürich, Switzerland.

**Arno Lukas,** Institute for Theoretical Chemistry and Radiation Chemistry, University of Vienna, UZAII, Althanstrasse 14, A-1090 Vienna, Austria.

**Jean-Baptise Manneville,** Physico-Chimie Curie UMR 168, Section Recherche, Institut Curie, 11 Rue Pierre et Marie Curie, F-75231 Paris CEDEX 05, France.

**Gerard Marriott,** Biomolekulare und Zelluläre Dynamik, Max-Planck-Institut für Biochemie, D-82152 Martinsried bei München, Germany.

**Bernd Mayer,** Institute for Theoretical Chemistry and Radiation Chemistry, University of Vienna, UZAII, Althanstrasse 14, A-1090 Vienna, Austria.

**Peter Mayrhofer,** Institute for Theoretical Chemistry and Radiation Chemistry, University of Vienna, UZAII, Althanstrasse 14, A-1090 Vienna, Austria.

**Philippe Méléard,** Centre de Recherche Paul Pascal – CNRS, Av. Schweitzer, F-33600 Pessac, France.

**Christine Ménager,** Laboratoire LI2C, Université Pierre et Marie Curie (Paris 6), Case courrier 63, 4 place Jussieu, F-75252 Paris CEDEX 05, France.

**Marin D. Mitov,** Institute of Solid State Physics, Bulgarian Academy of Sciences, Laboratory of Liquid Crystals, 72 Tzarigradsko Chaussee blvd., 1784 Sofia, Bulgaria.

**Hidetake Miyata,** Physics Department, Graduate School of Science, Tohoku University, Aramaki-Aoba, Aoba-ku, Sendai, Miyagi, 980-8578, Japan.

**Kazuhito Nagayama,** Department of Materials Science and Engineering, Kochi National College of Technology, 200-1 Monobe, Nankoku, Kochi 783-8508, Japan.

**Yoichi Nakatani,** Laboratoire de Chimie Organique des Substances Naturelles, associé au CNRS, Centre de Neurochimie, Université Louis Pasteur, 5 rue Blaise Pascal, F-67084 Strasbourg, France.

**David Needham,** Department of Mechanical Engineering and Materials Science, Duke University, Durham, NC 27708-0300, USA.

**Shuji Nishiyama,** Products R. & D., Laboratory, TOTO Ltd., 2-8-1 Kimura, Chigasaki, Kanagawa, 253-8577, Japan.

**Shin-ichirou Nomura,** Department of Physics, Graduate School of Science, Kyoto University, Kyoto 606-8502, Japan.

**Thomas Oberholzer,** Institut für Polymere, ETH-Zürich, Universitätstrasse 6, CH-8092 Zürich, Switzerland.

**Kazuo Ohki,** Physics Department, Graduate School of Science, Tohoku University, Aramaki-Aoba, Aoba-ku, Sendai, Miyagi, 980-8578, Japan.

**Nobukzau Osaki,** Department of Materials Science and Engineering, Kochi National College of Technology, 200-1 Monobe, Nankoku, Kochi 783-8508, Japan.

**Guy Ourisson,** Laboratoire de Chimie Organique des Substances Naturelles, associé au CNRS, Centre de Neurochimie, Université Louis Pasteur, 5 rue Blaise Pascal, F-67084 Strasbourg, France.

**Meghan Perkins,** Membrane Biophysics Laboratory, Roswell Park Cancer Institute, Buffalo, NY 14263, USA.

**Peter G. Petrov,** Max-Planck-Institut für Kolloid- und Grenzflächenforschung, Am Mühlenberg 2, D-14476 Golm, Germany.

**Tanja Pott,** Centre de Recherche Paul Pascal - CNRS, Av. Schweitzer, F-33600 Pessac, France.

**Bernard Pouligny,** Centre de Recherche Paul Pascal - CNRS, Av. Schweitzer, F-33600 Pessac, France.

**Jacques Prost,** Physico-Chimie Curie UMR 168, Section Recherche, Institut Curie, 11 Rue Pierre et Marie Curie, F-75231 Paris CEDEX 05, France.

**Jerôme Prost,** Laboratoire LI2C, Université Pierre et Marie Curie (Paris 6), Case courrier 63, 4 place Jussieu, F-75252 Paris CEDEX 05, France.

**Gert Rapp**, EMBL c/o DESY, Notkestrasse 85, D-22603 Hamburg, Germany.

**Dominik Rünzler,** Institute for Theoretical Chemistry and Radiation Chemistry, University of Vienna, UZAII, Althanstrasse 14, A-1090 Vienna, Austria.

**Olivier Sandre,** Laboratoire Physicochimie Curie, Institut Curie – Section de Recherche, 26 rue d'Ulm, F-75248 Paris CEDEX 05, France.

**Udo Seifert,** Max-Planck-Institut für Kolloid- und Grenzflächenforschung, Am Mühlenberg 2, D-14476 Golm, Germany.

**Toshinori Shimanouchi,** Department of Chemical Science and Engineering, Graduate School of Engineering Science, Osaka University, 1-3 Machikaneyama-cho, Toyonaka, Osaka 560-8531, Japan.

**Alok Singh,** Center for Bio/Molecular Science & Engineering, Naval Research Laboratory, Washington, DC 20375, USA.

**Susanne Stark,** Institute for Theoretical Chemistry and Radiation Chemistry, University of Vienna, UZAII, Althanstrasse 14, A-1090 Vienna, Austria.

**Akihiro Suzuki,** Graduate School of Nutritional and Environmental Sciences, University of Shizuoka, Yada 52-1, Shizuoka-shi 422-8526, Japan.

**Hikaru Tanaka,** Department of Materials Science and Engineering, Kochi National College of Technology, 200-1 Monobe, Nankoku, Kochi 783-8508, Japan.

**Johannes Thimmel,** Freie Universität Berlin, Fachbereich Physik, Institut für Experimentalphysik, Arnimallee 14, D-14195 Berlin, Germany.

**Caroline Thurner,** Institute for Theoretical Chemistry and Radiation Chemistry, University of Vienna, UZAII, Althanstrasse 14, A-1090 Vienna, Austria.

**Hiroshi Umakoshi,** Department of Chemical Science and Engineering, Graduate School of Engineering Science, Osaka University, 1-3 Machikaneyama-cho, Toyonaka, Osaka 560-8531, Japan.

**Victoria Vitkova,** Institute of Solid State Physics, Bulgarian Academy of Sciences, Laboratory of Liquid Crystals, 72 Tzarigradsko Chaussee blvd., 1784 Sofia, Bulgaria.

**Peter Walde,** Institut für Polymere, ETH-Zürich, Universitätstrasse 6, CH-8092 Zürich, Switzerland.

**Mathias Winterhalter,** Biozentrum, Biophysikalische Chemie, Klingelbergstrasse 70, CH-4056 Basel, Switzerland.

**Liyu Xu,** Max-Planck-Institut für Kolloid- und Grenzflächenforschung, Am Mühlenberg 2, D-14476 Golm, Germany.

**Keiji Yamazaki,** Otuka Electronics Company, 1-6 Higashi-cho, Hachiouji-shi, Tokyo-to 192-0082, Japan.

**Kenichi Yoshikawa,** Department of Physics, Graduate School of Science, Kyoto University, Kyoto 606-8502, Japan.

**Hisashi Yoshioka,** Graduate School of Nutritional and Environmental Sciences, University of Shizuoka, Yada 52-1, Shizuoka-shi 422-8526, Japan.

**Doncho Zhelev,** Department of Mechanical Engineering and Materials Science, Duke University, Durham, NC 27708-0300, USA.

# Preface

Liposomes, also called lipid vesicles or simply vesicles, are supramolecular assemblies of amphiphiles, surface active substances, that normally contain two hydrophobic tails and one hydrophilic head group. The investigation of vesicles belongs to the field of supramolecular chemistry. Due to the large size, giant vesicles are particularly attractive to investigate because their formation and the behaviour of one single supramolecular entity can be observed by light microscopy. The size of giant vesicles is typically in the range 10–100 μm, corresponding to an amphiphile aggregation number of typically $8 \times 10^8$–$8 \times 10^{10}$ per vesicle.

This volume has been prepared as the proceedings of the Workshop 'Giant Vesicles', which was held in Ascona, Switzerland, between June 21 and 25, 1998. The workshop was financially supported by the Centro Stefano Franscini of the Swiss Federal Institute of Technology (ETH) Zürich and the Swiss National Science Foundation. The main aim of this workshop was to bring together scientists working in different disciplines on giant vesicles and to promote an exchange of basic notions, new ideas, and new results. This volume provides a collection of ideas and results obtained from experimental investigations as well as from theoretical approaches, covering a variety of aspects ranging from pure mathematics or physics to biology.

Although the book contains contributions from several research groups currently working in the field of giant vesicles, it is certainly not complete. Therefore, in addition to what has been covered by this book, the interested reader should consult also the earlier work on giant vesicles by Ringsdorf *et al.* [1] and the current work of Menger and co-workers [2] on a variety of chemical and biochemical aspects of giant vesicles. Furthermore, much is known today about the physicochemical properties of giant vesicles and biomembranes at large thanks to the studies of Sackmann and his co-workers [3].

In order to eliminate possible confusions arising from the nomenclature of lipids and lipid assemblies, this volume uses the terms 'vesicles' and 'liposomes' synonymously. Furthermore, 'lecithin' is used as trivial name for any type of phosphatidylcholine, and most authors in the book have applied stereospecific numbering, *sn*, [4] and widely accepted abbreviations for the phospholipids used, which are, however, not recommended by the IUPAC-IUB Commission on Biochemical Nomenclature. POPC stands for example for 1-palmitoyl-2-oleoyl-*sn*-glycero-3-phosphocholine, which is the same as $\beta$-oleoyl-$\gamma$-palmitoyl-L-$\alpha$-phosphatidylcholine.

All chapters have been reviewed by us or by others and we would like to thank in particular Brian H. Robinson, Udo Seifert, and Mathias Winterhalter and, for their technical help, Nathalie Berclaz and Stefano Piotto.

Zürich, February 3, 1999

## References

1. H. Ringsdorf, B. Schlarb and J. Venzmer, *Angew. Chem.*, **100**, 117 (1988); *Angew. Chem., Int. Ed. Engl.*, **27**, 113 (1988).
2. (a) F. M. Menger and K. D. Gabrielson, *Angew. Chem.*, **107**, 2260 (1995); *Angew. Chem., Int. Ed. Engl.*, **34**, 2091 (1995); (b) F. M. Menger and J. S. Keiper, *Adv. Mater.*, **10**, 888 (1998); (c) F. M. Menger and J. S. Keiper, *Curr. Opinion Chem. Biol.*, **2**, 726 (1998).
3. E. Sackmann, in *Structure and Dynamics of Membranes*, (eds) R. Lipowsky and E. Sackmann, Handbook of Biological Physics, Vol. I, Elsevier, Amsterdam, 1995, p. 213.
4. (a) IUPAC-IUB, *Eur. J. Biochem.*, **2**, 127 (1967); (b) IUPAC-IUB, *Eur. J. Biochem.*, **79**, 11 (1977); (c) IUPAC-IUB, *Biochem. J.*, **171**, 21 (1978); (d) J. R. Silvius, in *Phospholipids Handbook*, (ed.) G. Cevc, Marcel Dekker, New York, 1993, p. 1.

# Part One
# Introduction

# Chapter 1

# Why Giant Vesicles?

*Pier Luigi Luisi*
*ETH, Zürich, Switzerland*

Why should giant vesicles constitute a field of inquiry distinct from normal vesicles? Size alone is not a sufficient good reason. These other reasons are the focus of the following discussion.

## 1. SELF-ORGANIZATION

A primary reason of interest and fascination is at the level of self-assembly and self-organization. This phenomenon, which we already know well for other structures, acquires in giant vesicles a particular dimension. One of the simplest giant vesicles was produced simply by dispersing oleic acid in water at pH 8.5 [1]. This was an oleic acid – oleate giant vesicle having a diameter of about 70 μm, as can be seen in Figure 1.1.

In this example of spontaneous vesiculation*, a total of about $10^{11}$ oleic acid – oleate molecules are involved in the formation of one vesicle. It is self-assembly and self-organization in the highest form, and unprecedented in chemistry – unless we consider crystallization. It is useful to remember that this creation of order operates under thermodynamic control: ordered structures are built by a process which is

*The term 'spontaneous' is used in the common sense use of the word, to signify that the process takes place by itself, without addition of external energy. More precisely the process is characterized by a negative free energy change *and*, simultaneously, by a low activation energy under the operational conditions. This qualification of the terminology is necessary because the strict thermodynamic definition of the term 'spontaneous' refers only to the thermodynamic quality (negative free energy change) without taking account of the kinetics, namely whether the process occurs or not. In fact, the strict thermodynamic meaning of the term 'spontaneous process' is not a very useful one.

*Giant Vesicles* Edited by P. L. Luisi and P. Walde

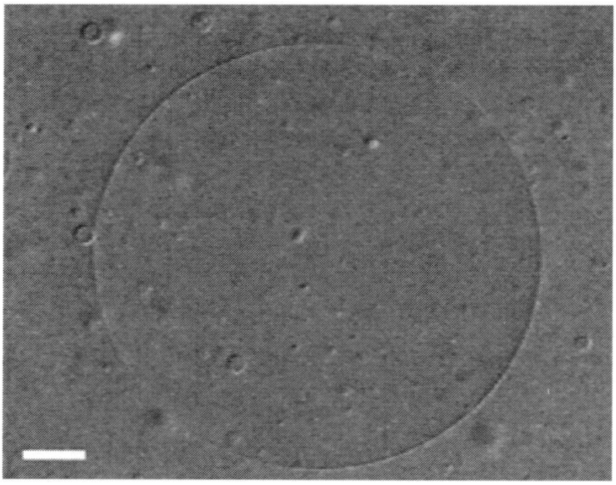

**Figure 1.1** Giant oleic acid–oleate vesicle prepared by dispersing oleic acid in 0.2 M bicine (*N,N*-bis[2-hydroxyethyl]glycine) buffer, pH 8.5. Length of the bar 10 μm.

characterized by a negative free energy change. The fact that order and thermodynamic control come together also has important implications in the origin of life – a matter that is returned to later.

Why and how do these molecules manage to assemble, within seconds, perfectly spherical double-layered aggregates? The 'why' is associated with thermodynamics, and addresses the question of why these gigantic structures, and not others, are energetically favored; the 'how' has more to do with the kinetic pathway of formation – through which intermediates and at which rates the giant vesicles form.

These two categories of question are yet to be answered satisfactorily for micelles and vesicles. Although some models and some ideas exist, a proper understanding remains an important challenge for physical chemists.

These questions also have a chemical component: it would be helpful to understand the relationship between the chemical structure of the monomeric surfactant and the propensity to assemble to giant vesicles. It has been observed that, generally, a small amount of giant vesicles accompanies the formation of normal vesicles. However, this tendency varies strongly from surfactant to surfactant. This situation also occurs in the case of electroformation [2]: some surfactants (e.g. fatty acids/soaps, phosphatidyl nucleosides) fail to give giant vesicles by the electroformation method [3], and in fact the method seems to be restricted to phosphatidylcholine or to lipid mixtures containing phosphatidylcholine. It is fair to say that the relationship between the chemical structure and the propensity to form vesicles is still poorly understood.

Concerning their structure, one positive feature is the stability of giant vesicles. Once formed, they remain stable, usually for a few days. They are also characterized by a good mechanical stability – one can puncture POPC(1-palmitoyl-2-oleoyl-*sn*-glycero-3-phosphocholine)-giant vesicles with needles, make microinjections and micromanipulate them, without destroying or transforming them. This stability is a great help when studying them.

## 2. GIANT VESICLES AS INDIVIDUAL STRUCTURES

The most obvious advantage of giant vesicles with respect to conventional ones, is that they can be observed by optical methods. Thus, for the first time, the research object can be seen in an optical microscope, as an individual macromolecular (poymolecular) entity. For chemists familiar with Avogadro numbers of molecules, this is quite a breakthrough, and one which implies new ways of thinking and operating.

For example, working with individual species means having to step back from the often comfortable statistics of large numbers. In fact, each individual tends to behave differently – one giant vesicle is large, another is smaller, one is stable to rupture, the other is not. Such individuality generally means a deviation from reproducibility. Thus, one research aim is to produce equal individuals, that behave reproducibly.

This difficulty does not seem to affect the main physical properties of giant vesicles, and for several years physical studies on bending elasticity have been analyzed by direct photographic observation; also, movements and shape transformations can be recorded in real time. Several reports on this type of physicochemical investigation are given in this book.

One important question relates to what extent the visual information obtained for giant vesicles can be transferred or used in the field of normal small vesicles. This question will become particularly relevant once the intermediates in the formation of giant vesicles – or in the mechanisms of budding, and growth and disintegration – have been observed.

With individuality comes compartmentalization. Vesicular self-assembly produces a closed species with an internal water core. There is a geometrical analogy here with the living cell, and in fact liposomes have been viewed for a long time as precursors of the protocell [4] – as soon as one has an individual compartment, one has the basic structural element for a life model. All life on Earth is based on cellular life, and all biological cells are closed compartments; in addition to cells, there are several organelles in eukaryotic cells (nucleus, lysosomes, Golgi complex, biological vesicles) which are based on compartments. A closed compartment has the possibility of defining an internal core that differs from the outside bulk medium – for example through a concentration gradient of special reagents, or through differences in dielectric constant, pH, salinity and so on between the internal and external domains. The notion of ‘self’, often used by people working in the area of artificial life, starts from the realization of this discrimination between an inside and an outside chemical world.

This closeness between vesicle compartmentalization and the living cell is one of the central interests of the research group in Zürich. This field was entered almost by accident, as a result of interest as to whether compartmentalized structures, such as micelles and vesicles, would be able to self-reproduce, namely to increase their population number by an autocatalytic process. A way of doing this was found, by using the hydrolysis reaction of a water-insoluble surfactant precursor: hydrolysis produces the surfactant which yields aggregates, which by micellar or vesicle catalysis accelerate in a remarkable way the hydrolysis rate and therefore the very production of vesicles, or micelles [5,6]. More recently, this concept of self-reproduction has been extended to giant vesicles. A further step in this research direction will be to try to increase the molecular complexity and specificity of vesicles, for example by hydrophobic binding with lipophilic catalytically active peptides and possibly to internalize them – so as to constitute a primitive model of a metabolic cell. This is seen as a bottom-up approach to the origin of life in the sense that, starting very basic thermodynamically driven reactions, ways are sought to escalate the complexity of the vesicular system.

Giant vesicles are used to pursue the top-down approach. This is based on microinjection techniques, namely the possibility of transforming the compartment of the giant vesicles into a bioreactor.

## 3. MICROINJECTION AND THE NOTION OF THE BIOENGINEERED MICROREACTOR

One further interest in giant vesicles lies in the fact that, due to their size, they permit the application of microinjection techniques. Chemicals can be injected inside giant vesicles by using microneedles, as biologists working with single cells have done for several years. In this way, the giant vesicle compartment becomes a microreactor and this can, in principle, be engineered to a state of considerable complexity. It is now possible to perform multiple puncturing on the same giant vesicle, to permit the addition of a series of different reagents in a temporal succession [3].

Considering the top-down approach, the use of giant vesicles in the field of origin of life is based on the following reasoning. The starting point is the recognition that the simplest living cells on Earth consist of no less than a few hundred proteins and a few hundred different nucleic acid families. There are good evolutionary reasons for this extreme complexity, but the question arises as to whether it would not be conceivable to have a living cell with, say, ten or so enzymes and only a few DNA and RNA families.

This question has implication for the origin of life, as it is likely that the first-formed cells were much simpler than those presently found; and there might also be interesting applications of this approach. Known as the minimal cell project (Figure 1.2) [7], this is a top-down approach in the origin of life because it utilizes enzymes and nucleic acids, namely compounds that are already operating in modern cellular

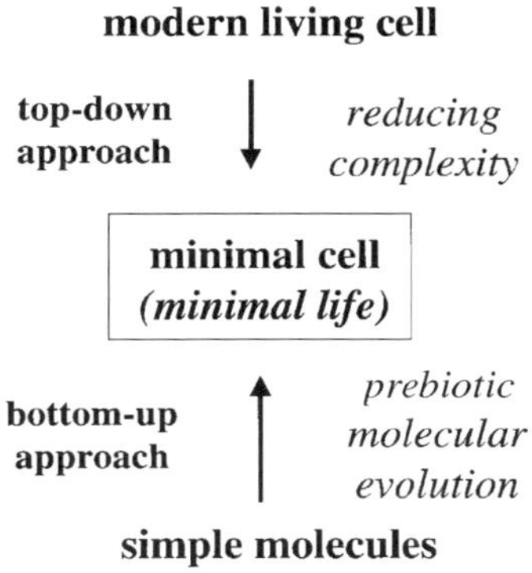


**Figure 1.2** Bottom-up and top-down approaches in research in the origin of life (minimal life).

life. The general idea is to create in the laboratory the simplest cellular form by microinjecting the minimal number of components into the giant vesicle. In particular, it would be interesting to construct a system that creates its own boundary by enzymatic reactions (Figure 1.3) – a system in which the enzymes are produced endogenously by a ribosomal system which has been introduced inside the giant vesicles by microinjection techniques.

Preliminary experiments, aimed at testing the stability of enzymes inside giant vesicles, and at testing the visualization of the reactivity of these enzymes, have already been reported, and this book contains contributions by Peter Walde (Chapter 22) and Thomas Oberholzer and Aline Fischer (Chapter 21) related to this subject.

This is a potentially very promising field for the chemical and biochemical applications of giant vesicles, to use them as microreactors for enzymatic reactions

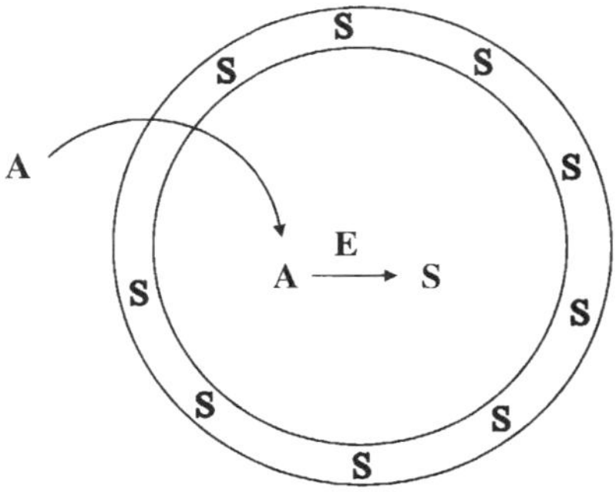


**Figure 1.3** Schematic representation of a hypothetical minimal cell, a cell that creates its own boundary. In or on the cell, the precursor molecule A is transformed into S which constitutes the boundary of the cell. The transformation of A into S is catalyzed by the enzyme E.

and bioreactors in general. It is at present largely undeveloped and it is hoped that its popularity will increase in the future.

## 4. SOME OF THE DIFFICULTIES

Having given some of the reasons for the fascination of giant vesicles, it is appropriate to mention a few possible difficulties.

One is the problem of reproducibility. Events witnessed with one giant vesicle are sometimes not easily observed again. For example, self-reproduction, budding and growth of oleic acid – oleate giant vesicles in one long series of experiments [1], could be video recorded in real time, but it was not possible to repeat such behavior subsequently. A second drawback, already mentioned, is that the formation of giant vesicles seems to be restricted to a small class of surfactants.

Another difficulty is in connection with the use of giant vesicles as chemical reactors. This has to do with the relatively small volume of a single compartment, involving volumes in the picoliter range, which severely limits the quantity of reagents that can be added by microinjection. This severely limits the amount of material obtainable from a chemical reaction occurring inside giant vesicles.

This drawback must be seen in connection with a general analytical difficulty, in that running reactions in giant vesicles has sense only if the reaction products are active under optical microscopy, i.e. if the products are colored. To run a reaction in giant vesicles and then destroy the giant vesicles to analyse for the product, for example by radioactivity or chromatography, generally does not make sense because one might just as well carry out the reaction in normal vesicles. Limiting reactions to those having a coloured product, together with the limitation of a small amount of product, makes many potentially interesting reactions difficult or impossible from an analytical point of view. Direct analysis of the products will be the major difficulty in the use of giant vesicles as microreactors.

## 5. CONCLUDING REMARKS

The workshop on giant vesicles [8] was probably the first occasion in which researchers in this new field had met and compared notes. It is hoped that a network of collaboration, expanding from Europe to Japan to USA to Australia, and consequent rapid scientific progress will be established. The field is already well established with a strong component in physical chemistry. Comparatively under-developed are the areas of pure chemistry and biology.

The positive aspects of research on giant vesicles and the drawbacks have been outlined. By and large, the points of challenge and interest largely overcome the misgivings arising from the disadvantages, and the field of giant vesicles is destined to grow considerably in the near future. One of the driving forces is likely to be the esthetic pleasure experienced when working with giant vesicles; that is, the great visual satisfaction seen in these beautiful ordered structures, the excitement in seeing them move about under the microscope lens, and the curiosity of spying on them to see what these individuals are going to do next.

## 6. REFERENCES

1. R. Wick, P. Walde and P. L. Luisi, *J. Am. Chem. Soc.*, **117**, 1435 (1995).
2. D. S. Dimitrov and M. I. Angelova, *Bioelectrochem. Bioenerg.*, **19**, 323 (1988).
3. P. Bucher, A. Fischer, P. L. Luisi, T. Oberholzer and P. Walde, *Langmuir*, **14**, 2712 (1998).
4. D. W. Deamer, *Origins Life*, **17**, 3 (1986).
5. P. A. Bachmann, P. L. Luisi and J. Lang, *Nature*, **357**, 57 (1992).
6. P. L. Luisi, *Adv. Chem. Phys.*, **92**, 425 (1996).
7. T. Oberholzer, M. Albrizio and P. L. Luisi, *Chem. Biol.*, **2**, 677 (1995).
8. Workshop on "Giant Vesicles", Centro Stefano Franscini, Monte Verità Ascona, Switzerland, June 21–25, 1998.

# Chapter 2

# Giant Vesicles: a Historical Introduction

*Danilo D. Lasic*

*Liposome Consultations, Newark, CA, USA*

## 1. INTRODUCTION

Although the discovery of lipid vesicles or liposomes is credited to Alec Bangham in the mid 1960s [1], scientific literature dating back to the mid-1800s contains articles on the colloidal behavior of lecithins and some other phospholipids. The chemical origin of polar lipids dates back to 1811 when Vauquelin described the binding of phosphorus to fatty acids in the material isolated by hot ethanol from brain [2]. Similar substances were obtained by others [3,4]. Resembling material was later isolated by Gobley from egg yolks and brain. It was a phosphorus-containing lipid, which he called lecithin [5]. He also showed that it contains glycerophosphoric acid; the whole molecular structure was elucidated by Diaconow and by Strecker in 1860s [6,7]. Hydration of dried extracts of such lipids and the formation of myelin figures have been described by Virchow [8] and Neubauer [9], and Thudichum analyzed various fractions of brain lipid extracts and realized the importance of phospholipids in life processes in the 1880s [10]. He wrote:

> *Phosphatides are the centre, life and chemical soul of all bioplasm whatsoever, that of plants as well of animals. Their chemical stability is greatly due to the fact that their fundamental radicle is a mineral acid of strong and manifold dynamicities. Their varied functions are the result of a collusion of radicles of strongly contrasting properties. Their physical properties are, viewed from a teleological point of standing, eminently adapted to their functions. Amongst these properties none are more deserving of further enquiry than those which*

*Giant Vesicles* Edited by P. L. Luisi and P. Walde

> *may be described as their power of colloidation. Without this power no brain as an organ would be possible, as indeed the existence of all bioplasm is dependent on the colloid state.*

It is certain that liposomes had to be present in these experiments [5,8–11]. For instance, Lehmann's book shows an optical micrograph of what is today known as large multilamellar vesicle; in 1911 the author called them 'Künstliche Zellen', artificial cells [12–14].

Optical microscopy observations of the swelling of various dry lipid extracts in water was, and probably still is, a popular experiment in many biological classes in high schools with students observing the fast-growing thin, cylindrical structures that give rise to beautiful colors between crossed nicols. The lipid used in these experiments was an extract from brain nerve tissue and, because these structures were first observed from dried myelin material, they were named myelin figures. Due to mechanical stresses under the cover slip and crystal defects in the lipid film, some of the cylinders, which were growing from the original lipid mass, probably detached and sealed the exposed edges forming suspended colloidal particles – liposomes. Although so obvious now, it took more than 100 years and an electron microscope to realize that these particles are self-closed and that they encapsulate the liquid medium, in which they freely diffuse, in their interior.

The early experiments established that the swelling behavior of lipids was due to the fast uptake of water in the lipid crystal lattice [8,9]. The amphiphilic nature and structure of the lipid bilayer and of biological membranes was established already in 1920s [15]. This work was followed by studies of colloidal behavior of lipid dispersions. The results lacked reproducibility and, because people at that time were not aware of the biological importance of polar lipid water systems, the work was largely unnoticed. It is now known that the poor reproducibility was due to impure lipids. Minor contamination with charged lipids or chemical degradation of lecithins, for instance, can drastically change their swelling and other colloidal properties. Therefore, researchers in colloid science worked mostly with more reliable metal oxide or silver iodide colloids. However, experiments with large multilamellar vesicles, or lecithin (or other phospholipid) sols, as they were referred to at the time, did not yield much useful information [23], untill their use in studies of blood clotting. That eventually led to the characterization of liposomes as self-enclosed artificial membranes.

Bangham used electron microscopy as well as osmotic and permeability experiments to show that the observed lecithin particles encapsulate a fraction of the aqueous phase and that the phospholipid bilayers represent a permeability barrier [16]. This realization profoundly changed the scientific evolution of liposomes. It must be recognized that liposomes were present in many experiments at that time, but they were simply considered as yet another colloidal dispersion. In the 1950s researchers published papers on the formation of lecithin micelles by sonication of lecithin sols [17], and accurately measured the micelle (i.e. liposome!) molecular weight [18]. They even used these suspensions in various experiments, including for the reduction of cholesterol levels in blood by administering sonicated oral lipid dispersions [19], but failed to realize their self-closed nature and resemblance with cell membranes. Bangham's work established many physicochemical properties including size distribution, osmotic behavior, and surface charge of self-closed colloidal smectic phases. Due to the large heterogeneity of the size distribution, however, these dispersions were not very useful in delicate experiments, such as those involving the binding of various blood proteins. It was soon realized that by sonicating suspensions of large liposomes, smaller vesicles are generated [20] which give rise to better-defined reagents (especially more reproducible surface area/mass of lipid ratio for preparations) and, thereby more reproducible results. Since then mostly smaller liposomes have been used in scientific research and especially in applications, where small unilamellar vesicles are almost exclusively used [21]. Giant liposomes [14,22,23] made a comeback with Helfrich's studies of swelling of lecithins [24] and vesicle shapes [25] and with Evans' measurements of mechanical properties of various lipid bilayers [26]. Theories of vesicle shapes followed from Helfrich [25], Svetina and Zeks [27,28], Lipowsky and Seifert [29,30], and others. Optical microscopic observations of the predicted shapes became very popular in the 1980s [31]. Theoretical and experimental studies of vesicle shapes were prevalent in late 1980s and early 1990s. In contrast to other endeavors in theoretical physics, experimental verification of the theory and predicted vesicle shapes is possible by monitoring giant unilamellar vesicles.

From the mathematical modeling of two-dimensional surfaces in three-dimensional space, the applications of giant vesicles has spread into several other disciplines. They can be used as microreactors into which various reagents, substrates, proteins, nucleic acids, channel-forming molecules and other substances can be reconstituted or microinjected. Reactions, as well as molecular transmembrane transport, can be studied by a variety of techniques, including patch clamp [32]. Such systems are being used in fundamental studies of muscle activity [33], signal transduction [34], for unimolecular DNA condensation [35], as well as in studies of the self-replication and origin of life, as practiced by the Luisi group.

## 2. DEFINITIONS AND BASIC THERMODYNAMICS

Morphologically, with respect to the vesicle size and number of lamellae, it is possible to distinguish between small (S), large (L), and giant (G) uni-, oligo-, or multilamellar [U, O, ML, respectively] vesicles {V}. Combination of these letters, ()[]{}, gives rise to several widely used abbreviations for various vesicles, such as SUV, LUV, LOV, GOV, and MLV. In the case when the encapsulated vesicles are not concentric, multivesicular liposomes are defined (MVL). Similar structure characterizes also DepoFoam, which consists of ca. 10 μm suspended particles with multiple interior compartments made of bifurcated lipid bilayers. Structurally, they resemble frozen foam. The high curvature at the bifurcation is accommodated by triolein molecules. These particles are used as a controlled release system for encapsulated hydrophilic drugs [36]. Another example of very large lipid vesicles are vesosomes, very large unilamellar vesicles or GUV encapsulating SUV, which are tethered together with ligand receptor bonds [37].

It is now well established, that in most cases vesicles are not a thermodynamically stable state, because energy, such as sonication, pressure (extrusion through small pores), homogenization, or similar are needed to produce liposomes [14]. In contrast, giant liposomes are typically referred to as being formed spontaneously. In addition to semantics and differences in definitions of spontaneity, this contradiction is also due to numerous crystal defects in the thin lipid film as well as the minute energy input during handling and observation of these diluted systems. During the swelling of dry lipid bilayers into hydrated myelin figures, crystal defects, such as disclinations and dislocations, can probably interrupt the sliding of the lipid molecules from the dry mass into myelin figures and cause buckling of bilayer fragments, resulting in its self-closure. However, portraying the complexity of swelling in unperturbed systems, the giant vesicles formed are often interconnected with invisible tethers, which can be broken by mild agitation. Similarly, the induction of crystal defects into the dry lipid film has been used to produce vesicles directly upon hydration of lipid film. Homogeneous LUV have been produced by hydration of very thin lipid films deposited on a micropatterned surface (etching for size and ion implantation for the charge density, i.e. polarity) of the silicon wafer in order to induce crystal defects in the deposited lipid film and control the size of swelling and peeling bilayers and, correspondingly, vesicle size [38]. With current rapid developments in the lab-on-a-chip technologies, similar approaches in the preparation of liposomes or liposome DNA complexes may soon come of age.

A typical characteristic of thermodynamically stable colloidal systems is narrow size distribution [39]. Therefore, strictly speaking, equilibrium analysis using the shape factor model [40] cannot be applied for kinetically trapped systems, such as liposomes [14]. However, in most practical examples, it is a good approximation. Although several reports describe thermodynamically stable vesicles with broad size distributions [41,42], truly spontaneous vesicles with a narrow size distribution have been produced only recently [43]. Only such vesicles can be referred to as a thermodynamic phase, analogously to micelles. Normal vesicles can be referred to as thermodynamic states with the same symmetry and phase transitions as the smectic lamellar phase, but not as thermodynamic phases which, as in the case of lamellar phase, require a periodic structure. Nonthermodynamically stable liposomes are kinetic traps [14] in those regions of phase diagrams where smectic precipitates (with infinite or large curvature radii) coexist in excess aqueous phase as the thermodynamically stable phase.

However, systems of giant vesicles cannot be referred to as colloidal suspensions because the particles are too large and the system is too heterogeneous. Therefore, giant vesicles should be treated as individual entities [44] and characterized by the size, shape and composition of the bilayer, and by the internal and external media.

## 3. PREPARATION OF GIANT VESICLES

Large and giant vesicles can be byproducts in the preparation of smaller vesicles. To increase the fraction of larger and giant vesicles several preparation procedures have been developed. Typically two different routes can be taken: the swelling of ultrathin and large defect-free lipid bilayers at minimal mechanical agitation, or the fusion of preformed smaller vesicles. Details of these techniques are described in several articles in this volume.

One of the first methods for the preparation of giant liposomes was described by Reeves and Dowben [45]. They deposited a thin lipid film from chloroform:methanol 2:1 (v/v). After drying, the film was hydrated without agitation or shaking. This method results in the preparation of a heterogeneous dispersion, and when giant vesicles became of interest other methods and improvements were developed. Many researchers deposit lipid film on a roughened (with sandpaper) Teflon surface and, prior to hydration, equilibrate the dry lipid film in a moist atmosphere [46,47]. The use of spinners allows very even film deposition [38] and it is very likely that other tools from microelectronics will be used to determine the shape and hydrophilicity/hydrophobicity of surfaces for the deposition of thin lipid films.

Other preparation methods involve dialysis, freeze thawing, or dehydration–rehydration of small unilamellar vesicles or their controlled fusion [14,48].

Giant vesicles (up to 50 μm) were produced from egg lecithin SUV upon freezing and thawing at very high concentrations of some alkali halides (1 – 3 M) followed by dialysis to lower concentrations [49]. The largest vesicles were formed when KCl and RbCl were used, whereas the use of LiCl did not yield GUV. The liposome size roughly correlates with eutectic temperature of frozen alkali metal chloride solutions, K > Rb > Na > Cs > Li. However, binding of these ions to the lipid surface may also have an effect. The presence of $Ca^{2+}$ increases vesicle size.

A novel approach is the use of electroswelling [50], as described in Chapter 3. In brief, dry lipid films are hydrated in an alternating electric field, typically 10 Hz at 0.3 – 3 V, which has been shown to affect the swelling of lipid, probably due to electroosmotic vibrations of the medium.

Vesicles in the size range from 0.5 to several micrometers can be prepared by other methods, such as detergent or other agent induced fusion, freeze-thaw cycles, the cochleate cylinder method, and the fusion of large interdigitated liposomes [14].

## 4. GIANT VESICLES AND MODELS OF VESICLE FORMATION

It seems now that the self-closure of fragments of lipid bilayers, and the budding off mechanism due to an induced membrane asymmetry, can explain the majority of the vesicle formation techniques [14,22]. Only in methods that involve the removal of an organic phase from lipid emulsions (water-in-oil), might the mechanism be different. In this case it resembles the inversion of an emulsion with consequent fragmentation and self-closure or spontaneous entropy-driven dissociation from specific cubic phases, such as $L_3$ ((normal) sponge phase).

Optical microscopy experiments with giant unilamellar vesicles confirm the hypotheses derived from working with smaller vesicles, in which unperturbed observations are practically impossible. The budding off model has been shown for a variety of systems [14,22,23,47], as were the observations of very large lipid flakes in various stages of self-closing, or self-opening, in a reversible process of liposome dissolution with the detergent [14]. The microscopic model [51] of vesicle formation by detergent depletion (or reversible process of preformed vesicle dissolution with detergent), in which the detergent shields the open bilayers against aqueous environment, was predicted in the vesicle formation by detergent depletion [33]. After initial skepticism, the model was confirmed in the case of SUV and LUV [14,52]. Recently, however, an almost identical scheme has appeared for the case of GUV [53] in which transition structures can be monitored in an unperturbed, artifact-free environment.

The analogies from small (<100 nm), large (0.1 to few μm) and giant (>few 50 μm) vesicles show the generality of behavior of biological membranes, from vesicle formation to vesicle shapes and their changes, and might be important in the understanding of the function of cell membranes. Although at higher curvature radii nonelastic bending elasticity limits this analogy, membrane behavior seems to scale at least three orders of magnitude. That is, from sizes where the bilayer thickness effects disappear (50–100 nm, depending on the bilayer composition, i.e. stretching and bending elastic moduli), to the 0.1–1 mm range, where vesicles become too unstable and simply too large for observation.

## 5. VESICLE SHAPES

Harvested after controlled swelling, vesicles have a broad spectrum of different sizes and shapes. Obviously, they originate in local microconditions (crystal defects) during swelling. Spherical shapes, however, are most frequently encountered because these shapes have minimal bending energy [25]. The richness of all the shapes clearly indicates that lipid swelling and liposome detachment is a dynamic process.

The changes of vesicle shapes are due to temperature or pressure changes, the addition of various amphiphiles or adsorbents, mechanical, electrical or magnetic treatments, or to adhesion [14,22,23,25,47]. Temperature changes induce area/volume differences due to the different expansion coefficients of lipid and water, as well as to the two- and three-dimensional response of the system to external stress, respectively. Budding, discocyte–stomatocyte, dumbbell–pear, and spontaneous tether formation transitions have been observed in lecithin GUV [47,54]. Since these studies in early 1990s, numerous papers on vesicle shapes and their transformations have been published. Most of the work concentrates on neutral lipids, despite the fact that charged liposomes might provide different phase diagrams and are more relevant for biological comparisons.

In these studies vesicle shape is influenced either by change in temperature or composition [22,29,47,48,54–57]. Similar topological changes of GUV under various physical and chemical stresses are extensively studied by Menger's group. They study hydration, adhesion, aggregation, fusion, fission, and disintegration of vesicles, which they refer to as cytomimetic supramolecular chemistry [58,59].

Shape changes of GUV were observed upon decompression of egg lecithin vesicles. Because water was pushed out during the compression (due to the fact that lecithin is more compressible than water), upon quick decompression, smaller vesicles budded off due to the excess of surface area. The compressibility of cholesterol-containing bilayers is smaller and no shape changes were observed [60].

Giant vesicles containing acidic lipids were shown to change shape upon the creation of a pH gradient across the membrane. The gradient caused flip-flop of acidic lipids, and dramatic shape changes resulted upon redistribution of only one lipid molecule in 1000 [61]. Such observations may have biological significance, because it is known that cell apoptosis (and senescence of red blood cells) begins with a flip-flop reorientation of phosphatidyl serine from the inner to the outer monolayer.

Based on the concepts of bending energy and broken symmetry of the two leaflets comprising the membrane, several models of vesicle shapes have been established [22,25,27–30].

Typically, models assume that the bilayer thickness is negligible with respect to the vesicle size and that the membrane is in a fluid state, i.e. it does not resist the shear forces within the bilayer. Each monolayer of the membrane can slide over the other one. Lipid molecules are insoluble in the aqueous phase and the flip-flop rate is zero. The spontaneous curvature model expands the bending energy in terms of the sum and product of the two curvatures, which can be defined on any plane, and assumes that the total area of vesicle is fixed. The proportionality constant for the first term is the bending elastic modulus, and for the second one is the Gaussian bending rigidity. As in the changes of vesicle shapes the topology does not change, the Gaussian term (the product of the curvatures) can be ignored because it is topologically invariant, as follows from the Gauss–Bonnet theorem. The area-difference–elasticity models superpose the architecture of the bilayer on the equation for the bending energy by splitting the bilayer into two monolayers and defining their stretching elastic energy. They therefore include a term for the stretching and compression of a monolayer. The assumption that the two monolayers are incompressible, yields the bilayer–couple model. Both models revert to the spontaneous curvature model for zero stretching energy and equivalence between the monolayers, respectively.

A variety of different shapes of predicted giant vesicles were indeed observed by optical microscopy. With the advent of cryo electron microscopy, similar shapes have also been observed in samples of SUV [22,35,62].

The analogy between small and giant vesicles is also reflected in the case of shapes of vesicles with encapsulated polymers. Gelation of a drug and counterion into fibrilar structures in SUV [63], as well as polymerization of encapsulated monomers in GUV [64] resulted in very similar vesicle elongation into oval shapes.

With respect to encapsulated or adsorbed molecules, such as annexins, vesicles can have very unusual shapes with very sharp edges. For example, perfect hexagons, squares and even triangles have been observed in samples with encapsulated doxorubicin sulfate or with adsorbed proteins [22,62].

In many different systems of positively or negatively charged lipid bilayers invaginated liposomes [35,65] have been observed that are scale invariant in SUV to GUV range. It is not known if these structures result from the presence of impurities and phase separation within the bilayer, local events (leaflet asymmetry during the growth of myelin figures for GUV or induced after extrusion in the case of SUV) before self-closure, or to the negative contribution of Gaussian curvature to the free energy, $kC_1C_2$ (where $k$ is the elastic modulus of Gaussian curvature, $C_1$ is curvature of the bilayer across the opening, and $C_2$ is the curvature of the opening).

Very interesting shapes can be observed in nonphospholipid diacyl surfactant systems [66]. In polymerizable lipids, for instance, tubular (with either circular or spiral cross-section) structures are often observed in the gel phase, which reversibly transform into liposomes above $T_c$ (the main gel-liquid crystalline phase transition temperature). Upon electrical pulses or the addition of polymers novel structures with novel morphologies, such a helices and strings-of-pearl, can be observed [67].

Although shape changes present a very interesting problem, their direct application to biological systems is not clear, because therein most changes occur by energy-driven enzymatic processes. Despite the fact that lecithin vesicles can exhibit the same shape changes as erythrocytes, it is still not known if shapes changes in red blood cells *in vivo* are lipid or protein driven. Nevertheless, there are many other useful applications of giant vesicles, as briefly described below and elsewhere [14,22,23,68].

## 6. APPLICATIONS OF GIANT VESICLES

Although vesicles that occur in nature and in drug delivery applications are much smaller, GUV can give valuable data on the behavior of SUV because most observations are universal and scale invariant (i.e. permeability coefficients, first-order elastic constants, etc. are liposome size independent) or can be scaled down to smaller dimensions (entrapped volume, encapsulation efficiency, surface area). For many applications, however, the large size is preferred due to easier observation, larger encapsulated volumes, and the ability to microinject various substances into these liposomes.

Because many of these applications are described in this volume, the wide spectrum, not the details, of these applications are given here.

### 6.1 Membrane mechanics and intermembrane interactions

Many membrane properties depend on the different elastic constants of the bilayer. Therefore their measurement can give valuable information on the characteristics of cells and the optimization of vesicles for various applications.

Bending stiffness of lipid bilayers and free energies of interactions between membranes of GUV have been measured by micropipette manipulation. Mechanical properties of the bilayers can be measured on a vesicle, which is controlled by micropipette suction via geometrical changes upon applied pressure: well-defined stresses give rise to measurable area and volume changes [26,69]. Interactions between vesicles require micromanipulation of two pipettes with attached vesicles [26,69–72]. More complex experiments involve lipid tethers with attached ligands, glass beads or (para)magnetic beads that can be manipulated magnetically, mechanically or chemically and various membrane parameters or interactions can be measured [73,74]. By using electropermeabilization, pores can be created and their time evolution in the vesicle membrane can be followed and the line tension measured [75]. The bending elasticity of the vesicle membrane can also be measured by shape fluctuations of tubular or spherical vesicles, as observed by optical microscopy [76–78].

Giant vesicles can be used to study adhesion onto various supports. In biological systems attractive forces are often specific, whereas repulsive forces are typically universal. Cell adhesion can be simulated by incorporating PEGylated lipids into the membrane to increase nonspecific repulsion [79–82]. Attractive interactions can be modeled by insertion of membrane-bound receptors and ligands or biotinylated lipids. Specific interactions between GUV and the solid support in which the attractive lock-and-key interactions have been modeled by incorporating biotinylated lipids into both surfaces and addition of coupler streptavidin in the medium have shown that at adhesion sites lateral phase separation occurs leading to domains of tight adhesion [83]. Using optical tweezers it is possible to measure the strength of individual chemical bonds. Because PEGylated liposomes have revolutionized liposomal drug delivery [84], it is possible that investigations of PEGylated GUV may be the next direction taken by studies of cell and bilayer interactions.

### 6.2 GUV in physiology and medicine

Giant vesicles are used to study the function, and to follow and characterize dysfunction of various organs, such as in investigations of muscle activity and signal transduction.

Sarcolemnal vesicles prepared from cell membranes of muscle tissue contain proteins and therefore information on glucose transport and activity of other membrane proteins. It has been shown that exercise increases the concentration of glucose transporter in the membrane of muscle cells, possibly via trafficking by protein-containing vesicles [85]. It has been shown that exercise doubles the glucose transport (to 5 nM per mg protein per second) due to the increased number of proteins and not to the increased pump rate [86]. By measuring the tensile strength and dilatational elasticity of such vesicles, which depend on the lipid–protein interactions, researchers hope to characterize muscle action [87].

Neurotransmission and the action of toxins has been investigated by using appropriate vesicles. GUV containing acetylcholine were prepared from lecithin and presynaptic plasma membranes and the effect of various toxins on the permeability of neurotransmitter has been studied. It has been shown that Botulinum toxin decreased membrane fluidity and inhibited $Ca^{2+}$-dependent transport of the neurotransmitter through the ion channel (ionophore A23187) [88]. GUV can be also patch-clamped and electrophysiological studies of ion channels can be performed. Giant sarcoplasmic reticulum vesicles have been prepared by fusion of smaller ones and single channel conductance has been measured [89]. All these data help in the understanding of membrane proteins and channels and depict cell function. Fusion of synaptic vesicles with plasma membrane and the influence of various factors on its rate were studied by the preparation of giant synaptic vesicles (by freeze thawing) and investigating their interactions with black lipid membranes (BLM) [90]. Fusion could be quantified by a release of the encapsulated fluorescent dyes on the *trans* side of the BLM.

This research is relatively new and it is very likely that the future will bring more discoveries in this area.

### 6.3 Vesicles and the origin of life

One of the basic questions in science is the origin of life on Earth. At some point the merging of self-replication of nucleic acids and boundary structures (self-association and self-compartmentalization of polar lipids) had to produce the first cell. Life probably started by GUV encapsulating appropriate molecules, such as nucleic acids and proteins or polypeptides [91] in the primordial hot soup or near the thermal vents. It has been also shown that vesicle-forming lipids could have been brought onto the planet from outer space [92]. Although such a scenario can explain the first cell, it has been demonstrated in the 1990s, that in the presence of appropriate substrates similar vesicles can grow and self-replicate [93]. Furthermore, replication of RNA and the synthesis of nucleic acid in GUV after microinjection of appropriate reagents has been demonstrated [94,95]. Evolution of life from the bilayer perspective, including the biosynthesis of cholesterol and its influence on increasing the complexity of the cell, has been argued by Bloom and Mouritsen [96].

Due to their unique characteristics, giant liposomes can be a useful model, as described above, for various mechanical and biological phenomena. Two slightly more philosophical comparisons can be made. Citing Bloom [97], giant vesicles and their behaviors have some similarities with more complex theories. After listening to a seminar on superstrings in which a phase transition in 11 dimensions induced by the increase of surface area/volume over a critical point was predicted, another attendee (E. Evans), mentioned that he could observe similar instability and transition in three dimensions using a normal optical microscope, by observing reversible protuberances from stressed giant vesicles.

To explain differences between some experimental results, Helfrich speculated that membranes have a superstructure [98]. The surface area of the membrane should be larger than that observed by optical microscopy due to the submicroscopic superstructure, which includes anomalous roughness and dispersivity [99]. Although no decisive experimental proof has been found yet, a curious similarity exists. In superstring theory new models deal with small membranes instead of strings. These multidimensional membranes can exist in three dimensions and have additional dimensions curled into and onto themselves, in a manner similar to the extra surface area of vesicles. Electron micrographs of large liposomes, formed in a nonagitating process, often bear tiny irregular blobs attached to their surface [14, page 95]. It is probable that these arise from unseen thin tethers that connect some of the liposomes and which rupture during sample preparation. Similarly, in optical microscopy of GUV, often during selecting and manipulating one liposome by a pipette many others feel the pull due to invisible tethers. However, such excess lipid on a spherical vesicle seems to be more a result of extremely rich structural polymorphism during lipid swelling and subsequent dynamic processes, rather than a general natural feature of lipid behavior. Similarly, upon partial polymerization or osmotic imbalance, wrinkled and flaccid vesicles can be produced [22].

## 6.4 Giant multilamellar vesicles

Giant vesicles also encompass multilamellar structures in the size range 1–50 μm. Although less interesting from a theoretical prospective (despite forming many similar shapes to GUV [100]), they have, due to their ease of preparation and ability to suspend large amounts of hydrophobic agents, found wide use in industry. They range from cosmetics and nutrition to drug delivery. MLV can serve as a drug carrier for hydrophobic molecules, acting as a matrix for their sustained release as well as a solubilizer for molecules with problematic solubility. Molecules that are not soluble in oils or in water tend to stick to polar – nonpolar boundaries, and liposomes have been shown to suspend many drugs with solubility problems monomolecularly. Such systems have been shown to be useful in high throughput screening of libraries of drug molecules [101].

Recent trends in cosmetics and nutrition have given rise to high quality products containing medicaments and natural substances with beneficial action, the so-called cosmoceuticals and neutraceuticals [102]. Often hydrophobic molecules are encapsulated in MLV which are also believed to increase permeation through the skin and absorption in the gastrointestinal tract. Although most of this work typically lacks rigorous scientific data, it is safe to say that lecithin – water-based vehicles are much safer than emulsions and gels based on oil, detergents or alcohols.

In conclusion, rapid development in the liposome field has been matched also in the field of giant vesicles, as this volume clearly testifies.

## 7. REFERENCES


1. A. D. Bangham and R. W. Horne, *J. Mol. Biol.*, **8**, 660 (1964).
2. L. N. Vauquelin, *Ann. Mus. Hist. Nat.*, **18**, 212 (1811).
3. J. P. Couerbe, *Ann. Chim. Phys.*, **56** (ser. 2), 160 (1834).
4. E. Frémy, *Ann. Chim. Phys.*, **56** (ser. 2), 463 (1841).
5. M. Gobley, *J. Pharm. Chim.* (Paris), **17** (ser. 3), 401 (1847).
6. C. Diaconow, *Zentr. Med. Wiss.*, **6**, 2 and 97 (1968).
7. A. Strecker, *Ann. Chem. Pharm.*, **148**, 77 (1868).
8. R. Virchow, *Virchows Archiv.*, **6**, 562, (1854).
9. C. Neubauer, *Zeit. Anal Chem.*, **6**, 189 (1867).
10. J. L. W. Thudichum, *A Treatise on the Chemical Constitution of the Brain*, Balliere, Tindall and Cox, London, 1884.
11. G. B. Ansell, in *Form and Function of Phospholipids*, (eds) G. B. Ansell, J. N. Hawthorne, R. M. C. Dawson, Elsevier, Amsterdam, 1, 1973, p. 1.
12. O. Lehmann: *Die scheinbar lebenden Kristalle*, Verlag J. F. Schreiber, Esslingen, München, 1907.
13. O. Lehmann: *Die neue Welt der flüssigen Kristalle*, Akademische Verlagsgesselschaft, Leipzig, 1911.
14. D. D. Lasic, *Liposomes: from Physics to Applications*, Elsevier, Amstersam, 1993.
15. E. Gordel and F. Grendel, *J. Exp. Med.*, **41**, 439 (1925).
16. A. D. Bangham, M. M. Standish and J. C. Watkins, *J. Mol. Biol.*, **13**, 238 (1965).
17. L. Saunders, J. Perrin, and D. Gammack, *J. Chem. Soc.*, **14**, 567 (1962).
18. N. Robinson, *J. Chem. Soc.*, **12**, 1260 (1960).
19. M. Friedman, S. O. Boyers and R. H. Roseman, *Proc. Soc. Exp. Biol. Med.*, **95**, 586 (1957).
20. D. Papahadjopoulos and J. C. Watkins, *Biochim. Biophys. Acta*, **135**, 639 (1967).
21. D. D. Lasic and D. Papahadjopoulos (eds), *Medical Applications of Liposomes*, Elsevier, Amsterdam, 1998.
22. R. Lipowsky and E. Sackmann (eds), *Structure and Dynamics of Membranes*, North Holland, Amsterdam, 1995.
23. D. D. Lasic and Y. Barenholz (eds), *Nonmedical Applications of Liposomes*, Vol. I-IV, CRC Press, Boca Raton, FL 1996.
24. M. Boroske, M. Elwenspoek and W. Helfrich, *Biophys. J.*, **34**, 95 (1981).
25. H. J. Deuling and W. Helfrich, *J. Physique*, **37**, 1335 (1976).
26. E. A. Evans, *Biophys. J.*, **14**, 923 (1974).
27. S. Svetina and B. Zeks, *Eur. Biophys. J.*, **17**, 101 (1989).

28. S. Svetina and B. Zeks, *in ref. 23*, Vol. I, pp. 13 – 42.
29. K. Berndl, J. Käs, R. Lipowsky, E. Sackmann and U. Seifert, *Europhys. Lett.*, **13**, 639 (1990).
30. R. Lipowsky and U.Seifert, *in ref. 23*, Vol. I., pp. 43 – 68.
31. H. P. Duwe, J. Käs and E. Sackmann, *J. Physique*, **51**, 945 (1990).
32. J. R. Monck, A. F. Oberhauser and J. M. Fernandez, *Mol. Membr. Biol.*, **12**, 151 (1996).
33. E. Neumann, J. Weber and T. Schurholz, *Arch. Physiol. & Biochem.*, **104**, 731 (1996).
34. N. Yakir and R. Rahamimoff, *J. Physiol.*, **485**, 683 (1995).
35. D. D. Lasic, *Liposomes in Gene Delivery*, CRC Press, Boca Raton, 1997.
36. J. Senior, *in ref. 23*, p. 733 – 750.
37. S. A. Walker, M. Kennedy and J. Zaszadinski, *Nature*, **387**, 61 (1997).
38. D. D. Lasic, A. Belic and T. Valentincic, *J. Am. Chem. Soc.*, **110**, 970 (1988).
39. C. Tanford, *The Hydrophobic Effect*, John Wiley & Sons, Chichester, 1980.
40. J. N. Israelachvili, *Intermolecular and Surface Forces*, Academic Press, 1991.
41. W. Hargreaves and D. W. Deamer, *Biochemistry*, **27**, 8261 (1988).
42. E. Kaler, A. Murthy, B. Rodriguez and J. N. Zaszadinski, *Science*, **245**, 1371 (1989).
43. R. Joannic, L. Auvray, and D. D. Lasic, *Phys. Rev. Lett.*, **78**, 3402 (1997).
44. IUPAC Committee on Liposome Nomenclature (H. Hauser, D. Crommelin, D. W. Deamer, D. D. Lasic, D. Marsh, D. Papahadjopoulos and T. E. Thompson), recommendation, in preparation.
45. J. P. Reeves and R. M. Dowben, *J. Cell Biol.*, **73**, 49 (1973).
46. D. Needham and E. A. Evans, *Biochemistry*, **27**, 8261 (1988).
47. J. Käs and E. Sackmann, *Biophys. J.*, **60**, 825 (1991).
48. F. M. Menger and K. D. Gabrielson, *Angew. Chem. Int. Ed. Eng.*, **34**, 2091 (1995).
49. N. Oku and R. C. MacDonald, *Biochemistry*, **22**, 855 (1983).
50. M. I. Angelova and D. S. Dimitrov, *Faraday Discuss. Chem. Soc.*, **81**, 303 (1986).
51. D. D. Lasic, *Biochim. Biophys. Acta*, **692**, 501 (1981)
52. D. H. Boal and M. Rao, *Phys. Rev.* A, **46**, 3037 (1992).
53. A. Saitoh, K. Takiguchi, Y. Tanaka and H. Hotani, *Proc. Natl. Acad. Sci. USA.*, **95**, 1026 (1998).
54. E. Sackmann, H. P. de Duwe and H. Engelhard, *Faraday Discuss. Chem. Soc.*, **81**, 281 (1986).
55. E. Sackmann, *FEBS Lett.*, **346**, 3 (1994).
56. L. Miao, B. Fourcade, M. Rao, M. Wortis and R. Zia, *Phys. Rev. A* **43**, 6843 (1991).
57. H. G. Döbereiner, J. Kas, D. Noppl, I. Sprenger and E. Sackmann, *Biophys. J.*, **65**, 1396 (1993).
58. F. Menger and K. Gabrielson, *J. Am. Chem. Soc.C,* ***116***, 1567 (1994).
59. F. Menger and S. J. Lee, *Langmuir*, **11**, 3865 (1995).
60. L. Beney, J. M. Perrier, M. Hayert and P. Gervais, *Biophys. J.*, **72**, 1258 (1997).
61. E. Farge and P. F. Devaux, *Biophys. J.*, **61**, 347 (1992).
62. P. Frederik, M. C. A. Stuart, S. Bomans and D. Lasic, *in ref. 23*, p. 309.
63. D. D. Lasic and F. Martin, Y. Barenholz, P. M. Frederik and T. J. MacIntosh, *FEBS Lett.*, **312**, 255 (1992).
64. H. Miyata and H. Hotani, *Proc. Natl. Acad. Sci.*, **89**, 11547 (1992).
65. N. S. Templeton, D. D. Lasic, H. Strey, D. D. Roberts and G. Pavlakis, *Nature BioTech*, **15**, 647 (1997).
66. M. A. Guedeau-Boudeville, *Compt. Rendues Acad. Sci.*, **541**, 1992.
67. J. C. Bradley, M. A. Guedeau-Boudeville, J. Jandeau and J. M. Lehn, *Langmuir*, **13**, 2457 (1997).
68. P. Walde and P. L. Luisi (eds), *Workshop on Giant Vesicles*, Book of Abstracts, Ascona, 1998.

69. E. A. Evans and D. Needham, *Faraday Discuss. Chem. Soc.*, **81**, 267 (1986).
70. D. Needham and R. Nunn, *Biophys. J.*, **58**, 997 (1990).
71. D. D. Lasic and D. Needham, *Chem. Rev.*, **95**, 2601 (1995)
72. E. A. Evans and M. Metcalfe, *Biophys. J.*, **46**, 423 (1984).
73. V. Heinrich and R. E. Waugh, *Ann. Biomed. Engin.*, **24**, 595 (1996).
74. L. Bo and R. E. Waugh, *Biophys. J.*, **55**, 509 (1989).
75. D. V. Zhelev and D. Needham, *Biochim. Biophys. Acta*, **1147**, 89 (1993).
76. R. M. Servuss, W. Harbich and W. Helfrich, *Biochim. Biophys. Acta*, **436**, 900 (1976).
77. H. P. Duwe and E. Sackmann, *Physica A*, **163**, 410 (1990).
78. J. Majhenc, F. Sevsek, S. Svetina and B. Zeks, *in ref. 68*, p. 43.
79. D. D. Lasic, *Angew. Chem. Int. Ed. Eng.*, **33**, 1785 (1994).
80. D. Needham, T. J. MacIntosh and D. D. Lasic, *Biochim. Biophys. Acta*, **1147**, 89 (1993).
81. T. Baekmark, S. Ellander, D. D. Lasic and E. Sackmann *Langmuir*, **11**, 3865 (1995).
82. T. Kuhl, D. Leckband, D. D. Lasic and J. Israelachvili, *Biophys. J.*, **55**, 509 (1994).
83. A. Albersdorfer, T. Feder and E. Sackmann, *Biophys. J.*, **73**, 245 (1997).
84. D. D. Lasic, *Nature*, **380**, 561 (1996).
85. S. Kristiansen, M. Hargraves and E. A. Richter, *Am. J. Physiol.*, **270**, E197 (1996).
86. T. Plough, J. Wojtaszewski, S. Kristiansen, P. Hespel, H. Galbo and E. A. Richter, *Am. J. Physiol.*, **264**, E270 (1993).
87. J. A. Nichol and O. F. Hunter, *J. Physiol.*, **493**, 187 (1996).
88. J. Lopez-Alonso, J. Canaves, M. Arribas, A. Casanova, J. Marsal and C. Solsona, *Neurosci. Lett.*, **196**, 37 (1995).
89. N. Hirashima, H. Ishibashi and Y. Kirino, *Biochim. Biophys. Acta*, **1067**, 325 (1991).
90. M. S. Perrin and R. C. McDonald, *Biophys. J.*, **55**, 973 (1989).
91. D. W. Deamer, *Microbiol. Mol. Biol. Rev.*, **61**, 237 (1997).
92. D. W. Deamer, *Nature*, **317**, 792 (1985).
93. K. Morigaki, S. Dallavalle, P. Walde, S. Colonna and P. L. Luisi, *J. Am. Chem. Soc.*, **119**, 292 (1997).
94. T. Oberholzer, M. Albrizio and P. L. Luisi, *Chem. & Biol.*, **2**, 677 (1995).
95. R. Wick, M. I. Angelova, P. Walde and P. L. Luisi, *Chem. & Biol.*, **3**, 105 (1996).
96. M. Bloom and O. Mouritsen, *in ref. 22*, p. 65.
97. M. Bloom, *Physics in Canada/La Physique au Canada*, **48**, 7 (1992).
98. W. Helfrich, *in ref. 22*, p. 691.
99. B. Klösgen and W. Helfrich, *in ref. 68*, p. 25.
100. D. D. Lasic, *in ref. 22*, p. 491.
101. D. D. Lasic, *Trends in Biotechnology*, **16**, 307 (1998).
102. D. D. Lasic, M. Brandl and T. Alfredson, unpublished results.

# Part Two

# Preparation Methods

# Chapter 3

# Liposome Electroformation

*MIGLENA I. ANGELOVA*
*Bulgarian Academy of Sciences, Sofia, Bulgaria*

## 1. INTRODUCTION

Bangham *et al.* [1] demonstrated in their pioneering work the capture effect of phospholipid liquid crystalline membrane structures formed by the swelling of lipid films on solid surfaces in excess aqueous solution. Since then, a number of investigators have used this procedure to prepare liposomes (lipid vesicles). These liposomes were: (a) predominantly multilamellar vesicles (MLV); (b) of wide size distribution (typically, from 0.2 to 20 μm); and (c) spontaneous formation of MLV from a variety of lipids or lipid mixtures was very slow (taking tens of hours), and in some cases even practically impossible. To solve these problems, several different modifications and additional procedures have been suggested. For example, external mechanical perturbations, as ultrasonication and extrusion through polycarbonate membranes, proved to be helpful for faster and thinner-membrane liposome formation, and to a certain extent, for narrowing the size distribution. The resulting vesicles – SUV (small unilamellar vesicles) or LUV (large unilamellar vesicles) – being about 30–50 nm in diameter (SUV) and 100–200 nm in diameter (LUV), respectively, were thoroughly investigated and in some cases used in medicine [2–6]. SUV and LUV have commanded the majority of attention in the past but, for certain cell-mimicking investigations, they have one crucial disadvantage in that they are submicroscopic. In addition, being of small diameter, they possess much higher curvature than cell membranes. As curvature affects lipid packing, SUVs and LUVs are poor models for a biological membrane. Although it is possible to uniformly bathe a sample of SUV or LUV in the active substance solution of interest, it is not

*Giant Vesicles* Edited by P. L. Luisi and P. Walde

possible to expose just one section of a single vesicle to the active substance, thereby, creating a localized asymmetry or lateral gradient on the vesicle membrane.

Cell-sized unilamellar vesicles (about 5–50 μm in diameter) have attracted much attention as they serve as a simple model of biological cells and membranes. However, the preparation of such giant vesicles, has always been a problem. To solve it, some other methods (not based on the process of swelling of lipid films on solid surfaces) have been used [3,5]. However, these methods bring their own disadvantages, such as formation in non-aqueous solutions.

Giant unilamellar vesicles (GUV), about 10–200 μm in diameter, represent a convenient subject for direct optical microscopy investigation of the mechanical and electric properties of membranes. Membrane shape and morphology transformation studies are conveniently carried out with GUVs of size 50–200 μm. The interactions of individual GUV with other colloidal particles, the effects of active substances *locally* microinjected to a small part of the membane of an individual giant vesicle, the creation of *local* lateral and transmembrane gradients of membrane composition, and the kinetics of the resulting membrane shape changes, can be followed directly by different optical microscopy regimes (amplitude or phase contrast, fluorescence, interferential contrast). Preparation of a good sample of such vesicles is a problem for any of the conventional methods [2,3].

The need for a non-trivial approach to study liposome formation, aiming for an efficient method for cell-size and GUV preparation was clear. Evidently, liposome formation requires membrane formation, separation, and bending. The starting idea was that external electric fields could affect these processes, because they can influence the formation of supramolecular structures, change intermembrane forces, induce bending, and affect the line tension of lipid domain boundaries. In addition, they can be precisely controlled. Therefore, external electric fields can help to elucidate mechanisms of liposome formation. Understanding the mechanisms of liposome formation helps in the preparation of liposomes with predetermined properties. External electric field effects in liposome formation may be also important when considering fundamental physicochemical, biophysical and biological problems [7], and origin of life [8].

The liposome electroformation method, originally published in [9] was developed [10–13] for plane-parallel as well as for cylindrical electroformation chambers, using both alternating current (AC), and direct current (DC) external electric fields. Many developments and applications have been described since then [8,14–30]. It was established that vesicle formation from a large variety of zwitterionic, as well as charged lipid mixtures can be induced and controlled by external electric fields applied during the swelling of lipid films on electrode surfaces in aqueous media. The sizes, lamellarity and yields of the resulting unilamellar vesicles can be controlled by the thickness of the initial lipid film, and the parameters of the applied external electric field (AC or DC).

## 2. LIPOSOME ELECTROFORMATION METHOD

The set-up for electroformation consists of the electroformation device and an external AC or DC supply. An optical phase contrast microscope (with objective 40×, phase contrast, with long working distance), equipped with video-camera and video-recording system is normally used for observation of GUV. The AC supply is a low-frequency generator providing, for example, 0.1–50 Hz at up to 7 V. An oscilloscope or multimeter monitors the applied electric field parameters. The electroformation device (chamber) can be of two different geometries (Figure 3.1).

Liposome electroformation protocols have the following steps: (i) depositing a dry lipid film on the electrode surface; (ii) application of the external electric field and filling the working chamber with water or aqueous solution; (iii) guiding vesicle

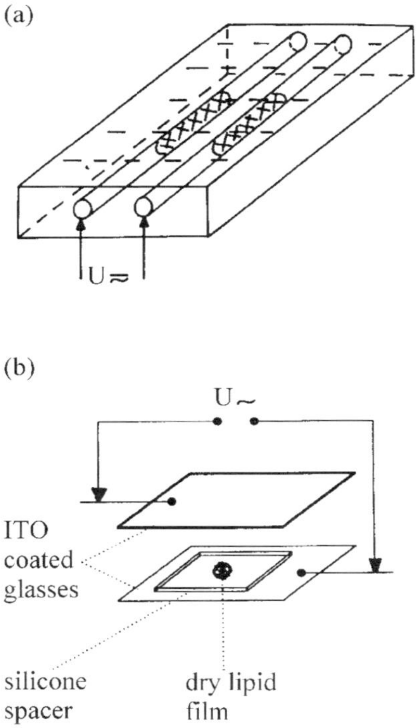


**Figure 3.1** The liposome electroformation device can be of two different geometries: (a) two parallel cylindrical wires for observation at the electrode edge in a direction perpendicular to the vesicle growth [13] (Reproduced from *Prog. Colloid Polymer Sci.*, **76**, 59 (1988) by permission of Springer-Verlag GmbH & Co. KG); and (b) two plane parallel transparent electrodes (indium tin oxide (ITO) coated glass places) for observation in the direction of vesicle growth [14] (Reproduced from *Prog. Colloid Polymer Sci.*, **89**, 127 (1992) by permission of Springer-Verlag GmbH & Co. KsG). $U_{\simeq}$ or $U_{\sim}$ indicates the point of external DC or AC electric field application.

growth by the applied electric field parameters (voltage and frequency). Detailed studies on the effects of the initial lipid film thickness and structure, the external field parameters (voltage, frequency, duration), swelling medium ionic strength, osmolarity and viscosity, are described elsewhere [9–13].

Most universal is the AC electroformation protocol. It has proved to be efficient for GUV preparation of a large variety of zwitterionic as well as charged or neutral lipid mixtures: phosphatidylcholines (PC), egg yolk L-$\alpha$-phosphatidylcholine (eggPC), dilauroyl-L-$\alpha$-phosphatidylcholine (DLPC), dimyristoyl-L-$\alpha$-phosphatidylcholine (DMPC), dipalmitoyl-L-$\alpha$-phosphatidylcholine (DPPC), $\beta$-oleoyl-$\gamma$-palmitoyl-L-$\alpha$-phosphatidylcholine (POPC), $\beta$-oleoyl-$\gamma$-stearoyl-L-$\alpha$-phosphatidylcholine (SOPC), diphytanoyl-L-$\alpha$-phosphatidylcholine (DPhPC), and some mixtures with L-$\alpha$-phosphatidylethanolamines (PE), L-$\alpha$-phosphatidylserines ($PS^-$), L-$\alpha$-phosphatidylglycerols ($PG^-$), sphingomyelin (SM), L-$\alpha$-phosphatidic acid ($PA^-$), cholesterol, D-sphingosine ($Sph^+$), dioleoyl-trimethylammonium-propane ($DOTAP^+$). Electroformed GUV were prepared in water, or 50 mM sucrose, or buffers such as 2 mM TRIS, phosphate, HEPES or BICINE (N,N-bis[2-hydroxyethyl]glycine). GUV electroformation is limited to low ionic strengths (below 10 mM NaCl). Presence in the swelling medium of even 1 mM $Ca^{++}$ proved sufficient to prevent GUV formation.

## 2.1. AC electroformation protocol

The choice of organic solvent from which the dried lipid film is deposited affects the GUV quality. Normally, one of the following mixtures is used: (i) chloroform : methanol 9 : 1 (v/v); (ii) diethyl ether : methanol 9 : 1 (v/v); (iii) chloroform : diethyl ether : methanol 2 : 7 : 1 (v/v/v); or 4 : 5 : 1 (v/v/v). The mixtures of two solvents are efficient for zwitterionic and neutral lipid film deposition. When a charged lipid is involved, the three-solvent mixture is preferable. The total lipid concentration in the organic solvents solution should be 0.2–1.6 mg ml$^{-1}$. It is up to the experimenter to establish the best solvent mixture and total lipid concentration in each particular case, which might also depend on the temperature (the ambient temperature affects the kinetics of solvent evaporation and, thereby, the starting lipid film properties). It takes normally a few trials to determine the optimal conditions. For the device shown in Figure 3.1(a), a 1 µl droplet of lipid solution is deposited (avoid sliding!) on each of the two parallel platinum wires (diameter 0.8 mm, distance between the wire centers 3 to 5 mm) and dried under nitrogen (or vacuum) for 30 min. With the device shown in Figure 3.1(b), one or a few droplets of 2.5–5 µl can be placed on the electrode plate (separation distance between the plates 0.3–3 mm) but taking care that the droplets do not touch each other. AC electric field (10 Hz, 0.2 V) is applied to the electrodes and water (or aqueous solution) is added to the working chamber, taking care to avoid agitation. The voltage is then raised to 1–7 V for 15 min (again, a decision for the experimenter). GUV formation takes about 2 h. Then, the AC frequency usually is lowered to 0.5 Hz for 1 min. This helps the mature vesicles to become spherical and to separate from the electrodes (if desirable). The voltage is lowered gradually to a minimum of about 0.2 V before the electric field is switched off.

Although each electroformation case needs to be considered individually, this is not difficult because the experimenter can observe the vesicle formation directly and control the processes.

## 2.2. Electroformed GUV properties

Electroformation is useful for preparing GUVs for studies involving individual vesicles (not demanding liposome suspensions of high lipid concentration). Any sample provides at least a few tens of "good" vesicles, (Figure 3.2). Bearing in mind that only about a few micrograms of lipid per preparation are used, it is clear that the method consumes very small amounts of lipids.

It is not possible to judge from the vesicle contour thickness or contrast in a phase contrast microscope whether an electroformed giant vesicle is really unilamellar. This is because the vesicle contour thickness or contrast in a phase contrast microscope image depends not only on the vesicle membrane thickness but also on the size of the observed vesicle. The fact that most of electroformed GUVs are unilamellar was proved by lipid membrane bending elasticity measurements [14,15,23,27] and by electron microscopy [25].

Electroformed GUVs are normally 5 – 200 μm in diameter, depending on the lipid composition, swelling medium, and external AC field parameters. At their birth, GUVs are connected by a filament to the rest of the lipid deposition. The membrane tension of a GUV just after its formation is dependent on the kinetics of formation. This can be controlled to a certain extent by the parameters of the external AC field (which establish the amplitude of the electroosmotic vibration). Larger electro-osmotic vibrations during GUV growth might result in excess membrane area, i.e.,

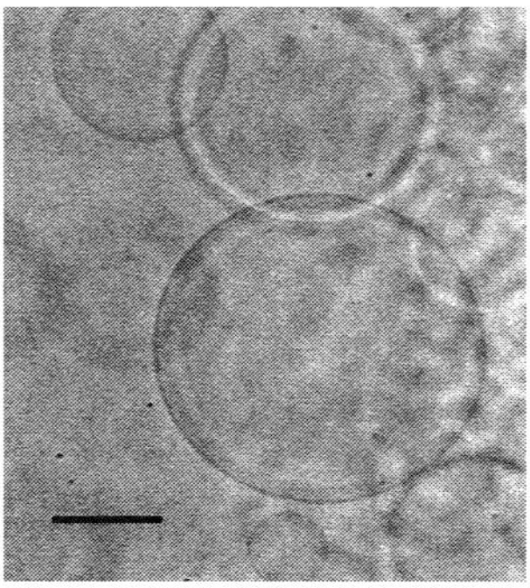

**Figure 3.2** Electroformed GUVs made from DMPC in distilled water at 45°C, bar = 30 μm.

the formation of a flacid, fluctuating GUV. If no particular AC regime is followed, GUVs are born stressed; it takes a few hours to get them relaxed and fluctuating. If well isolated GUVs are required, the suspension should be taken (gently!) out of the preparation chamber. Individual GUVs can be also picked up from the electrode with a sucking micropipette (thereby breaking the interconnecting filaments). Another efficient way of breaking the interconnecting filaments is to go from the liquid-crystalline to the gel state of the lipid (if possible).

Electroformed GUVs are ideal model objects for performing micromanipulation and *local* microinjection, thereby *creating localized and temporal membrane asymmetry and lateral gradients of the physico-chemical properties of the lipid membrane.*

## 2.3. Underlying Mechanisms

At least seven possible mechanisms for the effects of external electric field on the process of lipid swelling and vesicle formation have been suggested [9–13]:

(1) direct electrostatic interactions between electrode and bilayers;
(2) electroosmotically induced mechanical stresses;
(3) redistribution of double layer counterions between bilayers;
(4) decreased surface, membrane and line tension;
(5) electrochemical reactions;
(6) injection of charges from electrodes;
(7) reorientation and lateral redistribution of lipid molecules, inverse flexoelectric effect.

It should be noted that the level of speculation increases down the list. The exact mechanisms of liposome electroformation are still not entirely understood.

The 10 Hz electroformation protocol is quite universal. A sketch of the vesicle formation in this case is shown in Figure 3.3. The predominating mechanism of electroformation here is the electroosmotic periodic movement of the water medium at the water–electrode interface. These vibrations are directed perpendicular to the electrode surface (where the initial lipid film is deposited, Figure 3.3(a)) thus, pulling lipid lamellae off the electrode making so that they separate, growing as mushrooms (Figure 3.3(b)). The vesicles continuously increase in size up to, but not more than, 10–20 μm. (These are the vesicle sizes normally achieved by spontaneous formation from swelling lipid films.) At this stage the mushrooms begin to touch each other (Figure 3.3(c)). The zone of contact increases, and at a certain point the AC-induced vibrations cause the contact zone to be destabilized, making neighboring mushrooms fuse together into a large one (Figure 3.3(d)). It takes a few minutes for the resulting mushroom to become spherical, close its neck, and eventually separate from the electrode (Figure 3.3(e)). In fact, the electroosmotic vibrations are manifested as mechanical vibrations (like those in the case of sonication of lipid–water disper-

(a)
(b)
(c)
(d)
(e)

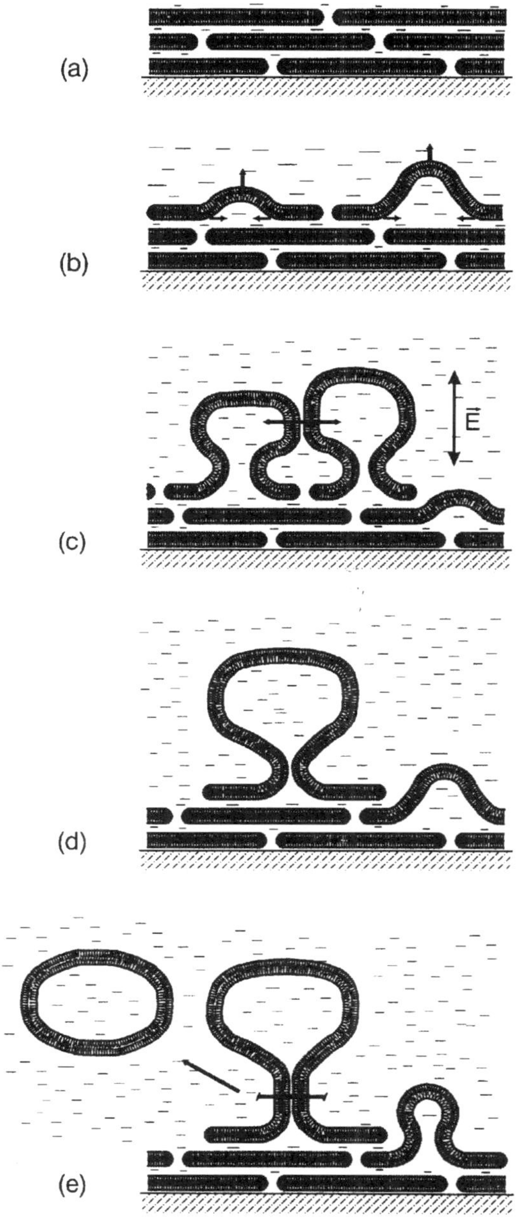


**Figure 3.3** A sketch of vesicle AC electroformation, see the text. $\vec{E}$ indicates the direction of the external electric field.

sions). However, electoosmotic agitation is much gentler and finer, and the process can be observed and precisely controlled.

## 3. RECENT APPLICATIONS OF GUV ELECTROFORMATION

One reason to develop liposome electroformation was to establish a method for the preparation of unilamellar cell-sized vesicles of a particular lipid composition, and thereby imitate biological cells with respect to their mechanical and electric properties. This could then be used for the direct optical microscopy model study of membrane interactions.

### 3.1. Bending elasticity measurements

The first application of liposome electroformation was in supplying the starting objects (isolated unilamellar vesicles of a particular lipid composition, size and membrane tension) for membrane bending elasticity measurements by image analysis of membrane thermal fluctuations [14,15,23,27]. See also Chapter 14 of this book.

The effects of external AC fields, when the system passes from a dielectric to conductive regime (in the particular case at 1 kHz, and 13 kHz) on the shape and thermal fluctuations of GUVs have been investigated [16].

### 3.2. Interactions of GUVs with colloidal particles – optical trapping studies

Membrane interactions with colloidal particles have been investigated by optical trapping and optical micromanipulation [17,18,20]. In these experiments, latex or glass spheres (2 – 20 μm in diameter) are manipulated by laser beams and brought into contact with GUVs. When doing so, it is possible to observe the fascinating phenomenon of physical endocytosis, in which the colloidal particle is engulfed by the GUV (made of SOPC, or DMPC) as a result of purely physicochemical mechanisms. These observations might possibly relate to events occurring in primitive living systems. One can assume that, at the beginning of life, a large variety of transports mechanisms through the primitive membrane (separating certain water space from the outer environment) was achieved by purely physicochemical mechanisms. Later, extremely complicated biochemical cell machinery was developed not only to ensure endocytic events, but also to prevent undesirable endocytic events, due to purely physicochemical mechanisms. Thus, *the primitive cell had to develop not only endocytosis ensuring, but endocytosis-preventing structures and mechanisms as well.*

Furthermore, small colloidal particles encapsulated by GUV membrane perform bidimensional Brownian motion on the membrane [17], and can be used as microviscosimeters for measuring the shear viscosity of lipid bilayers [21, 28]. See also Chapter 15 of this book.

### 3.3. Shape transformations of GUVs upon phase transitions of the constituting lipids

Electroformed GUVs have proved to be a unique system for visualizing the effects of lipid membrane phase transitions on a closed single bilayer. In addition, DMPC GUVs have been used in studies of interdigitated phase formation, induced by ethanol and temperature variations [22].

### 3.4. Morphology transformation induced by local microinjection of active substances to GUVs

One interesting application of electroformed GUVs is the study of the effect of active substances locally microinjected into GUVs. The resulting local membrane composition gradients induce local membrane mechanical and electrical property gradients. In some cases it is possible that membrane surface hydrodynamics phenomena take place as well. These experiments open a huge field of research concerning nonequilibrium phenomena in bidimensional lipid membranes (in the fluid or gel state). The biological relevance of such studies is unique bearing in mind the fact that important physiological processes (such as endo- and exocytosis, signal transduction, fertilization, cell defense, etc.) involve local delivery and interactions of biochemichally active substances with the cell membrane. Work in this field includes visualization of enzymatic reactions on GUVs (for example the injection of phospholipase $A_2$, phospholipase D [19,26] and sphingomyelinase) (see also Chapters 20 and 22). GUVs were applied for the first time as microreactors for the endogeneous production of membrane-constituting lipids [8,30].

The effects of adding DNA locally to cationic GUVs made of natural lipids have been investigated [29] and *DNA induced endo- and exocytosis* were observed. The phenomena depend on membrane composition, vesicle membrane tension, and DNA molecule length and conformation. The results support the idea that DNA–lipid interactions should be taken into account when considering problems of cell division and differentiation, as well as cell transfection.

## 4. CONCLUSION

The electroformation method is used in a number of laboratories for studies involving individual GUVs. In comparison with earlier methods for the preparation of giant vesicles, electroformation has advantages, leading to the formation of giant unilamellar vesicles from a variety of lipids.

## 5. ACKNOWLEDGMENTS

This work was supported by the French – Bulgarian Laboratory, Vesicles and Membranes, under the framework of international scientific cooperation between the C.N.R.S. (France) and the Bulgarian Academy of Sciences.

## 6. REFERENCES


1. A. D. Bangham, M. M. Standish and J. C. Watkins, *J. Mol. Biol.*, **13**, 238 (1965).
2. A. D. Bangham (ed.) *Liposome Letters*, Academic Press, London, 1983.
3. G. Gregoriadis (ed.) *Liposome Technology*, CRC Press, Boca Raton, 1984.
4. K. H. Schmidt (ed.) *Liposomes as Drug Carriers*, Georg Thieme Verlag, Stuttgart, 1986.
5. M. Rosoff (ed.) *Vesicles*, Marcel Dekker, New York, 1996.
6. D. D. Lasic and D. Papahadjopoulos (eds) *Medical Applications of Liposomes*, Elsevier Science B.V., Amsterdam, 1998.
7. F. M. Menger and M. I. Angelova, *Acc. Chem. Res.*, **31**, 789 (1998).
8. R. Wick and P. L. Luisi, *Chem. Biol.*, **3**, 277 (1996).
9. M. I. Angelova and D. S. Dimitrov, *Faraday Discuss. Chem. Soc.*, **81**, 303; 345 (1986).
10. M. I. Angelova and D. S. Dimitrov, *Mol. Cryst. Liq. Cryst.*, **152**, 89 (1987).
11. D. S. Dimitrov and M. I. Angelova, *Progr. Colloid Polym. Sci.*, **73**, 48 (1987).
12. D. S. Dimitrov and M. I. Angelova, *Bioelectrochem. Bioenerg.*, **19**, 323 (1988).
13. M. I. Angelova and D. S. Dimitrov, *Prog. Colloid. Polym. Sci.*, **76**, 59 (1988).
14. M. I. Angelova, S. Soléau, P. Méléard, J.-F. Faucon and P. Bothorel, *Progr. Colloid Polym. Sci.*, **89**, 127 (1992).
15. M. I. Angelova, S. Soléau, P. Méléard, J.-F. Faucon and P. Bothorel, *Springer Proc. Phys.*, **66**, 178 (1992).
16. M. D. Mitov, P. Méléard, M. Winterhalter, M. I. Angelova and P. Bothorel, *Phys. Rev. E*, **48**, 628 (1993).
17. M. I. Angelova, B. Pouligny, G. Martinot-Lagarde , G. Gréhan and G. Gouesbet, *Progr. Colloid Polym. Sci.*, **97**, 293 (1994).
18. B. Pouligny, G. Martinot-Lagarde and M. I. Angelova, *Progr. Colloid Polym. Sci.*, **98**, 280 (1995).
19. R. Wick, M. I. Angelova, P. Walde and P. L. Luisi, *Chem. Biol.*, **3**, 105 (1996).
20. C. Dietrich, M. I. Angelova and B. Pouligny, *J. Phys. II France*, **7**, 1651 (1997).
21. K. Velikov, K. Danov, M. I. Angelova, C. Dietrich and B. Pouligny, *Coll. Surf. A*, **149**, 245 (1999).
22. M. I. Angelova, R. Mutafchieva, R. Dimova and B. Tenchov, *Coll. Surf. A*, **149**, 201, (1999).
23. L. Fernandez-Puente, I. Bivas, M. D. Mitov and P. Méléard, *Europhys. Lett.*, **28**, 181 (1994).
24. M. A. Guedeau-Boudeville, L. Jullien and J.-M. di Meglio, *Proc. Natl. Acad. Sci. USA*, **92**, 9590 (1995).
25. L. Mathivet, S. Cribier and P. Devaux, *Biophys. J.*, **70**, 1112 (1996).
26. V. Dorovska-Taran, R. Wick and P. Walde, *Anal. Biochem.*, **240**, 37 (1996).
27. P. Méléard, G. Gerbeaud, T. Pott, L. Fernandez-Puente, I. Bivas, M. D. Mitov, J. Dufourcq and P. Bothorel, *Biophys. J.*, **72**, 2616 (1997).
28. K. Velikov, C. Dietrich, A. Hadjiisky, K. Danov and B. Pouligny, *Europhys. Lett.*, **40**, 405 (1997).
29. M. I. Angelova, N. Hristova and I. Tsoneva, Eur. *Biophys. J.*, **28** (2), 142 (1999).
30. P. Bucher, A. Fischer, P. L. Luisi, T. Oberholzer and P. Walde, *Langmuir*, **14**, 2712 (1998).

# Chapter 4

# Formation of Giant Vesicles from Different Kinds of Lipids using the Electroformation Method

*Aline Fischer, Pier Luigi Luisi, Thomas Oberholzer, and Peter Walde*
*ETH-Zürich, Switzerland*

## 1. INTRODUCTION

Since the late 1960s, work with giant vesicles (diameter >1 μm) has aroused increasing interest. A number of methods for the preparation of giant vesicles are described in literature. Reeves and Dowben [1] were among the first to describe the controlled formation of giant vesicles. Their method, the gentle hydration method, consists of the gentle swelling of a lipid film in water or in an aqueous solution of nonelectrolytes. By this approach, they obtained vesicles with diameters up to 10 μm. Deamer and Bangham [2] introduced the ether evaporation method, which involves the injection of an etheric solution of lipids into a hot aqueous solution. Szoka and Paphadjopoulos [3] described a similar method, known as reverse-phase evaporation. In this method the lipids are dissolved in an organic solvent and an aqueous solution is added. After sonication, giant vesicles are obtained by evaporation of the organic solvent. Giant vesicles are also obtained by dialysis [4,5], freeze-and-thaw [6,7] or titration [8].

The method used in this work is the so-called electroformation method, originally developed by Angelova and Dimitrov [9]. This method is of particular interest because it allows the preparation of giant vesicles with diameters around 100 μm, which can easily be used for microinjection experiments. The vesicles grow on a platinum wire under the influence of an alternating or direct electric field.

*Giant Vesicles* Edited by P. L. Luisi and P. Walde

Until recently the electroformation method has been used mainly with POPC (1-palmitoyl-2-oleoyl-*sn*-glycero-3-phosphocholine). The aim of the work described here is to enlarge the field of this application to other kinds of lipids, such as phosphatidylglycerol, phosphatidylethanolamine, phosphatidylserine, didodecyldimethylammonium bromide, and phospholiponucleosides. The formation of giant vesicles with phospholiponucleosides is of particular interest, as their potential molecular recognition properties are currently being investigated with conventional vesicles having diameters below 1 μm [10–15].

## 2. MATERIALS AND METHODS

### 2.1 Chemicals

The materials and sources used are as follows. 1-Palmitoyl-2-oleoyl-*sn*-glycero-3-phosphocholine (POPC) and 1-palmitoyl-2-oleoyl-*sn*-glycero-3 [phospho-rac-(1-glycerol)] (POPG) were from Avanti Polar Lipids, USA. Bovine brain phosphatidylserine (PS) was from SERVA, Germany. 5′-(1-Palmitoyl-2-oleoyl-*sn*-glycero-3-phospho)-uridine (POP-U), 5′-(1-palmitoyl-2-oleoyl-*sn*-glycero-3-phospho)-adenosine (POP-A), 5′-(1,2-dioleoyl-*sn*-glycero-3-phospho)-uridine (DOP-U), 5′-(1,2-dioleoyl-*sn*-glycero-3-phospho)-cytidine (DOP-C), 5′-(1,2-dioleoyl-*sn*-glycero-3-phospho)-adenosine (DOP-A), 5′-(1,2-dioleoyl-*sn*-glycero-3-phospho)-inosine (DOP-I) were synthesized by Baglioni and co-workers [10]. 5′-(*n*-Hexadecylphospho)-uridine (HDP-U) and 5′-(*n*-hexadecylphospho)-cytidine (HDP-C) were synthesized by Heiz *et al.* [14]. *N*-(TexasRed® sulfonyl)-1,2-dioleoyl-*sn*-glycero-3-phosphoethanolamine triethylammonium salt (TexasRed-DOPE, product T-647) [17] and YO-PRO™-1 iodide (YO-PRO-1, product Y-3603) [18] were obtained from Molecular Probes, USA. Didodecyldimethylammonium bromide (DDAB) was purchased from Fluka, Switzerland; 1,2-dipalmitoyl-*rac*-glycero-3-phosphoethanolamine (DPPE) was purchased from Sigma, USA.

### 2.2 Preparation of Giant Vesicles

Giant vesicles were prepared by electroformation. The investigation chamber was constructed at our institute. It consisted of two platinum wires (99.99 % from Möller AG, Switzerland) fixed parallel in a poly(vinylidene fluoride) casing. The bottom of the chamber contained a 0.5 mm piece of borosilicate glass (from Guinchard SA, Yverdon, Switzerland) glued to the casing. The top of the chamber was open to enable micromanipulation.

The lipids were dissolved in methanol:diethylether (1:9, v/v) at a concentration of 0.2 mg ml$^{-1}$. A drop of approximately 3 μl was carefully deposited in two different areas of each wire with a Hamilton syringe. The lipid film was then dried under a nitrogen stream for about one minute and afterwards dried under vacuum in a desiccator for at least 15 h. The investigation chamber was then put on the stage of the light microscope (Axiovert 135 TV from Zeiss, Germany) and connected to the frequency generator (Conrad Electronic, Germany) generating an AC field. About 1 ml of water was added to the chamber and the growing vesicles were directly observed by light microscopy.

## 3. RESULTS AND DISCUSSION

From earlier work [9] it is known that electroformation can be used to prepare giant vesicles from several lipids and lipid mixtures: L-α-phosphatidylcholine [9e], 1,2-dimyristoyl-*sn*-glycero-3-phosphocholine (DMPC) [9a,b,c], bovine phosphatidylserine [9e], and digalactosyl diacylglyceride (DGDG) [9e]; negatively charged mixtures containing phosphatidylcholine (71 %), phosphatidylethanolamine (21.5 %) and phosphatidylserine (7.5 %) [9a–e]; positively charged mixtures composed of dodecyl amine and L-α-phosphatidylcholine (1:10 and 1:100, molar ratios) [9b,c] and neutral mixtures of egg phosphatidylethanolamine and egg phosphatidylcholine [9e] or egg phosphatidylethanolamine and egg lysophosphatidylethanolamine [9e]; the use of mixtures of egg phosphatidylcholine and cholesterol is also known [9d,e].

In this work experimental conditions were sought for the preparation of giant vesicles with several other lipids. The results are summarized in Table 1 and are discussed below. Giant vesicle formation with negatively charged POPG was much more difficult than with POPC. Pure POPG vesicles formed rapidly after addition of water but were not very stable. After 1–2 h only small vesicles with diameters up to 40 μm could be observed. The number of giant vesicles was much smaller than in the case of POPC and the vesicles often contained additional smaller vesicles in their aqueous interior. When POPG was mixed with POPC (1:4) vesicle formation was much more successful.

In the literature the formation of giant vesicles from PS [9e] is described, but our studies indicated that giant vesicles do not form from pure PS, at least under the conditions of 10 Hz and 8 V, using bovine brain PS.

Mixtures of POPC and DPPE or TexasRed-DOPE formed giant vesicles in the same way as pure POPC; this is confirmed by earlier observations [19].

Positively charged mixtures of DDAB and POPC led to adhesion to neighboring vesicles (Figure 4.1A). It seems that the vesicles share the membrane over a certain contact area. The number of these contact areas increases with the amount of DDAB.

Vesicle formation with pure phosphonucleosides was difficult. With POP-U and DOP-U only a few very small vesicles formed, whereas with POP-A, DOP-A, DOP-C, and DOP-I, vesicles were not formed. However when the phosphonucleosides were mixed with POPC (1:1 for POP-U and DOP-U or 1:4 for POP-A and DOP-I) vesicle formation was successful and the vesicles could also be stained with YO-PRO-1 (Figure 4.1(B–E)). YO-PRO-1 is a cationic cyanine dye containing benzox-

**Table 1** Preparation of giant vesicles by the electroformation method using other lipids than POPC

| Lipid mixture | Frequency[a] (Hz) | Field[a] (V) | Observations |
|---|---|---|---|
| POPG | 10 | 8 or 17 | directly after the addition of water, mushrooms[b] with diameters up to 90 μm formed; after 1–2 h only spheres[c] with diameters around 40 μm could be observed |
| POPG:POPC (1 : 4, wt %) | 10 | 5 | within 1 h vesicles with diameters up to 80 μm formed |
| Bovine brain PS | 10 | 8 | locally a swelling of the lipid film could be observed, but no vesicles formed |
| DPPE:POPC (1 : 400, wt %) | 10 | 2 | giant vesicles formed in the same way as with pure POPC |
| TexasRed-DOPE : POPC (1:50, wt %) | 10 | 2 | giant vesicles with diameters not larger than 50 μm formed |
| DDAB:POPC (10 : 90, wt %) | 10 | 2 | a lot of vesicles shared membranes with neighboring vesicles, the vesicles were not always well formed |
| DDAB:POPC (2.5 : 97.5, wt %) | 10 | 2 | a few vesicles shared membranes with neighboring vesicles; the shape of the vesicles was the same as with pure POPC |
| POP-U | 10 | 8 | within 2 h only very few vesicles with diameters up to 20 μm formed |
| POP-U:POPC (1:1, molar ratio) | 10 | 2–10 | Within 1–2 h vesicles with diameters up to 100 μm formed |
| POP-A:POPC (1:1, molar ratio) | 10 | 8 | vesicle formation was not successful |
| POP-A:POPC (1:4, molar ratio) | 10 | 3 or 8 | within 2 h a few giant vesicles formed |
| DOP-U | 10 | 8 | within 2 h only very few vesicles with diameters up to 20 μm formed |
| DOP-U:POPC (1:1, molar ratio) | 10 | 16 | within 2 h a few giant vesicles formed |
| DOP-A | 10 | 2.5 | vesicle formation was not successful |
| DOP-C | 10 | 2.5 | vesicle formation was not successful |
| DOP-I:POPC (1:1, molar ratio) | 10 | 5 | vesicle formation was not successful |

**Table 1** (*continued*)

| Lipid mixture | Frequency[a] (Hz) | Field[a] (V) | Observations |
|---|---|---|---|
| DOP-I:POPC (1:4, molar ratio) | 10 | 2 | vesicle formation was successful, giant vesicles were formed |
| HDP-C | 10 | 4 | vesicle formation was not successful |
| HDP-C:POPC (1:1, molar ratio) | 10 | 4 | after a few minutes vesicles, comparable to pure POPC vesicles, formed; several minutes later, only different forms of aggregates could be observed |
| HDP-U:POPC (1:1, molar ratio) | 10 | 4 | giant vesicles formed quickly, followed by reorganization to different forms of aggregates |

[a]The values for the frequency and voltage are values for which vesicle formation was best.
[b]The term mushroom is used for open vesicles that look like mushrooms [20].
[c]The term sphere is only used for closed vesicles.

azole and a quinoline ring and bearing two positive charges. It is essentially nonfluorescent but becomes green fluorescent when bound to DNA or RNA [18]. The fact that vesicles containing phosphonucleosides were stained by YO-PRO-1 and pure POPC vesicles were not, clearly shows (i) that the membrane of the formed vesicles contained the phosphonucleosides used and (ii) that YO-PRO-1 can also bind to single nucleosides. With pure HDP-C or HDP-U, vesicle formation was not successful. Giant vesicles were only formed on mixing with POPC (1:1); they were stained with YO-PRO-1. The vesicles however, were not stable. They became quickly transformed into giant vesicles that contained some kinds of aggregates, still stained by YO-PRO-1.

## 4. CONCLUSIONS

The results show that vesicle formation by electroformation for pure lipids other than POPC (or other phosphatidylcholines) is difficult. Vesicles can only form when the lipids are mixed with POPC. It seems that small changes in the chemical structure of the lipid strongly influence the lipid swelling in the electric field and the whole electroosmosis-supported formation process.

Among the phospholipids used, the phospholiponucleosides are of particular interest because their potential molecular recognition properties are currently being investigated with small vesicles having diameters below 1 μm [10–15]. The

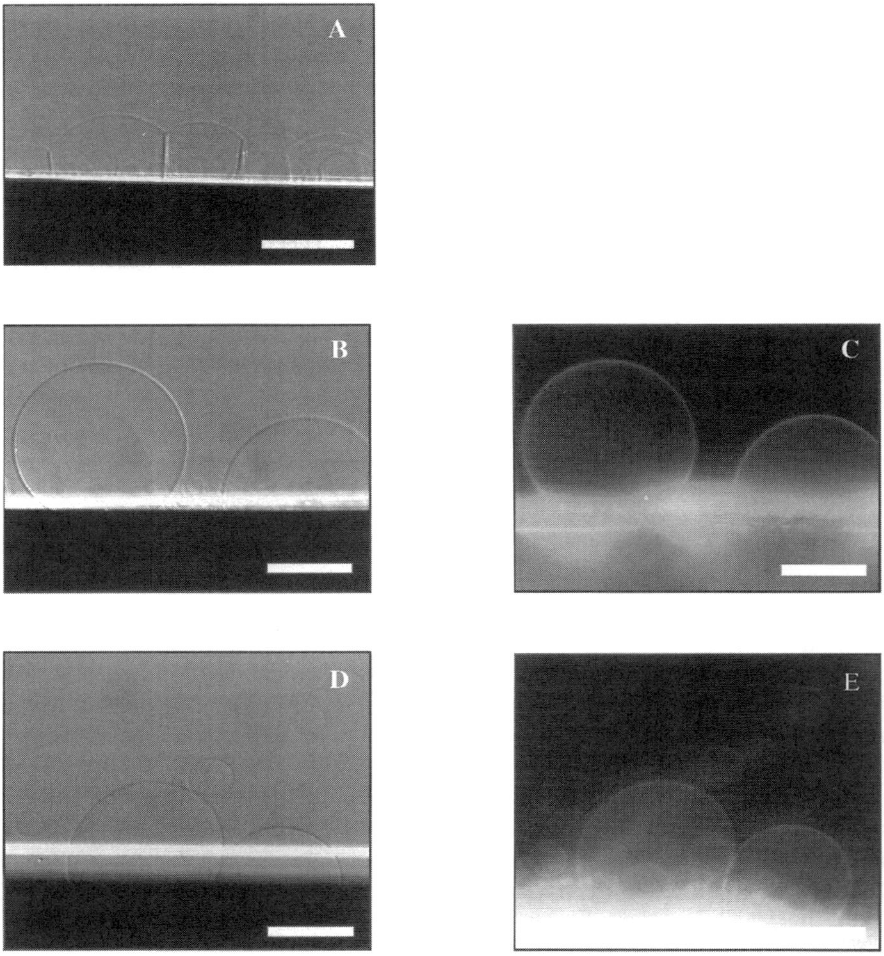


**Figure 4.1** Giant vesicles from different kinds of lipids, prepared by the electroformation method.
A: DDAB: POPC (1 : 9, wt%);
B, C: POP-U: POPC (1 : 1, molar ratio);
D, E: DOP-I: POPC (1 : 4, molar ratio);
A, B, D: Differential interference contrast;
C, E: fluorescence mode. Length of the bar: 50$\mu$m.

possibility of forming giant vesicles with phospholiponucleosides could enable visualization, by light microscopy, of the molecular recognition between two vesicles bearing two complementary bases.

## 5. REFERENCES


1. J. P. Reeves and R. M. Dowben, *J. Cell. Physiol.*, **73**, 49–60 (1969).
2. D. W. Deamer and A. D. Bangham, *Biochim. Biophys. Acta*, **443**, 629–634 (1976).
3. F. Szoka and D. Paphadjopoulos, *Ann. Rev. Biophys. Bioeng.*, **9**, 467–508 (1980).
4. N. Oku, J. F. Scheerer and R. C. MacDonald, *Biochim. Biophys. Acta*, **692**, 384–388 (1982).
5. N. Oku and R. C. MacDonald, *Biochim. Biophys. Acta*, **734**, 54–61 (1983).
6. N. Oku and R. C. MacDonald, *Biochemistry*, **22**, 855–863 (1983).
7. K. Higashi, S. Suzuki, H. Fujii and Y. Kirino, *J. Biochem.*, **101**, 433–440 (1987).
8. W. R. Hargreaves and D. W. Deamer, *Biochemistry*, **17**, 3759–3763 (1978).
9. (a) M. I. Angelova and D. S. Dimitrov, *Faraday Discuss. Chem. Soc.*, **81**, 303–311 (1986); (b) D. S. Dimitrov and M. I. Angelova, *Progr. Colloid Polym. Sci.*, **73**, 48–56 (1987); (c) D. S. Dimitrov and M. I. Angelova, *Bioelectrochem. Bioenerg.*, **19**, 323–336 (1988); (d) M. I. Angelova and D. S. Dimitrov, *Progr. Colloid Polym. Sci*, **76**, 59–67 (1988); (e) M. I. Angelova, S. Soléau, P. Méléard, J. F. Faucon and P. Bothorel, *Progr. Colloid Polym. Sci.*, **89**, 127–131 (1992).
10. D. Berti, P. Baglioni, S. Bonacchio, G. Barsacchi-Bo and P. L. Luisi, *J. Phys. Chem. B*, **102**, 303–308 (1998).
11. S. Bonaccio, P. Walde and P. L. Luisi, *J. Phys. Chem.*, **98**, 6661–6663 (1994).
12. S. Bonaccio, M. Wessiken, D. Berti, P. Walde and P. L. Luisi, *Langmuir*, **12**, 4976–4978 (1996).
13. S. Bonaccio, D. Capitani, A. L. Segre, P. Walde and P. L. Luisi, *Langmuir*, **13**, 1952–1956 (1997).
14. C. Heiz, U. Rädler and P. L. Luisi, *J. Phys. Chem. B*, **102**, 8686–8691 (1998).
15. U. Rädler, C. Heiz, P. L. Luisi and R. Tampé, *Langmuir*, **14**, 6620–6624 (1998).
16. D. Berti, L. Franchi, P. Baglioni and P. L. Luisi, *Langmuir*, **13**, 3438–3444 (1997).
17. R. P. Haugland, *Handbook of Fluorescent Probes and Research Chemicals*, 5th edn, Molecular Probes: Eugene, OR, 1992.
18. R. P. Haugland, *Handbook of Fluorescent Probes and Research Chemicals*, 6th edn, Molecular Probes: Eugene, OR, 1996.
19. V. Dorovska-Taran, R. Wick and P. Walde, *Anal. Biochem.*, **240**, 37–47 (1996).
20. P. Bucher, A. Fischer, P. L. Luisi, T. Oberholzer and P. Walde, *Langmuir*, **14**, 2712–2721 (1998).

# Chapter 5

# Observations of a Variety of Giant Vesicles under an Optical Microscope

*KEN-ICHIROU AKASHI AND KAZUHIKO KINOSITA, JR.*

*Keio University, Yokohama, Japan; Core Research for Evolutional Science and Technology, Genetic Programming Team 13, Kawasaki, Japan*

*HIDETAKE MIYATA*

*Tohoku University, Sendai, Japan*

*HIROYASU ITOH*

*Tsukuba Research Laboratory, Hamamatsu Photonics K.K., Tsukuba, Japan; Core Research for Evolutional Science and Technology, Genetic Programming Team 13, Kawasaki, Japan*

## 1. INTRODUCTION

Giant vesicles are useful in the study of dynamics of lipid membranes. For example, the shape changes of spherical vesicles, such as formation of projections, budding, or invagination, which resemble cell deformation or endocytosis [1,2], have been observed by light microscopy. It has been found that the yield of unilamellar giant vesicles greatly increases when the lipid membranes are electrically charged. This is because the electrostatic repulsion prevents the membranes from adhering to one another. Under these favorable conditions, many nonspherical vesicles having branched projections or concavities are also formed. In multilamellar vesicles, the

*Giant Vesicles* Edited by P. L. Luisi and P. Walde

lamellar spacing becomes large, and then complicated structures, such as membranous pores (similar to fusion pores), become distinguishable by light microscopy.

## 2. MATERIALS AND METHODS

A detailed protocol for the preparation of giant vesicles has been described elsewhere [3,4]. Briefly, electrically neutral phosphatidylcholine (PC) was used as the major component, with negatively charged phosphatidylglycerol (PG); both were obtained from Avanti Polar Lipids Inc. (Alabaster, AL, USA). Each of these lipids was dissolved at 7.5 mg $ml^{-1}$ in chloroform : methanol (2 : 1 by volume) and mixed at an appropriate ratio (see below). An 80 μl aliquot of the lipid solution was dried at 45°C with a rotary evaporator to form a thin film on the bottom of a test tube. The film was completely dried in vacuo and then prehydrated at 45°C with water-saturated $N_2$ gas. About 5 ml of an aqueous solution containing 0.1 M sucrose and appropriate salts (internal solution) was gently added to the tube. The tube was incubated at 37°C for about 2 h, and then gently rocked to disperse the lipid film uniformly in the solution. (It was found that giant unilamellar vesicles had already formed at this stage.) After further overnight incubation at room temperature, an almost transparent white cloud containing giant vesicles formed in the tube.

The yield of giant unilamellar vesicles is highly dependent upon the lipid compositions and ionic conditions of the aqueous solution. With pure PC, vesicle formation is promoted in a solution of divalent cations, such as $Ca^{2+}$, at 1 – 30 mM, compared to low yields in nonelectrolyte or monovalent cation solutions. If PG was added to PC (PC : PG by mole ratio 9 : 1), high yields are obtained even in physiological ionic conditions.

The vesicles were observed on an inverted microscope in phase-contrast and confocal fluorescence modes, as follows. The white cloud of lipid was gently dispersed in the tube and introduced into an observation chamber. The chamber was filled with the same solution as the internal solution except that 0.1 M sucrose was replaced with 0.1 M glucose (external solution). Because of the difference in the refractivity of the internal and external solutions, the contrast of the vesicle's images was enhanced in the phase-contrast mode. For the confocal fluorescence microscopy, a confocal scanner unit (CSU 10, Yokogawa, Japan) was used, and the lipophilic fluorescent dye Nile Red (Molecular Probes, Inc., OR, USA) was added to the starting lipid solution at 0.3 wt% of the lipid.

## 3. RESULTS AND DISCUSSION

In a successful preparation, giant unilamellar vesicles (diameter >20 μm) were present at several hundreds per 50 μl aliquot of vesicle suspension (0.12 mg lipid per mililiter) [4]. Although most of the vesicles appeared spherical or ellipsoidal, vesicles with irregular shapes were also observed.

Figure 5.1 shows the images of vesicles having tube-like projections, (A) and (B), or concavities (C) and (D). The tube-like structures had a round tip and their positions changed during observation, indicating that they were not pulled out by any external force. The vesicles having concavities were isolated from surrounding objects (confirmed by changing the focus), indicating that these vesicles maintained their unique shapes by themselves. The vesicles with projections or concavities consisted of a single continuous membrane, which vigorously fluctuated (undulated). In spite of the fluctuation, the overall shapes did not change, unless a mechanical perturbation such as convection in the chamber occurred. These results indicate that the shapes observed here were dynamically stable (or at least metastable) for single membrane vesicles. In this regard, it is believed that intermembrane repulsion (electrostatic in the present case) contributes to the stability of these irregular shapes. Without the repulsion, inherent attractive interactions (such as van der Waals forces) would cause adhesion between adjacent membranes within a vesicle during fluctuation, leading to the destruction of tubes or concavities consisting of single bilayers.

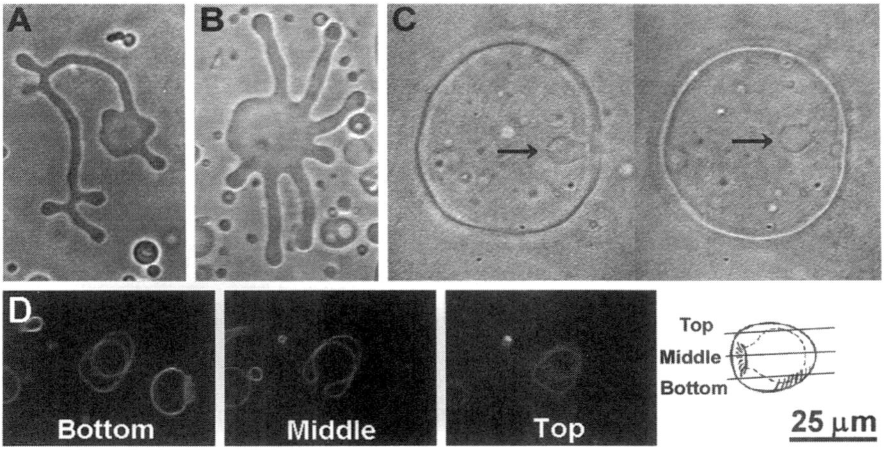


**Figure 5.1** (A–C) Phase-contrast and (D) confocal images of nonspherical vesicles. The lipid composition and ionic conditions were: (A), 1-palmitoyl-2-oleoyl-*sn*-glycero-3-phosphocholine (POPC) and 1-palmitoyl-2-oleoyl-*sn*-3-[phospho-rac-(1-glycerol)] (POPG) (9 : 1 by weight) without salt; (B) and (D), 1,2-dioleoyl-*sn*-glycero-3-phosphocholine (DOPC) in 3 mM $CaCl_2$; (C) egg-PC and egg-PG (9 : 1) in 10 mM KCl. In (A) and (B), the inside of the vesicles appears dark because of the difference in the refractivity of the internal and external solutions (see text). In (C), the internal and external solutions were the same. The arrows indicate concavities. (D) shows sections of a stomatocyte, which is schematically illustrated on the right-hand side. Temperature, $22 \pm 2°C$.

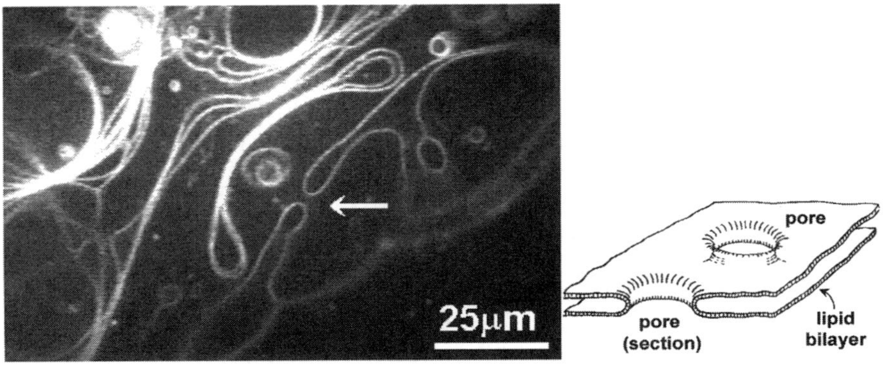


**Figure 5.2** A confocal light microscopic image of a part of a multilamellar vesicle. The lipid was POPC, and the ionic condition was 3 mM $CaCl_2$. The arrow indicates a membranous pore, which is schematically illustrated on the right-hand side. Temperature, $22 \pm 2°C$.

Figure 5.2 shows the inside structure of a multilamellar vesicle observed under a confocal fluorescence microscope. Examination of sections at different focul levels revealed a structure consisting of two membranes linked to each other with a cylindrical membrane (see the area pointed out by the arrow, and the schematic drawing in Figure 5.2). Because the lamellar spacings were large enough to distinguish each of the lamellae, as a result of the electrostatic repulsion, the complicated structure was more clearly resolved than in a previous study [5]. This linked structure shown here may represent an intermediate leading to the formation of vesicles of nonspherical topology.

In conclusion, the introduction of electrostatic repulsion between lipid membranes resulted in the formation of giant vesicles with complicated structures and enabled the observation of the structural details under an optical microscope. This strategy is useful to the investigation of the dynamics and structure of lipid membrane vesicles.

## 4. REFERENCES

1. H. Hotani, *J. Mol. Biol.*, **178**, 113 – 120 (1984).
2. J. Käs and E. Sackmann, *Biophys. J.*, **60**, 825 – 844 (1991).
3. K. Akashi, H. Miyata, H. Itoh and K. Kinosita, Jr., *Biophys. J.*, **71**, 3242 – 3250 (1996).
4. K. Akashi, H. Miyata, H. Itoh and K. Kinosita, Jr., *Biophys. J.*, **74**, 2973 – 2982 (1998)
5. W. Harbich, R. M. Servuss and W. Helfrich, *Z. Naturforsch.*, **33a**, 1013 – 1017 (1978).

# Part Three
# Basic Theoretical Aspects

# Chapter 6

# Bending Elasticity of Fluid Membranes

*WOLFGANG HELFRICH*

*Freie Universität, Berlin, Germany*

## 1. INTRODUCTION

Bending elasticity is a long-standing concept of continuum mechanics and has been used mostly to deal with solid rods and plates. More recently, it has been applied to fluid membranes, especially the lipid bilayers of giant vesicles, to understand their equilibrium shapes and shape fluctuations. For continuum theory to be applicable, the membranes should be reasonably smooth or, in other words, not fluctuate too much.

Strictly speaking, fluid membranes such as amphiphilic monolayers and bilayers are continua only in the two lateral dimensions, whereas in the normal direction their extension is molecular. The simplest model of them is a mathematical surface without any thickness. However, thin as it is, the third dimension of the membranes determines their elastic properties of stretching and bending.

The bending elasticity of fluid membranes is closely related to the director field elasticity of liquid crystals. Of the three elastic deformations in nematics, which are splay, bend, and twist, only splay remains as it does in the case of smectics. In fact, a membrane is like an isolated smectic layer and this is why membrane curvature is sometimes expressed in terms of splay and saddle splay.

The following is a brief introduction to fluid membrane bending elasticity. The emphasis is on some basic ideas and not on particular models or applications. The theory of vesicle shapes is treated in Chapter 7. A subject to be included in the following is non-Hookean bending elasticity, i.e. energy terms of higher than

*Giant Vesicles* Edited by P. L. Luisi and P. Walde

quadratic order in the principal curvatures or dependent on the derivatives of curvatures.

Higher-order bending elasticity comes into play only at very high or rapidly changing curvatures where continuum theories are on the border of their validity. Nevertheless, such theories seem to supply plausible explanations for the dispersive phase and other types of superstructure found with some lipid membranes. Higher order bending elasticity may also be instrumental in the separation of single lipid membranes from multilayer systems. Recent studies of these puzzling phenomena are described in Chapters 17 and 18.

## 2. HOOKEAN ELASTICITY

In order to define the curvatures of a membrane, it is convenient and often sufficient to think of a mathematical surface. In a purely formal way, a formula for the Hookean bending energy of such a surface is derived, after finding suitable expressions that are linear or quadratic in the principal curvatures. These preparations are followed by three-dimensional descriptions of monolayers and bilayers. Subsequently, stress profiles and thermal undulations are discussed in terms of Hookean bending elasticity.

### 2.1 Mathematical surfaces

A mathematical surface is a function of two coordinates. In the so-called Monge representation, it is simply expressed by $z = z(x, y)$, with $x$, $y$, and $z$ being Cartesian coordinates. A more general description of a surface would be $\vec{r} = \vec{r}(x, y)$ with $\vec{r}$ being an arbitrary position in three-dimensional space. The surfaces are assumed to be continuous and differentiable, which implies that they are free of breaks and abrupt bends.

Consider a particular point on such a surface and its tangent plane. In a vicinity of the point, the surface can be approximated by a paraboloid. For a convenient description, use local Cartesian coordinates $\xi$, $\eta$, and $\zeta$ whose origin is at the point and whose $\xi$, $\eta$ plane coincides with the tangent plane. The paraboloid is the second-order expansion of the surface function $\zeta = \zeta(\xi, \eta)$. The first derivatives vanish by definition in this local Monge representation.

For clarity of notation, we replace $\xi$ and $\eta$ by $x$ and $y$ and express the height of the surface above the tangent plane by $u(x, y)$. Accordingly, the second-order expansion takes the form

$$u = \tfrac{1}{2} u_{xx} x^2 + u_{xy} xy + \tfrac{1}{2} u_{yy} y^2 \tag{1}$$

where the subscripts denote derivatives. This can be rewritten as

$$u = -\tfrac{1}{2} \vec{\rho} \cdot \vec{\vec{\Omega}} \cdot \vec{\rho}$$

with the radius vector

$$\vec{\rho} = (x, y)$$

and the curvature tensor

$$\vec{\vec{\Omega}} = -\begin{pmatrix} u_{xx} & u_{xy} \\ u_{xy} & u_{yy} \end{pmatrix} \tag{2}$$

The direction of the $u$ axis and the sign of $\vec{\vec{\Omega}}$ are matters of convention. Let us stipulate that on a sphere $u$ points outwards so that $u_{xx} = u_{yy} < 0$.

The components of the curvature tensor depend in general on the orientation of the $x$, $y$ coordinate axes in the tangent plane. The mixed derivatives $u_{xy}$ vanish when the coordinate axes coincide with the directions of the principal curvatures $c_1 = 1/R_1$, and $c_2 = 1/R_2$, where $R_1$ and $R_2$ are the associated radii. In these special cases, separated from each other by 90° rotations, the principal curvatures are $u_{xx}$ and $u_{yy}$. However, there are linear and quadratic expressions in the components of $\vec{\vec{\Omega}}$ that are scalar invariants under rotations of the $xy$ cross. They are

$$\mathrm{Tr}\,\vec{\vec{\Omega}} = c_1 + c_2 \tag{3}$$

$$(\mathrm{Tr}\,\vec{\vec{\Omega}})^2 = (c_1 + c_2)^2 \tag{4}$$

and

$$\det \vec{\vec{\Omega}} = \tfrac{1}{2}[(\mathrm{Tr}\,\vec{\vec{\Omega}})^2 - \mathrm{Tr}\,(\vec{\vec{\Omega}})^2] = c_1 c_1. \tag{5}$$

The simple forms on the right-hand sides are obtained directly in the special cases where $u_{xy} = 0$. Note that $c_1 + c_2$ is twice the mean curvature and $c_1 c_2$ is the Gaussian curvature in differential geometry. In liquid crystal terminology, $c_1 + c_2$ is splay and $-c_1 c_2$ is saddle splay. The sign of the curvatures has been defined such that $c_1 + c_2$ is positive for a sphere or cylinder.

The invariants are needed to express the bending energy of membranes. This is because the bending energy density of a fluid, laterally isotropic membrane can consist only of terms that do not depend on the orientation of the $xy$ cross. Multiplying each of them with a coefficient leads to one of the standard formulas for the bending energy surface density,

$$g_\mathrm{b} = \tfrac{1}{2}\kappa(c_1 + c_1)^2 - \kappa c_0(c_1 + c_2) + \bar{\kappa} c_1 c_2 \tag{6}$$

where $\kappa$ is the bending rigidity and $\bar{\kappa}$ is the elastic modulus of Gaussian curvature. The factor $c_0$ in the linear term is the spontaneous curvature of the membrane. Its meaning becomes obvious from another standard form,

$$g_\mathrm{b} = \tfrac{1}{2}\kappa(c_1 + c_2 - c_0)^2 + \bar{\kappa} c_1 c_2 \tag{7}$$

which differs from eq. (6) only by a constant.

Both formulas for the bending energy density are regarded as expansions up to quadratic order about the *flat* state of the membrane. This elasticity is considered to be Hookean, despite the possible presence of a linear term. Only for purely cylindrical curvature (e.g. $c_2 \equiv 0$) would the first term of eq. (7) suffice to describe an expansion about a spontaneously curved state $c_1 = c_0$. However, both $\kappa$ and $c_0$ could depend on the origin of the expansion.

For those unfamiliar with tensors and traces there is a shortcut to eqs (6) and (7). Accepting the existence of two principal curvatures along orthogonal directions, one can argue that any expression that does not change when $c_1$ and $c_2$ are interchanged is a suitable invariant. In fact, $c_1 + c_2$, $(c_1 + c_2)^2$ and $c_1 c_2$ satisfy this requirement. There are two additional quadratic invariants, $(c_1 - c_2)^2$ and $c_1^2 + c_2^2$, which have been used by some authors. They may be preferable in certain cases, but are not independent of the standard set, $(c_1 + c_2)^2$ and $c_1 c_2$. (A general procedure of finding the invariants, well-known from liquid crystals [4], starts with a different elastic modulus for each contributing term, $u_{xx}^2$, $u_{xx}u_{xy}$, etc. in the quadratic case, and reduces the number of independent moduli by equating the energies for different rotation angles of the $xy$ cross.)

## 2.2 Monolayers

Now consider physical membranes with a finite thickness, beginning with amphiphilic monolayers that occur in microemulsions at the boundaries of oil and water. Their up – down asymmetry is usually associated with a spontaneous curvature. A schematic cross-section including the sign conventions is given in Figure 6.1.

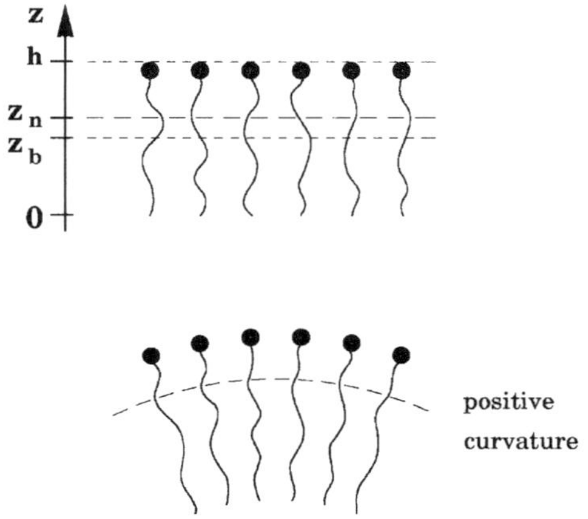


**Figure 6.1** Schematic cross-section through monolayer of amphiphilic molecules consisting of polar head (solid dot) and hydrocarbon chain. (PCs, PEs and DGDGs are two-chain molecules with zwitterionic polar heads.) The thickness of the monolayer $h$, as well as the positions of the neutral surface $z_n$, and the actual bending surface, $z_b$ are indicated in the flat state for $c_0 < 0$ (see text). The sign of curvature is defined by the lower diagram.

The curvatures of a monolayer have to be defined in terms of a reference surface that will be placed at some depth within the layer. Although the position may not be of practical interest, it is crucial in theoretical models. A frequent choice is the so-called neutral surface for which bending and stretching are energetically decoupled membrane deformations. Although the neutral surface is a useful concept in model calculations, it is in general not the surface of equal molecular density in nonuniformly curved monolayers, as will be seen below.

The theory of plate bending shows that the elasticities of bending and stretching are closely related because both derive from the bulk elasticity of the plate [2]. In principle, this theory can be adapted to fluid membranes so that it accounts for their elasticities of lateral and transverse dilations and of molecular tilt. The elastic modulus of lateral shear is made zero to obtain fluidity.

Only a simplified version of the fluid plate model is presented here to establish a crude relationship between the stretching modulus and the bending rigidity of fluid monolayers. The resulting formulas, frequently used in the literature, permit some estimates and serve to illustrate the role of the reference surface.

Hooke's law for the stretching energy per unit area of fluid monolayer, $g_{\mathrm{m,s}}$, may be written as

$$g_{\mathrm{m,s}} = \tfrac{1}{2}\lambda_{\mathrm{m}}\left(\frac{\Delta A}{A}\right)^2 \tag{8}$$

where $\Delta A/A$ is the (relative) dilation and $\lambda_{\mathrm{m}}$ is the monolayer stretching modulus. The latter should be half the bilayer stretching modulus $\lambda$. Values of the order of $\lambda = 200\,\mathrm{mN\,m^{-1}}$ have been measured on typical lipids such as PCs, PEs, and DGDGs [3]

The laterally isotropic monolayer is idealized as a homogeneous fluid plate of thickness $h$ and subjected to cylindrical curvature $c_1 = 1/R$. It is then intuitively clear that the neutral surface coincides with the middle surface of the plate. If the middle surface is at $z = 0$, the dilation of the monolayer on one side of the middle and its compression on the other obey

$$\frac{\Delta A(z)}{A} = \frac{z}{R}$$

Integrating the lateral stress $\sigma(z) = (\lambda_{\mathrm{m}}/h)\Delta A(z)/A$ over the thickness $h$ of the monolayer, one finds the lateral tension produced by the bend,

$$\gamma_{\mathrm{m}} = \int_{-h/2}^{h/2} \frac{\lambda_{\mathrm{m}}}{h}\frac{z}{R}\,\mathrm{d}z$$

to be indeed zero.

For such a monolayer the bending rigidity $\kappa_m$ may be obtained from the identity

$$\tfrac{1}{2}\kappa_m \frac{1}{R^2} = \frac{1}{2}\int_{-h/2}^{h/2} \frac{\lambda_m}{h}\left(\frac{z}{R}\right)^2 \mathrm{d}z \tag{9}$$

which immediately yields

$$\kappa_m = \lambda_m \frac{h^2}{12} \tag{10}$$

Substituting $\lambda_m = 100\,\mathrm{mN\,m^{-1}}$ and $h = 2\,\mathrm{nm}$ gives $\kappa_m = 1/3 \times 10^{-19}$ J. Doubling this value results in nearly $10^{-19}$ J, the order of magnitude of bending rigidities measured on lipid bilayers [3–5].

A more complex situation arises if up – down asymmetry of the monolayer produces a spontaneous curvature. Using again the neutral surface as reference, let us first write down the total elastic energy up to quadratic order in the deformations:

$$g_m = \tfrac{1}{2}\kappa_m(c_1 + c_2)^2 - \kappa_m c_0(c_1 + c_2) + \bar{\kappa}_m c_1 c_2 + \tfrac{1}{2}\lambda_m\left(\frac{\Delta A}{A}\right)^2 \tag{11}$$

However, the neutral surface cannot be the surface of equal molecular density if the membrane curvature is nonuniform [6]. This is because in mechanical equilibrium the monolayer lateral tension contains the bending energy density $g_{m,b}$ as an extra term,

$$\gamma_m = \gamma_{m,0} + g_{m,b}, \quad \text{where } \gamma_{m,0} = \lambda_m \Delta A/A \tag{12}$$

so that the gradient of lateral tension balances the gradient of curvature elastic potential energy. The curvature-induced part of the lateral tension affects the Hookean bending energy when to lowest order it is a linear function of $c_1 + c_2$, i.e. in the presence of spontaneous curvature, as will be seen shortly.

To calculate the position of the surface of equal molecular density relative to the neutral surface, consider again cylindrical curvature $c_1$ and assume $\gamma_m = 0$ for $c_1 = 0$. Accordingly, the change in area of the neutral surface due to the lateral tension of bending, $-\kappa_m c_0 c_1$, is

$$\frac{\Delta A}{A} = -\frac{\kappa_m c_0 c_1}{\lambda_m} \tag{13}$$

In addition to this uniform effect, the curvature produces the dilation

$$\frac{\Delta A(z)}{A} = c_1(z - z_n)$$

depending on the distance $z - z_n$ from the neutral surface at $z = z_n$. By definition, the sum of the two dilations vanishes at the position $z = z_b$ of the surface of equal molecular density. Therefore, the distance $\delta = z_b - z_n$ between the two reference surfaces is

$$\delta = \frac{\kappa_m c_0}{\lambda_m} \tag{14}$$

and thus independent of the curvature $c_1$. Because of the sign convention for $u$, a positive $\delta$ means that the surface of equal molecular density is outside the neutral surface in cylindrical and spherical membranes.

As a rule, the distance $\delta$ should be very small in comparison with the thickness $h$ of the monolayer. Inserting eq. (10) in eq. (14), yields $\delta = h^2 c_0/12$. Even for $c_0 = |1/h|$, which should be near the upper limit, $\delta$ comes out to be only a small fraction of the monolayer thickness. In most regards this shift is of little interest.

A possibly important effect of the linear dependence of $\lambda_m$ on curvature for $c_0 \neq 0$ is a lowering of the effective bending rigidity. Let us calculate it in terms of the neutral surface. For cylindrical curvature the total elastic energy density $g_m$ referring to this surface may be expressed by

$$g_m = \tfrac{1}{2}\kappa_m c_1^2 - \kappa_m c_0 c_1 - \frac{\Delta A}{A}\kappa_m c_0 c_1 + \tfrac{1}{2}\lambda_m \left(\frac{\Delta A}{A}\right)^2 \tag{15}$$

The dilation $\Delta A/A$ of the neutral surface to be inserted in eq. (15) is

$$\frac{\Delta A}{A} = -c_1 \delta$$

as follows from eqs (13) and (14). The first two terms of eq. (15) represent the regular bending energy. The third term corrects the second, $-\kappa_m c_0 c_1$, for the change of molecular density on the neutral surface. It is always negative, resulting from the compression of the monolayer in regions of negative bending energy and vice versa. The fourth term stands for the stretching energy; it is always positive and half as large as the third. The last two terms can be absorbed into the first, the sum of all three terms being

$$\tfrac{1}{2}\kappa_m \left(1 - \frac{\kappa_m c_0^2}{\lambda_m}\right) c_1^2 \tag{16}$$

From eq. (16) it is concluded that the effective bending rigidity is [6]

$$\kappa'_m = \kappa_m \left(1 - \frac{\kappa_m c_0^2}{\lambda_m}\right) \tag{17}$$

The correction $\kappa_m c_0^2/\lambda_m$ should in most cases be negligible. Taking $\kappa_m/\lambda_m = h^2/12$ from eq. (10), it is seen that $|c_0|$ has to be about $1/h$ for a 10 % reduction of the bending rigidity. However, spontaneous curvatures of this magnitude do occur in lipid monolayers forming an inverse hexagonal phase and, unlike that of $\delta$, the correction of $\kappa_m$ is quadratic in $c_0$. The complete formula for the Hookean elastic energy density of the monolayer may now be written as

$$g_m = \tfrac{1}{2}\kappa'_m(c_1 + c_2)^2 - \kappa'_m c'_0(c_1 + c_2) + \bar{\kappa}_m c_1 c_2 + \tfrac{1}{2}\lambda_m\left(\frac{\Delta A}{A}\right)^2 \tag{18}$$

where the substitution of $\kappa'_m c'_0$ for $\kappa_m c_0$ defines the effective spontaneous curvature

$$c'_0 = c_0\left(1 - \frac{\kappa_m c_0^2}{\lambda_m}\right)^{-1}$$

and $\Delta A/A$ is now the relative change in area in addition to that caused by the lateral tension $-\kappa_m c_0 c_1$.

Formula (18) has been derived for the neutral surface as a reference. In the presence of spontaneous curvature it is a disadvantage of this reference surface that its molecular density depends on curvature. For the most part, formula (18) remains valid for the surface of equal molecular density. The first term and the last keep their forms because any corrections of the simple squares are of higher than quadratic order in the deformations. As will be seen in Section 2.4, $\kappa_m c_0$ is independent of the position of the reference surfaces, but $\bar{\kappa}_m$ slightly depends on it.

In general, it is sufficient to know that there is a surface in the monolayer for which stretching and bending deformations are in effect decoupled. This will be the classical neutral surface (for $c_0 = 0$) or a new surface of equal molecular density at a slightly shifted position (for $c_0 \neq 0$). In the latter case, nonuniform curvature $c_1 + c_2$ is associated with an inherent lateral tension that is proportional to it. Henceforth, the term 'bending surface' is used in both cases.

## 2.3 Bilayers

A bilayer is composed of two monolayers that are free to slide upon each other in the fluid state. It is assumed in the following that their immediate contact does not affect their elastic properties. Bilayers with equal monolayers on both sides are called symmetric. Full up–down symmetry, possible only in the flat state, requires that the molecular densities and the aqueous media are identical on both sides. Symmetry is facilitated if there is flip-flop, i.e. local exchange, of the amphiphiles between the monolayers. As sketched in Figure 6.2, the sign convention for the outer monolayer will be applied to the bilayer as a whole.

Let us begin with the symmetric bilayer, considering first the case of rapid exchange. It does not matter whether the exchange is through flip-flop or via bilayer pores and edges connecting the monolayers. Often lateral flow in the monolayers substitutes for real exchange. There will be a difference in area and, thus, the number of molecules between the monolayers wherever the bilayer is bent with a curvature $c_1 + c_2 \neq 0$. The bending surface of the bilayer is its middle surface, the total

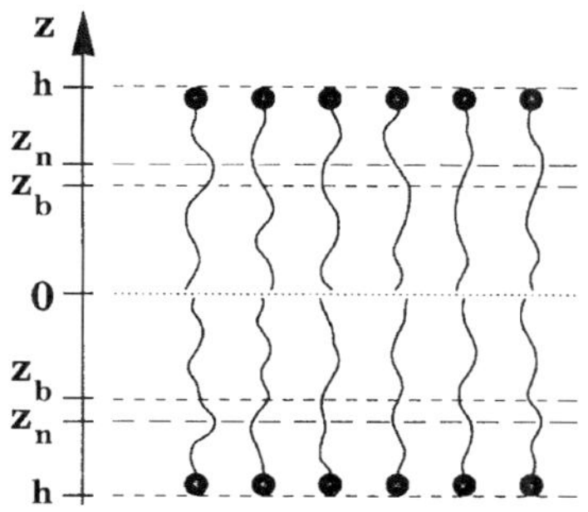


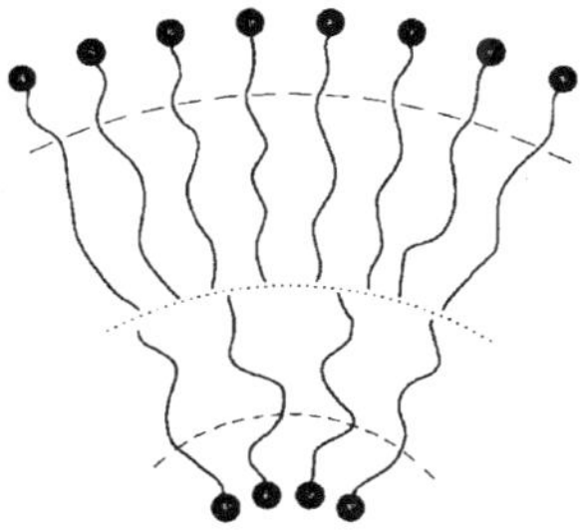


**Figure 6.2** Schematic cross-section through bilayer of amphiphilic molecules. For the meaning of $h, z_n$ and $z_b$ see Figure 6.1. The lower diagram defines positive curvature for a vesicle membrane facing inside and outside water.

number of molecules per unit area of this surface being, to the first order, independent of $c_1 + c_2$.

The total elastic energy of a symmetric bilayer may be written as

$$g_b = \tfrac{1}{2}\kappa(c_1 + c_2)^2 + \bar{\kappa}c_1c_2 + \tfrac{1}{2}\lambda\left(\frac{\Delta A}{A}\right)^2 \tag{19}$$

with

$$\kappa = 2\kappa'_m; \quad \lambda = 2\lambda_m$$

as the full up–down symmetry rules out spontaneous curvature. The bilayer modulus of Gaussian curvature, $\bar{\kappa}$, is not simply $2\bar{\kappa}_m$, but in addition contains a term proportional to monolayer spontaneous curvature. This can be seen very directly by considering a symmetric bilayer with saddle curvature. In such a case,

$c_1 + c_2$ vanishes for the bilayer because of $c_2 = -c_1$. However, it does not vanish for the bending surfaces of the monolayers. For the outer monolayer,

$$(c_1 + c_2)_{\text{out}} = \frac{1}{R + z_{\text{b}}} = \frac{1}{R - z_{\text{b}}} = 2z_{\text{b}} c_1 c_2 \tag{20}$$

where $R$ is the radius of the two principal curvatures and $z_{\text{b}}$ is the distance of the bending surface from the bilayer middle surface. The resulting contribution to the energy density of bilayer saddle curvature is

$$-\kappa'_{\text{m}} c'_0 (c_1 + c_2)_{\text{m}} = -2\kappa'_{\text{m}} z_b c'_0 c_1 c_2.$$

As for symmetry reasons the inner monolayer makes the same contribution, this leads to [7]

$$\bar{\kappa} = 2\bar{\kappa}_{\text{m}} - 2\kappa' z_b c'_0 \tag{21}$$

Very slow flip–flop, taking hours or days, is frequent in pure one-component lipid bilayers. A lipid imbalance between the monolayers can thus persist and produce a spontaneous curvature of the bilayer, $c_{0,\text{b}}$. In order to deal with this effect, let us first write the bending energy in its complete form:

$$g = \tfrac{1}{2}\kappa (c_1 + c_2)^2 - \kappa c_{0,\text{b}} (c_1 + c_2) + \bar{\kappa} c_1 c_2$$

The forces conjugate to the deformations $c_1$ and $c_2$ are torques per unit length, $\tau$. For instance, along a hypothetical cut through the bilayer perpendicular to the direction of $c_1$ one has

$$\tau_1 = \partial g / \partial c_1 = \kappa (c_1 + c_2) - \kappa c_{0,\text{b}} + \bar{\kappa} c_2 \tag{22}$$

In the flat state (or a flat region) the torque per unit length is simply $-\kappa c_{0,\text{b}}$. In the case considered, it is due to a dilation of one bilayer and an equal compression of the other, if the bilayer as a whole has no lateral tension. Employing the homogeneous plate model for the monolayers ($c_0 = 0$), the line torque balance may be written as

$$\kappa c_{0,\text{b}} = -z_{\text{n}} \lambda_{\text{m}} \left[ \left( \frac{\Delta A}{A} \right)_{\text{out}} - \left( \frac{\Delta A}{A} \right)_{\text{in}} \right] \tag{23}$$

As the monolayer dilations of the outer and inner monolayers need not cancel each other, this formula allows for nonvanishing bilayer lateral tension. The height $z_{\text{n}}$ is the distance of the monolayer neutral surfaces from the bilayer middle. It enters as the average lever arm of the elastic forces due to monolayer dilation and compression. In the closed bilayer of a vesicle, the difference of the monolayer dilations can change with the shape of the vesicle. Simple geometric arguments lead, in this particular model, to

$$\left( \frac{\Delta A}{A} \right)_{\text{out}} - \left( \frac{\Delta A}{A} \right)_{\text{in}} = 2z_{\text{n}} \frac{1}{A} \oint (c_1 + c_2) \mathrm{d}A + \text{constant} \tag{24}$$

where the closed integral is taken over the middle surface of the bilayer. Combination of the last two equations results in a shape-dependent bilayer spontaneous curvature

$$c_{0,\mathrm{b}} = -\frac{2z_\mathrm{n}^2\lambda_\mathrm{m}}{\kappa}\frac{1}{A}\oint(c_1 + c_2)\mathrm{d}A + \text{constant}$$

For an estimate, use eq. (10) together with $z_\mathrm{n} = h/2$ and $\kappa = 2\kappa_\mathrm{m}$ to arrive at the simple relationship [8]

$$c_{0,\mathrm{b}} = -3\frac{1}{A}\oint(c_1 + c_2)\mathrm{d}A + \text{constant} \tag{25}$$

Of course, for real amphiphilic bilayers, the numerical factor relating spontaneous curvature to average curvature need not be exactly 3. In the general case with $c_0 \neq 0$, the situation is more complicated because the decoupling of bending and stretching implies that any monolayer spontaneous curvature $c_0$ diminishes on dilation of the neutral surface by the factor $(1 + \Delta A/A)^{-1}$.

The fact that the spontaneous curvature of a bilayer with arrested flip–flop depends on vesicle shape comes with many other names in the literature, such as nonlocal bending energy [9], bilayer couple model [10] and area difference elasticity (Chapter 7). The shape dependence of spontaneous curvature does not change the equilibrium shape equations of vesicles and their solutions. However, it affects the stability and relative energies of these shapes, thus being important in shape transformations of vesicles subjected to changes of the environment and the resulting phase diagrams of vesicles. Shape fluctuations, especially short-wave length undulation modes, generally have little or no effect on spontaneous curvature because they are weak and their positive and negative curvatures nearly cancel.

Spontaneous curvature of a symmetric bilayer can also arise from differences between the aqueous media inside and outside the vesicle such as different solutes and concentrations. (Any dependence of $c_0$ on local curvature $c_1 + c_2$ is absorbed in $\kappa$). This kind of spontaneous curvature does not depend on vesicle shape and is sometimes regarded as the true spontaneous curvature.

This section concludes with a few words on bilayers that are asymmetric because their monolayers consist of different molecules. In such cases a reasonable choice of the bilayer bending surface may be the middle between the monolayer bending surfaces. It ensures that bilayer bending without stretching conserves the sum of the monolayer bending surface areas for arbitrary cylindrical curvature. The bending rigidity and the spontaneous curvature of the bilayer may be expected to obey

$$\kappa = \kappa_\mathrm{out} + \kappa_\mathrm{in}$$

$$\kappa c_{0,\mathrm{b}} = \kappa_\mathrm{out} c_{0,\mathrm{out}} - \kappa_\mathrm{in} c_{0,\mathrm{in}}$$

if both monolayers are free of lateral tension where they are flat. (A justification of the latter equation is provided in the next section.) An estimate of the shape dependence of spontaneous curvature has to take into account that the stretching moduli will be different for the two monolayers.

## 2.4 Stress profiles

The calculation of the bending rigidity from the homogeneous plate model has shown the lateral stress $\sigma(z)$ to be a useful quantity in dealing with the elastic response of membranes. In amphiphilic membranes there is a nonvanishing profile of isotropic lateral stress $\sigma(z)$ even in the flat state at zero lateral tension. The bending rigidity $\kappa$ of the monolayer depends on the derivatives of the stresses with respect to $c_1 + c_2$ which in general are difficult to calculate. However, the two other bending elastic parameters, $-\kappa c_0$ and $\bar{\kappa}_m$, can be computed from the stress profile of the flat state [11]. Also needed is the change in membrane area as a function of the height $z$ and of the principal curvatures.

The stress profile expected for a flat monolayer of zero lateral tension is sketched in Figure 6.3, with a positive stress at the water–hydrocarbon interface and a compensatory negative stress in the region of the hydrocarbon chains. In the case of charged membranes, there will be additional negative stresses in the water at $z > h$.

Employing the surface of equal molecular density as the bending surface, the curvature-dependent dilation of a bent monolayer is

$$\frac{\Delta A(z)}{A} = (z - z_b)(c_1 + c_2) + (z - z_b)^2 c_1 c_2. \tag{26}$$

Because of

$$g_b = \int \sigma(z) \frac{\Delta A(z)}{A} \mathrm{d}z$$

for fixed $\sigma(z)$, this leads immediately to

$$-\kappa_m c_0 = \int \sigma(z)(z - z_b)\mathrm{d}z \tag{27}$$

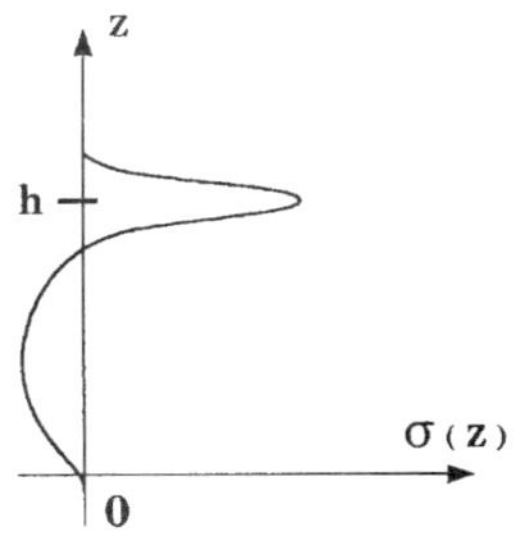


**Figure 6.3** Schematic stress profile of the flat monolayer shown in Figure 6.1.

and

$$\bar{\kappa}_{\rm m} = \int \sigma(z)(z - z_{\rm b})^2 {\rm d}z$$

In the case of vanishing lateral tension, ie. for

$$\gamma_{\rm m} = \int \sigma(z){\rm d}z = 0$$

eq. (27) reduces to the useful result

$$-\kappa_{\rm m} c_0 = \int \sigma(z) z \, {\rm d}z \tag{28}$$

It does not depend on the position of the reference surface as claimed above. No limits are given for these integrals to allow for an electrostatic double layer on the side of the water.

These considerations are easily extended to bilayers of zero lateral tension ($\gamma = 0$), resulting in

$$-\kappa c_{0,{\rm b}} = \int \sigma(z) z \, {\rm d}z \tag{29}$$

The formula applies to symmetric and asymmetric bilayers, including the case of equal but opposite lateral tensions of the monolayers ($\gamma_{\rm out} = -\gamma_{\rm in} \neq 0$). An abbreviated version of it has been used above to calculate in the homogeneous plate model the spontaneous curvature induced by lipid imbalance. For the symmetric bilayer with $\gamma_{\rm out} = \gamma_{\rm in} = 0$ one finds in addition

$$\kappa_{\rm m} = \int \sigma(z) z^2 \, {\rm d}z \tag{30}$$

Any dilation of the bilayer bending surface due to Gaussian curvature is proportional to $c_1 c_2$ and therefore does not affect Hookean elasticity. Substituting $(z - z_{\rm b}) + z_{\rm b}$ for $z$ in eq. (30) leads to

$$\bar{\kappa} = 2\left[\int_0^\infty \sigma(z)(z - z_{\rm b})^2 + 2z_{\rm b} \int_0^\infty \sigma(z) z\right] {\rm d}z$$

where the factor 2 in front compensates for the fact that the integration is only over the outer monolayer. Obviously, not only has a formula for $\bar{\kappa}$ been obtained, but also eq. (21) has been recovered in a very different way. The last equation is equally valid with $z_{\rm n}$ in place of $z_{\rm b}$. Comparison of the two alternatives shows that, unlike the bilayer modulus $\bar{\kappa}$, the monolayer modulus $\bar{\kappa}_m$ depends on the position of the monolayer bending surface.

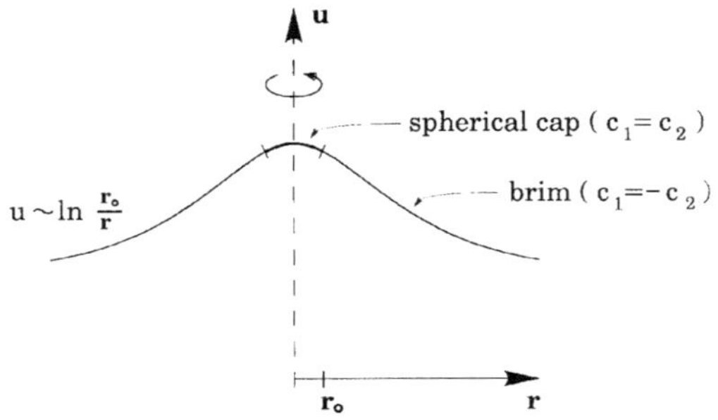


**Figure 6.4** Cross-section through membrane deformed by a single local bending deformation (called a hat), containing the axis of rotational symmetry

## 2.5 Fluctuations

The shape fluctuations of giant lipid vesicles are easily visible in light microscopy, their strength being on the order of 10 % of the vesicle radius. Visible thermal membrane undulations are suppressed by lateral tensions $\gamma \geq 10^{-3}\,\mathrm{mN\,m^{-1}}$. Usually, these bending fluctuations are described in terms of global modes of membrane normal displacement. Examples are spherical harmonics for fluctuating spheres and sine waves for fluctuating flat membranes with periodic boundary conditions. The latter is the favorite model system of theoretical studies.

The mean-square fluctuation amplitudes are obtained from the equipartition theorem. For a sine wave $u = u_0 \sin(\vec{q} \cdot \vec{\rho})$ on a quadratic piece of membrane,

$$\langle u_0^2 \rangle = \frac{2kT}{A\kappa q^4} \tag{31}$$

where $u_0$ is the displacement amplitude, $q$ is the wave vector of the mode, and $A$ is the area of the membrane. The mean-square curvature amplitude obeys

$$\langle (c_1 + c_2)_0^2 \rangle = \frac{2kT}{A\kappa} \tag{32}$$

as $c_1 + c_2 = q^2 u$ to linear order. In the presence of a lateral tension $\gamma$, $\kappa q^4$ has to be replaced by $\kappa q^4 + \gamma q^2$ in eq. (31). Accordingly, lateral tension suppresses most strongly long-wavelength modes with $q^2 < \gamma/\kappa$.

Alternatively, the bending fluctuations can be expanded into local bending modes, the so-called hats [11,12]. In the hat model the membrane is divided into practically circular pieces, which may be chosen as small as a molecular cross-section. The disks fluctuate independently in their curvatures $c_1 + c_2$. If a single hat is excited, it creates in a flat membrane a cap of spherical curvature $c_1 = c_2$ with a brim of saddle curvature $c_1 = -c_2$, as sketched in Figure 6.4. The brim drops (or rises) logarithmically with the distance $\rho$ from the hat's center so that the principal curvatures go to zero with $1/\rho^2$. The hat model gives an idea of the membrane roughness on the molecular scale. The r.m.s. boundary angle of a fluctuating cap can be derived from eq. (32). With $\kappa = 10^{-19}$ J and $k \cdot T = 4 \cdot 10^{-21}$ J one obtains 3°, regardless of disk size. Near-flatness is an important assumption in almost all theoretical treatments of membrane fluctuations.

## 3. HIGHER-ORDER BENDING ELASTICITY

Bending energies that vary with higher than quadratic order in the principal curvatures have occasionally been invoked to explain why vesicles and other bilayer structures cannot become arbitrarily small. Gradient terms such as $[\vec{\nabla}(c_1 + c_2)]^2$ with $\vec{\nabla} = (\partial/\partial x, \partial/\partial y)$ are also counted among the non-Hookean terms. The order of an invariant of the bending energy surface density is its inverse power of length, e.g. two or quadratic with $(c_1 + c_2)^2$. Accordingly, the order of both $[\vec{\nabla}(c_1 + c_2)]^2$ and $(c_1 + c_2)^4$ is four or quartic. Inspecting these terms for a sine wave $u = u_0 \sin(\vec{q} \cdot \vec{\rho})$, one finds them to vary as $u_0^2 q^6$ and $u_0^4 q^8$, respectively. This suggests that gradient terms should be more important than others of the same order for $u_0^2 q^2 \ll 1$.

Only the even orders can occur in the bending energy density of a symmetric bilayer. The second order is therefore followed by the fourth. A complete set of quartic invariants under rotation of the $xy$ cross of local Cartesian coordinates has been given in partial tensor notation by Mitov [13]. They are

$$
\begin{gathered}
(c_1 + c_2)^4, \quad (c_1 c_2)^2, \quad (c_1 + c_2)^2 c_1 c_2 \\
[\vec{\nabla}(c_1 + c_2)]^2, \quad (\nabla_\gamma \Omega_{\alpha\beta})^2, \\
\vec{\nabla}^2 (c_1 + c_2)^2, \quad \vec{\nabla}^2 (c_1 c_2) \\
\varepsilon_{\alpha\beta} \Omega_{\beta\gamma} \nabla_\gamma \nabla_\alpha (c_1 + c_2)
\end{gathered}
\tag{33}
$$

with $\alpha$, $\beta$ and $\gamma$ being summed over 1 and 2, i.e. $x$ and $y$, and

$$\varepsilon_{\alpha\beta} = \begin{pmatrix} 0 & 1 \\ -1 & 0 \end{pmatrix}$$

The last term of eq. (33) represents a chiral deformation; it can be shown to vanish for a symmetric bilayer. In the case of periodic boundary conditions, the two terms starting with $\vec{\nabla}^2$ can also be dropped, their integrals being zero because of the divergence theorem.

There are reasons for considering higher-order bending elasticity. More and more evidence for anomalous behavior of the bilayers of zwitterionic lipids that are typical of biological membranes (PCs, PEs, and DGDG) has accumulated since the early 1980s in the author's experimental group at FUB [5]. Initially there were light microscopic observations that could not be understood in terms of Hookean bending elasticity. They were made in highly swollen samples with a large excess of pure or salted water. Two of them are reasonably well reproducible and have been investigated in detail. First, a mutual membrane adhesion induced by lateral tension, in particular its large contact angles, led to the postulate of an anomalous submicroscopic roughness which in turn calls for a superstructure of the membranes involved. Second, the reversible precipitation of so-called dark bodies from the aqueous phase in highly swollen PC–water systems pointed to the existence of a dispersive bilayer phase in water. In the meantime, these hypotheses have been checked by electron microscopic studies, as summarized in Chapter 17.

Let us focus here on the possible explanation as given by higher-order bending elasticity of two of the bilayer structures found in those studies: (a) a graininess of vesicle membranes with a grain spacing of 4–6 nm, and (b) irregular knotted sticks, or pearl strings, with occasional branchings and knot diameters of 20 nm or more. The graininess was observed in EYPC and DOPC membranes with cryo-transmission electron microscopy. The knotted sticks, which seem to represent the dispersive phase, were seen in freeze-etching electron microscopy of POPC, DMPC, and EYPC vesicle preparations.

The same primitive model using no more than two of the five relevant fourth-order invariants listed above seems to be sufficient to understand both graininess and knotted sticks. Its bilayer bending energy density is expressed by

$$g = \tfrac{1}{2}\kappa(c_1 + c_2)^2 + \bar{\kappa}c_1c_2 + \bar{\kappa}_2(c_1c_2)^2 + \kappa'[\vec{\nabla}(c_1 + c_2)]^2 + \bar{\kappa}_4(c_1c_2)^4 \qquad (34)$$

The moduli $\bar{\kappa}_2$ and $\kappa'$ happen to be the only fourth-order moduli that can be estimated rather straightforwardly in terms of a special stress profile of the monolayer and general features of its geometry, respectively [14,15]. The predicted values are

$$\bar{\kappa}_2 \approx -\gamma_{\text{int}}h^4 \quad \text{and} \quad \kappa' \approx (1/30)\gamma_{\text{int}}h^4$$

where $\gamma_{\text{int}} = 50\,\text{mN}\,\text{m}^{-1}$ is the interfacial tension at the hydrocarbon–water interface and $h = 2\,\text{nm}$ the monolayer thickness. Note that $\bar{\kappa}_2$ is negative. It is this modulus that makes the membrane prefer strongly bent states. The eighth-order term at the end of eq. (34) is added to prevent an unrealistic divergence of the curvature. Dimensional analysis provides an order-of-magnitude estimate of its modulus, $\bar{\kappa}_4 \approx \gamma_{\text{int}}h^8$.

Four of the five terms in eq. (34) were needed to reproduce the graininess in a basically flat membrane by Monte Carlo simulation [16]. The Gaussian curvature term, $\bar{\kappa}c_1c_2$, was left out as is common in many other calculations because its integral depends only on the topology of the surface. A graininess as obtained for a fluctuating circular region embedded in nonfluctuating surroundings is shown in Figure 6.5, together with the values of the elastic moduli and the temperature used in the calculation. Lowering the temperature leads to a transformation of the disordered graininess into a well-ordered quadratic texture resembling an egg carton. An energy

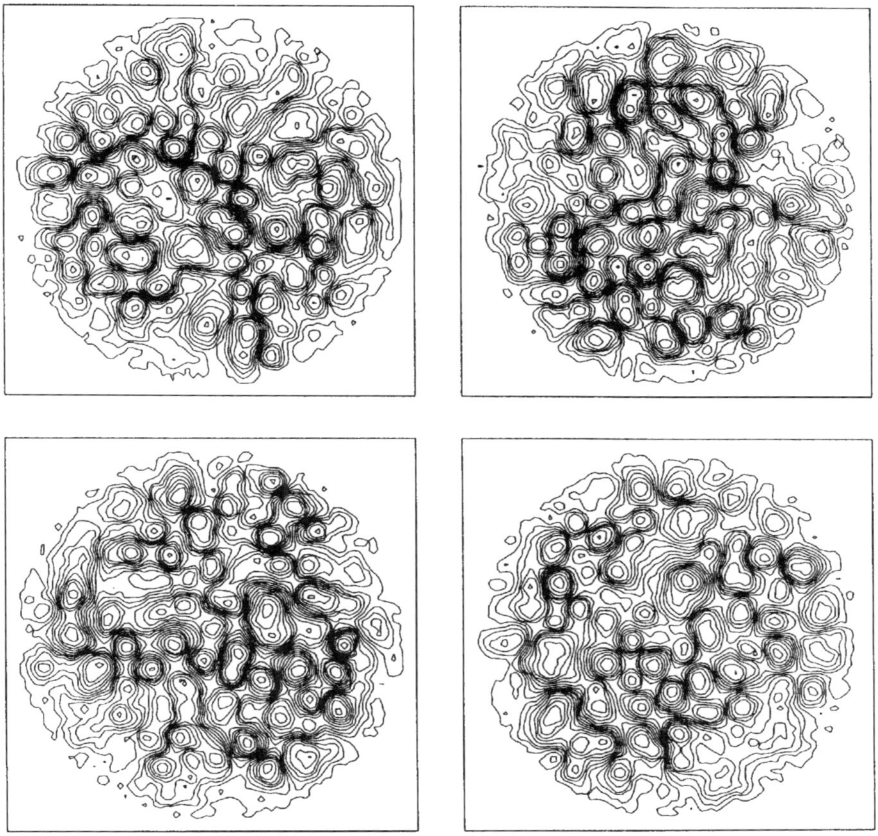

**Figure 6.5** Four snapshots of a grainy membrane superstructure as obtained from Monte Carlo simulation of a membrane disk (with zero height and slope along the circular circumference). Contour lines are shown. The sides of the square frames represent 66 nm. The bending elastic moduli and the temperature of the simulation are $\kappa = 0.5 \times 10^{-19}$ J, $\kappa' = 1 \times 10^{-38}$ J m$^2$, $\bar{\kappa}_2 = -0.9 \times 10^{-36}$ J m$^2$, $\bar{\kappa}_4 = 0.9 \times 10^{-71}$ J m$^6$, temperature = 470 K.

barrier separates the ordered and disordered superstructures from the flat state of the membrane. It is not possible with the present method of simulation to search for other features of the superstructure of vesicle membranes such as sharp bends and a coarse version of the graininess which have been seen in electron microscopy. Moreover, the nature of the postulated anomalous roughness is still an open question.

A negative modulus of the $(c_1c_2)^2$ term can also be responsible for the formation of knotted sticks. The narrow constrictions between more or less spherical knots are characterized by very strong saddle curvature, thus providing large negative contributions to the bending energy. The same term will lower the bending energies of the spherelike knots as it does not depend on the sign of Gaussian curvature. However, this effect is unlikely to neutralize the Hookean bending energy of the sphere, $4\pi(2\kappa + \bar{\kappa})$, since PC bilayers are not known to vesiculate spontaneously.

The fusion of two sticks, either end to end or by branching, is expected to lower their total bending energy as it generates an additional constriction. Linear addition should be favored over branching because the simple stick has two hemispherical endings while the singly branched stick has three. Moreover, accommodating three constricted connections instead of two on a knot may cost some extra energy. (Note, however, that constrictions may also branch.) Fusion changes the topology of the membranes. The change of bending energy per fusion event, $-4\pi\bar{\kappa}$, is most probably negative, which is another reason for fusion. Entropy seems to be the only factor favoring disintegration of the knotted sticks.

Higher-order bending energy may also be involved in the swelling of lipid multilayer systems in excess water, as described in Chapter 18. The process to be explained is the second stage of swelling, in which the PC membranes separate possibly one by one from a multilayer system at its so-called equilibrium spacing. Originally, this unbinding was thought to require the presence of a superstructure, such as the graininess. The energy barrier to be overcome for its formation should be lowest for the outermost membrane, which is least hindered by other membranes. The resulting roughness of the outermost membrane would decrease its van der Waals attraction to the stack, thus permitting it to unbind. However, the discussion of the micrographs recently obtained in cryo-transmission electron microscopy of DOPC vesicles casts doubt on this hypothesis, suggesting instead that even smooth vesicles do not adhere to each other [17]. In addition, the new experiments on multilayer swelling seem to rule out a waiting time of unbinding (Chapter 18), whereas a delay is expected if a superstructure has to be formed first.

Therefore, an alternative mechanism, involving higher-order bending elasticity but no superstructure, was proposed to resolve the conflict between an apparent equilibrium spacing in the multilayer system and separation of the outermost membranes. As the membranes obstruct each other in the stack with a period of less than 7 nm, the fluctuations of the stacked membranes are much weaker than those of a single membrane. This alone would not lift the contradiction as may be inferred from the current theory of the unbinding transition [18]. However, a different situation arises if higher-order terms significantly decrease the bending energy below its Hookean value at the stronger curvatures $c_1 + c_2$ reached in freely fluctuating membranes. A particularly suitable model to deal with such an effect is the hat model of bending fluctuations. The negative $(c_1c_2)^2$ term lowers the bending energy of the spherical cap and a similar amount of negative energy comes from the saddle curvature in the brim region next to the cap where saddle curvature is strong.

This increases the probability of finding hats with an anomalously strong cap curvature. Inspection shows that a small distortion of the Gaussian distribution of cap curvatures prescribed by Hookean elasticity is enough to produce a substantial decrease of the effective bending rigidity. The complete model remains to be worked out, but preliminary calculations indicate that the estimated value of the negative modulus of the $(c_1c_2)^2$ term is of the right order of magnitude to explain the metastability of the multilayer systems of lipid bilayers in water.

## 4. CONCLUDING REMARKS

The treatment given here disregards the possibility of tilt, i.e. a collective inclination of the hydrocarbon chains of the amphiphile with respect to the membrane surface normal [19]. Tilt may be expected to occur in regions of nonuniform membrane curvature, but it is usually so weak that its angle and elastic energy surface density can be neglected. This assumption may fail at the very high curvature gradients occurring in grainy bilayers and in the knotted sticks of the dispersed phase. A theoretical study of these problems remains to be done.

Also, mixed membranes were not discussed. They will be bumpy, even in the absence of thermal bending fluctuations, whenever the constituent molecules differ by their spontaneous curvatures. This holds for bilayers as well as monolayers, because different pairs of opposite molecules may have different bilayer spontaneous curvatures.

In the case of mixed membranes it is attractive to solve eq. (32) for the flexibility $1/\kappa$ and derive this macroscopic quantity from the strength of the mean-square local bending fluctuations. The fluctuations of $c_1 + c_2$ arise not only from the molecular bending fluctuations but also from the diffusion of the molecules (or pairs of molecules in bilayers) if their spontaneous curvatures are different. Therefore, mixing can lead to a dramatic decrease of the bending rigidity in comparison with one-component membranes [12].

The emphasis of this review of membrane elasticity is on the fluid plate model of the amphiphilic monolayer. The homogeneous fluid plate is used for some elementary considerations and estimates. Plates modeling actual monolayers will possess a nonvanishing stress profile even when undeformed, i.e. in the flat state at zero lateral tension.

Fluid membranes are molecularly thin two-dimensional objects in three-dimensional space. Their extreme deformability distinguishes them from liquid crystals and other three-dimensional systems. Moreover, their fluidity and complex internal structure sets them apart from polymers. The intervention of higher order bending elasticity and its consequences for membrane shape and fluctuations seem to be related to these characteristic properties.

## 5. REFERENCES


1. F. C. Frank, *Discuss. Faraday Soc.*, **25**,1 (1988).
2. L. D. Landau and E. M. Lifshitz, Theory of Elasticity, Pergamon, London, 1959.
3. E. Evans and D. Needham, *J. Phys. Chem.*, **91**, 4219 (1987).
4. E. Evans and B. Rawicz, *Phys. Rev. Lett.*, **64**, 2094 (1990); R. E. Waugh, J. Song, S. Svetina and B. Žekž, *Biophys. J.*, **61**, 974 (1992).
5. W. Helfrich, in: *Structure and Dynamics of Membranes*, R. Lipowsky and E. Sackmann (eds), Elsevier Science BV, Amsterdam, 1995.
6. W. Helfrich and M. M. Kozlov, *J. Phys. II France*, **3**, 287 (1993).
7. A. G. Petrov and I. Bivas, *Prog. in Surf. Sci.*, **16**, 389 (1984).
8. W. Helfrich, *Z. Naturforsch.*, **29c**, 510 (1974).
9. E. Evans, *Biophys. J.*, **14**, 923 (1974).
10. S. Svetina and B. Žekž, *Biochim. Biophys. Acta*, **42**, 86 (1983).
11. W. Helfrich, in: *Physics of Defects,* eds. R. Balian, M. Kleman and J.-P. Poirier, North Holland, 1980.
12. W. Helfrich, *Liq. Crystals*, **5**, 1647 (1989); W. Helfrich and M. M. Kozlov *J. Phys. II France*, **4**, 1427 (1989).
13. M. D. Mitov, *Comptes rendus de l'académie bulgare des Sciences*, **31**, 513 (1978).
14. R. Goetz and W. Helfrich, *J. Phys. II France*, **6**, 215 (1996).
15. M. M. Kozlov, *Langmuir*, **8**, 1541 (1992).
16. A. Jud and W. Helfrich, unpublished research.
17. B. Klösgen and W. Helfrich, *Biophys. J.*, **73**, 3016 (1997).
18. R. Lipowsky, in: *Structure and Dynamics of Membranes*, R. Lipowsky and E. Sackmann (eds), Elsevier Science BV, Amsterdam, 1995.
19. M. Hamm and M. M. Kozlov, *Eur. Phys. J. B*, **6**, 519 (1998).

# Chapter 7

# Giant Vesicles: A Theoretical Perspective

*UDO SEIFERT*

*Max-Planck-Institut für Kolloid- und Grenzflächenforschung, Golm, Germany*

## 1. INTRODUCTION

Giant vesicles constitute a paradigmatic soft-matter system. The characteristic interplay between energy and entropy, i.e. between deterministic forces and fluctuations that is one of the key features of soft matter, becomes visible in the microscope. Rarely in physics, are the relevant phenomena observable so directly, rather than requiring deconvolution of intricate data. Even though only a few parameters determine the state of such a vesicle, these vesicles can exhibit an amazing variety of phenomena. Therefore, their quantitative explanation becomes a real challenge because a simple model must contain and reflect this diversity. Due to the smallness of the parameter space of these systems, a fairly complete elucidation of the equilibrium properties of vesicle shapes has become possible.

As well as physics, both chemistry and biology contribute to the origins of and the continuing interest in this system. In fact, the quest for an understanding of the various shapes of red blood cells and their mutual transformations as external conditions are changed, is one of the original motivations to study vesicles. Even though a red blood cell's membrane is structurally more complex than a bilayer, a reductionistic approach focusing just on the lipid bilayer was expected to identify important physical mechanisms, present also in real cells. The chemist's interest in vesicles covers a vast range – from drug delivery to the use of vesicles as micro-reactors (as many of the contributions to this volume demonstrate).

*Giant Vesicles* Edited by P. L. Luisi and P. Walde

This article conveys the value and merits of a theoretical perspective. As theorists prefer simple well-defined systems, the focus is on the behavior of fluid single-component phospholipid bilayer vesicles. Generally speaking, this theory has several aims. It wants to create, define and clarify notions and concepts using simple model situations. It provides quantitative data through analytical or numerical calculations of existing or prospective experimental situations. Specifically, Section 2 describes briefly how to arrive at a fairly complete understanding of the equilibrium behavior of giant vesicles. Equilibrium refers here to a single vesicle under observation. It enjoys both mechanical equilibrium of its mean shape and thermal equilibrium of the shape fluctuations around this mean shape. Equilibration of a whole vesicle ensemble with respect to its size distribution would take much larger time-scales of days if not weeks. Section 3 deals with the interaction of a vesicle with a planar substrate. Section 4 describes dynamical phenomena in equilibrium and, in particular, the little that is known about nonequilibrium phenomena. The latter topic is one at the forefront of current research. This contribution is mostly nontechnical; a comprehensive list of references and more technical detail can be found in a review article [1].

## 2. EQUILIBRIUM SHAPES

### 2.1 Curvature models

From the very beginning of research in this field, the amazing variety of different shapes that fluid vesicles can acquire was simply stunning, [1–7]. Also see Chapter 10. Some of these shapes can transform into another if the external conditions, such as osmotic pressure or temperature, are changed. A natural approach for understanding the factors that determine these shapes and their transformations was to search for an appropriate (free) energy whose minimization would yield these shapes [8–15]. For a symmetrical membrane, whose chemical conditions on the two sides are identical, the flat conformation is locally the state of lowest energy. For a closed shape, which is necessarily nonflat, the selection of the correct energy has to be guided by the essential physical properties of closed bilayer membranes.

#### *(a) Length-scale separation and curvature model*

The bilayer is about 4 nm thick, a giant vesicle is about $10^4$ times larger, (Figure 7.1). This separation of length scales enables the membrane to be described as a two-dimensional surface embedded in three-dimensional space. Any such surface can locally be characterized by the *two* radii of curvature $R_1$ and $R_2$ from which are derived *two* quantities characterizing these curvatures. These are the mean curvature, $H \equiv (1/R_1 + 1/R_2)/2$, and the Gaussian curvature, $K \equiv 1/R_1R_2$. For cylindrical

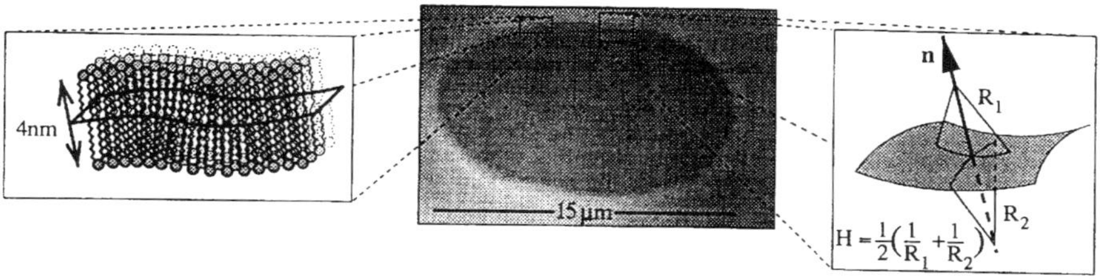


**Figure 7.1** Video microscopic image of a SOPC (1-stearoyl-2-oleoyl-*sn*-glycero-3-phosphocholine) vesicle (courtesy of H.-G. Döbereiner) and sketches of the bilayer (left) and a surface with mean curvature $H$ (right).

surfaces the Gaussian curvature vanishes, whereas for saddles the mean curvature is zero. Bending such a membrane costs local energy [8,9]

$$f = (\kappa/2)(2H)^2 + \kappa_G K, \tag{1}$$

where $\kappa$ and $\kappa_G$ are the ordinary and the Gaussian bending rigidity, respectively. For phospholipid membranes, a typical value for $\kappa$ is $10^{-12}$ erg.

### *(b) Fluidity, insolubility and incompressibility*

As the membrane is in the fluid state of matter (i.e. smectic A in liquid crystal terminology), it cannot withstand shear in its plane. Moreover, the solubility of the double chain phospholipids is extremely low. Therefore, there is practically no exchange of material between membrane and solution. This fact, together with the small compressibility of the membrane, implies that for almost all phenomena the membrane can be considered as locally incompressible.

### *(c) Large edge energy and fixed topology*

Any open edge of a membrane patch exposed to water costs considerable energy. This is why free membrane patches do not exist. Likewise, the topology of a vesicle does not change because this would imply the formation of transient edges. Therefore, only closed membrane configurations need to be considered, i.e. vesicles for which the energy

$$F_\kappa = (\kappa/2)\oint \mathrm{d}A(2H)^2 \tag{2}$$

is given by the integral of eq. (1) over its closed surface with total area $A \equiv 4\pi R^2$. From the Gauss–Bonnet theorem, the integral of the second term of eq. (1) depends only on the topology (i.e. the number of holes or handles) of a vesicle, not on its

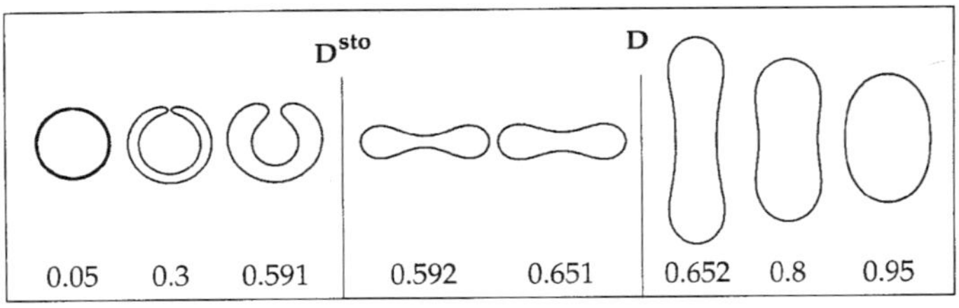


**Figure 7.2** Equilibrium shapes minimizing $F_\kappa$ in eq. (2) for given reduced volume $v$ between 0.05 and 0.95. With decreasing volume, the discontinuous transition $D$ separates the prolate/dumbbell branch from the oblate/discocyte branch. The latter are separated by the discontinuous transition $D^{sto}$ from the stomatocytes. The absolute size does not enter these shape calculations or those shown in Figure 3. Reproduced from *Phys. Rev. A*, **44**, 1182 (1991) by permission of the American Physical Society.

shape. Therefore, the Gaussian term can be discarded when considering vesicles of fixed topology. For example, all the shapes shown in Figure 7.2 are of spherical topology.

### *(d) Osmotic control of volume*

Experimentally, one fixes these systems to the osmotic conditions of the external solution. Even with pure water, there are necessarily osmotically active impurities around. As the membrane is permeable to water, the volume changes up to the point where practically no more osmotic pressure difference acts [9]. The bending moments that can be derived from eq. (2) are just too weak to balance any nonzero osmotic pressure. Therefore, vesicle configurations must be determined at a prescribed volume.

The constraints on area $A$ and volume $V$ are added with Lagrange multipliers $\bar{\Sigma}$ and $P$ to the energy in eq. (2). Minimization of the resulting expression must be performed numerically. Due to the scaling invariance of the bending energy, the shape corresponding to minimal energy depends only on one parameter, which is the reduced volume $v \equiv V/[(4\pi/3)(A/4\pi)^{3/2}]$ [13]. Neither the value of the bending rigidity $\kappa$ or the absolute size of the vesicle will affect the result at this stage. The shapes obtained from such a minimization for various values of $v$ are shown in Figure 7.2. They comprise three types: prolates/dumbbells, oblates/discocytes, and the cup-shaped stomatocytes. This model clearly captures the important aspects of vesicle shapes because shapes found previously in the laboratory are obtained theoretically. However, something important is missing as this minimization does not lead to pears or shapes with buds, let alone starfish vesicles. As these latter shapes have been observed experimentally, the model described so far is incomplete.

## 2.2 Bilayer aspect

The aspect missing so far concerns the bilayer feature of the membrane. It is not possible to describe the membrane just as a structureless two-dimensional surface. As one key signature of its molecular architecture, the bilayer architecture must be retained in the model. The two monolayers are tightly coupled but the exchange of lipid molecules between them is slow compared to the experimental time-scale of these observations. Therefore, the numbers $N^+$ and $N^-$ of molecules in the outer and inner monolayer, respectively, are conserved. The number difference $N^+ - N^-$ leads to the prefered area difference $\Delta A_0 = (N^+ - N^-)a_0$ between the two layers. Here, $a_0$ is the equilibrium area per lipid molecule. The actual area difference, $\Delta A = A^+ - A^-$, can be expressed as an integral over the closed surface as $\Delta A = 2D \oint \mathrm{d}AH$ where $D$ is the distance between the neutral surfaces of the two monolayers, i.e. roughly half the bilayer thickness. Deviations of the true area from the prefered area cost energy [15]:

$$F_{\mathrm{ADE}} = \frac{\alpha\pi\kappa}{2D^2A}(\Delta A - \Delta A_0)^2 \tag{3}$$

where $\alpha$ is somewhat model dependent, but close to 1. Even though appreciation of this bilayer feature had been around for a while [10,16,17], it required systematic theoretical and experimental work to show that eq. (3) represents this aspect correctly [15].

Minimizing the total energy, $F \equiv F_\kappa + F_{ADE}$, now leads to pears, budding and a multitude of starfish vesicles, some of which are shown in Figure 7.3 [7]. These shapes depend not just on the reduced volume $v$ but on one more parameter which basically is the number difference of lipid molecules in the two layers. These shapes of lowest energy can be arranged in a two-dimensional phase diagram [15,18]. As

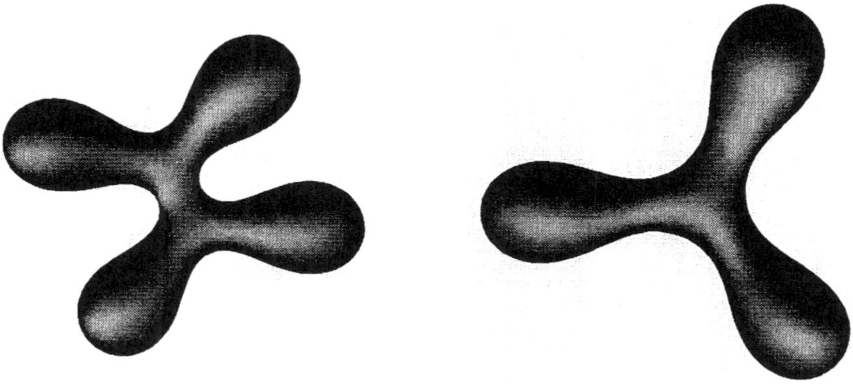

**Figure 7.3** Two calculated starfish vesicles obtained by minimizing $F_\kappa + F_{\mathrm{ADE}}$. Reproduced from *Europhys. Lett.* **33**, 403 (1996) by permission of the European Physical Society.

one changes temperature or the osmotic conditions, a trajectory in this phase diagram cuts across the different region of stability of the different shapes. Transitions from one shape to another, say from prolate to pear, can be abrupt (i.e. discontinuous, or first order) or smooth (i.e. continuous or second order). All experiments performed so far on the shapes of such vesicles seem to be compatible with this theoretical model.

The relevance of the bilayer feature can be tested noninvasively by observing a vesicle for a long time without changing its external conditions. Its shape should change slowly as an affect of flip–flop between the layers. The driving force for such a flip–flop can be derived from the gradient of the energy in eq. (3) with respect to number difference. After sufficient time, in a symmetric environment, the shape must settle into one of those shown in Figure 7.2, because by then the bilayer aspect has been relaxed. The time-scale for such a change should be related to the flip–flop time, which is believed to be in the range of days [20]. However, as the fraction $D/R \simeq 10^{-3}$ of flipped lipids would already lead to an observable shape change, such an experiment would yield results with in a much shorter observation time.

## 2.3 Spontaneous curvature

The discussion so far has been confined to membranes that are symmetric with respect to both the chemical composition of the two leaflets of the bilayer and the liquid environment. Originally, Helfrich suggested a spontaneous curvature model in which the term $2H$ in eqs (1) and (2) is replaced by the term $(2H - C_0)$ [9,11]. In this case, $C_0$ is the curvature that a free membrane patch cut from a vesicle in a Gedanken experiment would acquire spontaneously. One can show that minimizing such a modified energy eq. (2) without eq. (3) leads to the same multitude of shapes although with different stability properties [15,19]. Even for a vesicle under such asymmetric conditions the second energy term in eq. (3) must be retained if its two leaflets do not exchange molecules. In the latter case, the spontaneous curvature and the area difference term can then be combined into an effective spontaneous curvature.

A subtle refutation of the simple spontaneous curvature model without the bilayer aspect follows from the observation of vesicles of non-spherical topology, i.e. vesicles with holes (like anchor rings) or with handles. For vesicles with at least two holes or handles, the shape of lowest total energy is not unique but rather one fold continuously degenerate due to conformal invariance. This theoretical finding led to the prediction that such vesicles should permanently change their shape, a phenomenon termed conformal diffusion [21]. This was indeed verified experimentally somewhat later [22]. Apart from the esthetic pleasure when visualizing conformal transformations under the microscope, this observation also shows the spontaneous curvature model to be incomplete with out the additional term.

## 2.4 Fluctuations

The vesicle shapes can be obtained by minimizing the appropriate curvature energy subject to the geometric constraints. Video microscopy reveals that these shapes typically exhibit thermal fluctuations. How can these fluctuations be described?

It is instructive to start from a planar membrane. The local position of such a membrane with respect to the mean plane can be parametrized by a height variable $h(x, y)$. The energy given by eq. (2) then becomes

$$F = \frac{\kappa}{2}\int \mathrm{d}x\,\mathrm{d}y(\nabla^2 h)^2 = \frac{\kappa}{2}\int \frac{\mathrm{d}^2 q}{(2\pi)} q^4 h_{\mathrm{q}} h_{\mathrm{q}} \tag{4}$$

Here $\nabla^2 \equiv (\partial_x^2 + \partial_y^2)$ is the Laplace operator and $\mathbf{q}$ is a wave vector. The Fourier transformation is defined as $h(\mathbf{x}) \equiv \int h_{\mathbf{q}} e^{i\mathbf{q}\mathbf{x}} \mathrm{d}^2 q/(2\pi)^2$. Equipartition then tells us the mean amplitudes at temperature $T$:

$$\langle |h_{\mathbf{q}}|^2 \rangle = \frac{k_B T}{\kappa q^4} \tag{5}$$

where $\langle \ \ldots \ \rangle$ is the thermal expectation value taken with the Boltzmann weight, $\exp(-F/k_{\mathrm{B}}T)$, and $k_{\mathrm{B}}$ is the Boltzmann constant.

In principle, one could attempt to calculate the fluctuations around any of the mean shapes using a similar approach. In practice, the differential geometry involved is tricky and taking care of the geometrical constraints is not entirely trivial. The latter problem can be solved by introducing an effective tension and pressure difference whose numerical value are given by the Lagrange multiplier $\bar{\Sigma}$ and $P$, used to implement the area and volume constraints, respectively [19]. The former must be attacked numerically which is somewhat cumbersome. So far, only a few attempts have been made to calculate such fluctuation spectra around nontrivial mean shapes [23,24]. The notable exception is for quasispherical vesicles from whose fluctuation spectrum the value of the bending rigidity can be deduced by comparing data to theory [25–32].

For nonspherical vesicles progress has been made with in just one interesting mode such as the pear-like fluctuations in the prolate shape [33]. As the budding transitions approached, the amplitude of these fluctuations increases significantly, in accordance with the theoretical prediction.

## 2.5 More complex membranes

The theory described so far is complete for a single-component bilayer vesicle. Additional complexity arises from modifying the membrane. It is then important to consider first the properties of the corresponding free membrane without boundaries. In a second step, one should address the vesicle problem where additional complication may arise because of the closed topology and the associated constraints. There follows a subjective selection of such problems.

A slightly more complex system is a two-component membrane. If the two components mix locally and the compositions of the two monolayers are the same, a flat membrane still has lowest energy. However, if one of the components induces or prefers a certain spontaneous curvature, a coupling between local composition and local curvature arises [34,35]. Such an effect may become significant if, for instance budding happens. The daughter vesicle may then have a different composition than the mother vesicle [36]. For this type of more complex membrane the vesicle can still be described by a total energy $F$ of the form eq. (2) + eq. (3) where now parameters such as $\kappa$, $\kappa_G$, and $C_0$ depend on the properties of the two components.

Conceptually different is the situation where, due to the more complex membrane, additional energy terms occur which differ from eqs (2) or (3). An example is again given by a two-component membrane, now considered in a regime where fluid–fluid phase separation is supposed to occur. The additional energy in this case is the line energy associated with the domain boundary. For large enough domains, this energy can be relieved by forming a bud [37]. For domains on a vesicle, additional subtleties can arise because of the constraints [38].

Related systems are inclusions in membranes, which could model large proteins or rigid gel-like domains. Even if these inclusions are flat and symmetric they can affect the conformation and the fluctuations of the surrounding membrane via boundary conditions. This induces effective interactions between them [39,40]. Although most of these problems have been studied mainly for planar membranes, a recent study addresses inclusions on a spherical vesicle [41].

Another type of inclusion relates to polymers attached covalently to a fraction of the lipids of the bilayer. If these lipid anchors are far apart, the membrane tries to bend away from the polymer in order to increase the volume available to the fluctuating polymer. Therefore, the polymer creates a local spontaneous curvature [42]. If the polymers become denser, the membrane gets effectively thicker which leads to a strong increase of the bending rigidity. If such a decorated membrane forms a nonspherical vesicle, coupling between polymer composition and local curvature will lead to an inhomogeneous distribution of the anchored polymers.

## 3. ADHESION AND INTERACTION

### 3.1 Experimental details

From the point of view of having a well-defined system, an isolated vesicle freely floating in solution is almost perfect. For interesting applications or a higher level in complexity, interaction with other objects can be considered. The simplest case corresponds to adhesion to a rigid substrate [43]. For this case, a comprehensive set of experimental data can be obtained using reflection interference microscopy

[44,45]. This technique permits the study of the gross features of a bound vesicle, such as its contour radius or the size of the adhesion disc, and also of the fluctuation spectrum of the bound part. This experiment shows clearly that the problem of a vesicle adhering to a substrate has two complementary aspects, a macroscopic one (overall shape) and a mesoscopic one (fluctuation spectrum in the vicinity of the substrate). In a first step, recognizing this distinction is key for a successful theoretical modeling of this problem. In a second step, the two aspects must be reconciled self-consistently.

Although adhesion to a rigid substrate is the paradigmatic case, one can also consider adhesion of a vesicle to another vesicle or to another membrane. These situations have been studied experimentally using light microscopy [46] and micropipet aspiration [47,48].

## 3.2 Macroscopic model

From the macroscopic perspective, the simplest model for the adhesion of a vesicle to a substrate consists in adding an adhesion energy $F_a \equiv -WA^*$ to the bending energy given in eq. (2) [43]. Here, $W$ is the contact potential and $A^*$ the size of the adhesion patch with which the vesicle adheres to the substrate. Such a contact potential enters the boundary condition at the point of contact. First, the contact angle is necessarily zero as any sharp bend would have an infinite curvature energy. This implies that the membrane is curved only in one direction, and $1/R_2^* = 0$ along the line of contact. Second, the contact curvature obeys $1/R_1^* = (2W/\kappa)^{1/2}$, which follows from minimizing with respect to the area of contact [43]. This boundary condition, which does not depend on the area of the vesicle, holds for all variants of the curvature model. Depending on both the size of the vesicle, $R \equiv (A/4\pi)^{1/2}$, and the strength of the adhesion potential, $W$, adhesion can either be weak or strong. The crucial quantity is the scaled strength of the adhesion potential $w \equiv WR^2/\kappa$, (Figure 7.4).

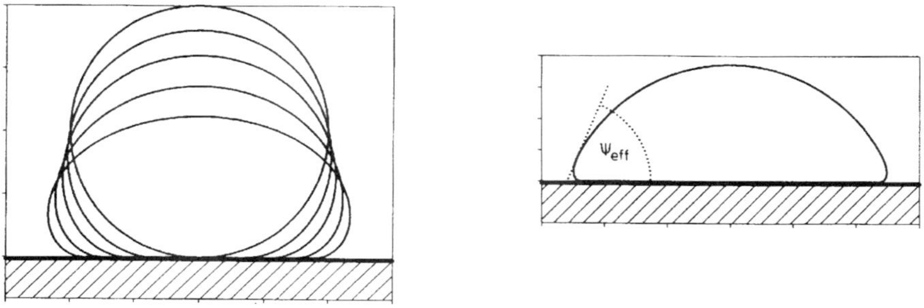


**Figure 7.4** Typical shapes of bound vesicles for weak adhesion (left) and strong adhesion (right). Only in the latter case can an effective contact angle $\Psi_{eff}$ be defined. Reproduced from *Phys. Rev. A*, **42**, 4768 (1990) by permission of the American Physical Society.

*(a) Weak adhesion (w of order 1)*

It turns out, somewhat surprisingly, that one needs a threshold value $w_c$ for adhesion to occur. If the adhesion strength is too small, the possible gain in adhesion energy is smaller than the cost in deforming its shape so that the vesicle will not adhere even in the presence of the weakly attractive substrate [43]. The specific value of $w_c$ depends on the reduced volume and the other parameters but is of order 1 [49]. For $w > w_c$, the vesicle is bound. Its contact area $A^*$ increases with increasing strength of the potential $W$. For $w \gg 1$ one enters the regime of strong adhesion.

*(b) Strong adhesion ($w \gg 1$)*

Physically, in this regime, the bending energy becomes irrelevant and the shape is determined solely by the desire to maximize the contact area $A^*$ for given total area and enclosed volume. The optimal shape is then a spherical cap with an effective contact angle $\Psi_{\text{eff}}$ determined by the geometrical constraints by an explicit relation $\Psi_{\text{eff}}(v)$ [50]. The limiting cases are $\Psi_{\text{eff}} = \pi$ for a sphere ($v = 1$) and $\Psi_{\text{eff}} = 0$ for a flat pancake ($v = 0$). Interestingly, this effective contact angle also obeys a Young equation [43]

$$W = \bar{\Sigma}(1 + \cos \Psi_{\text{eff}}). \tag{6}$$

In contrast to the adhesion of liquid droplets where $\bar{\Sigma}$ would be the surface tension, for vesicle adhesion $\bar{\Sigma}$ is not an independent quantity but rather given by the (numerical) value of the Lagrange multiplier used to implement the area constraint. As eq. (6) shows, $\bar{\Sigma}$ is a function of both the reduced volume through $\Psi_{\text{eff}}(v)$ and $W$.

Experimentally, the difference between strong and weak adhesion has been seen clearly. With light microscopy [46], the mutual interaction of vesicles has been studied and large rounded contact regions corresponding to weak adhesion were found. Spherical cap-like configurations are seen in micro-pipet experiments [48] as well as in freeze-fracture electron microscopy for smaller vesicles [51].

From the macroscopic point of view, the problem of a vesicle adhering to a substrate is thus solved if we know the strength of the contact potential $W$ which must be extracted from a more mesoscopic approach.

## 3.3 Mesoscopic model

The bound part of a closed vesicle is typically several nanometers away from the substrate and still shows thermal fluctuations [45]. These conditions require refinement of the model by paying due attention to the mesoscopic interactions between membrane and substrate. Various forces are involved in this interaction, such as van der Waals interaction, electrostatic contributions in the case of charged membranes, and short-range repulsive forces such as the hydration force or protrusion forces [52].

All these direct forces can be combined into an effective potential $V_d(l)$. This quantity gives the energy of a (fluctuationless) planar membrane at a distance $l$ from the substrate. Typically $V_d(l)$ increases dramatically for small $l$ and vanishes for large $l$. In between, there are either one or two minima. If these direct forces were the full story, the effective adhesion strength in the macroscopic approach could be taken as the value of the direct potential at the minimum $l_0$, i.e. $W = -V_d(l_0)$. However, the situation is more complex due to thermal fluctuations.

The presence of the substrate confines its fluctuations [53]. Naively, this effect can be described by a steric or fluctuation potential [50]:

$$V_f(\Sigma; l) = \left(\frac{6b^2(k_B T)^2}{\kappa l^2}\right)\left(\frac{y^2}{\sinh^2 y}\right) \quad \text{with} \quad y \equiv (\Sigma/bT)^{1/2} l/2. \tag{7}$$

If the lateral tension $\Sigma$ vanishes, the second bracket becomes unity. This form corresponds to the famous Helfrich potential [53]. The numerical prefactor $6b^2 \simeq 0.1$ can be estimated by Monte-Carlo simulations [54]. In the presence of a lateral tension $\Sigma$, which pulls at the membrane, the power law decay of the first bracket is cutoff by the second factor at a length scale $l_\Sigma \equiv (4bT/\Sigma)^{1/2}$ [55,52]. The best choice for $b$ in this case is $b = 1/2\pi$. The form of eq. (7) covering the full range of tension and separation values has been derived using a simple self-consistent picture [50].

Within a superposition approach, the fluctuation potential $V_f(\Sigma; l)$ should be added to the direct potential $V_d(l)$ to form the total potential $V(l) \equiv V_d(l) + V_f(\Sigma; l)$. The energy of the bound part of the vesicle, which is at a local distance $l(x, y)$ from the substrate, is then given by

$$F^a = \int dA \left\{\frac{\kappa}{2}(\nabla^2 l)^2 + V(l) + \bar{\Sigma}\right\}$$

$$\approx \frac{1}{2}\int \frac{d^2q}{(2\pi)^2}(\kappa q^4 + (\bar{\Sigma} - W)q^2 + V''(l_0))|h_{\mathbf{q}}|^2 \tag{8}$$

In the first equality we have to include $\bar{\Sigma}$ from the area constraint. The second equality follows after expanding for small fluctuations, $h(x, y) \equiv l(x, y) - l_0$, around the minimum at $l_0$. The area element d$A$ becomes $dA \approx dx\,dy(1 + (\nabla h)^2/2)$ and for the potential $V(l)$ holds $V(l) \approx -V''(l_0)h^2/2$.

The prefactor of the $q^2$ term in eq. (8) is the effective tension

$$\Sigma \equiv (\bar{\Sigma} - W) = -\frac{\cos \Psi_{\text{eff}}}{1 + \cos \Psi_{\text{eff}}} \tag{9}$$

which carries two contributions. The first arises from the area constraint, which acts like an external pulling force at the bound part of the membrane. The second one describes the fact that a fluctuating membrane exposes more area to the attractive potential minimum [56]. The final equality, which arises from using eq. (6), shows that for strong adhesion there is a universal relationship between the effective adhesion energy $W$ and the effective tension $\Sigma$ acting on the fluctuations of the bound part. This depends only on the reduced volume $v$ which determines $\Psi_{\text{eff}}$. This relationship states, in particular, that the effective tension becomes negative for those vesicles with a relatively small reduced volume for which $\Psi_{\text{eff}} < \pi/2$. Such a negative tension should show up in the measurable spectrum of fluctuations of the bound part [45], which can be obtained using equipartition on eq. (8) as

$$\langle |h_{\mathbf{q}}|^2 \rangle = \frac{k_B T}{\kappa q^4 + \Sigma q^2 + V''(l_0)} \tag{10}$$

The final step consists in determining not only the ratio between the tension $\Sigma$ and the depth of the adhesion potential $W$ as given by eq. (9) but also their individual value. As $W$ corresponds to the depth of the total potential at its minimum $l_0$,

$$W = -[V_{\text{d}}(l_0) + V_f(\Sigma; l_0)]. \tag{11}$$

With this equation, the problem of vesicle adhesion taking into account steric repulsion is closed and solvable for any given form of the direct potential $V_{\text{d}}(l)$. A specific example is shown in [50] where the distinction between $\Sigma$ and $\bar{\Sigma}$ was not made. This refinement is irrelevant for large reduced volume or tense vesicles, which were the focus in [50], but becomes important for small reduced volume, where the present picture predicts a negative $\Sigma$.

## 3.4 Adhesion and unbinding transition

Two mechanisms for an adhesion or unbinding transition have been described above. Ignoring fluctuations, the competition between adhesion energy and bending energy predicts a scale-dependent adhesion transition at $w_c(v)$. For a given $W$, this relationship implies that large vesicles with $R > R_c \equiv (w_c \kappa / W)^{1/2}$ are predicted to adhere, whereas smaller ones are free [43]. Physically, the large ones can take advantage of the adhesion energy without having to pay too much in bending.

The second mechanism happens via fluctuations. This unbinding transition has been studied mainly for free planar tensionless membranes [57,58]. The theory described above shows that the same threshold also holds for the closed topology except for quite tense vesicles [50]. In the latter case the fluctuations may be restricted so much that such a vesicle is still bound even though an open membrane patch would unbind due to fluctuations. This regime corresponds to genuine tension-induced adhesion [59]. The crossover between the two different mechanisms can be estimated to happen at a length scale in the 100 nm range [43]. So far, no clear experimental record of either type of transition has been given for vesicles.

# 4. NONEQUILIBRIUM DYNAMICS

## 4.1 Classification

Characteristic of giant vesicles is their slow dynamics, visible in the microscope. The typical time-scale for shape changes on the micron scale is about a second, which makes vesicles a perfect system for studying dynamical aspects directly. Dynamics in the micron world of vesicles is overdamped, i.e. inertial effects can safely be ignored [60]. This is somewhat counterintuitive to the experience of working with fluid phenomena, which is shaped largely by the so-called large Reynolds number regime, where inertia matters. For example, in the micron world, motion is reversible and objects come to rest immediately after the driving forces cease to act. For a classification of vesicle dynamics, one has to distinguish between the dynamics of equilibrium and nonequilibrium phenomena.

Equilibrium dynamics of vesicles comprises the dynamical fluctuations around locally stable mean shapes. Quantitatively, such fluctuations have been studied for quasi-spherical vesicles [29,61] and for prolate shapes in the vicinity of the budding transition [33]. A nontrivial example of dynamical equilibrium fluctuations has been observed at the prolate – oblate transition [62]. As the activation energy between the two locally stable shapes is just a few $k_B T$, occasionally thermal fluctuations are large enough to drive the vesicle into the other minimum. This system thus constitutes one of the few examples showing a thermally induced macroscopic bistability.

Nonequilibrium dynamics is the more challenging and less well explored regime. One should distinguish two classes. First, there is relaxation into a new equilibrium after a parameter change. This class comprises the decay of a metastable shape such as the budding process induced by raising the temperature [5,63]. The driving force in this case is the curvature energy which favors a different shape at the new external conditions. Conceptually related is the spectacular pearling instability of cylindrical vesicles, which develops upon the action of optical tweezers [64]. In this case, the laser generates a tension for which a homogeneous cylinder is no longer a stable configuration. The second class of nonequilibrium behavior refers to genuine nonequilibrium states of vesicles induced by external fields, such as hydrodynamic flow. The behavior of vesicles under such conditions has only recently been started to be explored theoretically [65,66] and by experiment [67].

## 4.2 Dissipative mechanisms and equation of motion

Over-damped dynamics are characterized by the competition between driving forces and dissipation. Driving forces for vesicles arise from bending energy and from external fields such as optical tweezers or hydrodynamic flow. Dissipation (or friction) takes place both in the surrounding liquid and in the membrane, in principle. For giant vesicles of micron size, the dominant dissipation is viscous dissipation in the embedding fluid [68,69]. Therefore, a full treatment of the hydrodynamics of this fluid using the Stokes approximation [60] is mandatory for a faithful description of the dynamics of membranes.

Dissipation in the membrane can be classified into three phenomena: drag between the two monolayers [70], shear viscosity within each layer, and permeation through the membrane. Calculation of the relaxation spectra of bending fluctuations involving the first two mechanisms show that on scales of microns and larger, hydrodynamic dissipation is dominant [71]. In the submicron range, friction between the layers becomes relevant. On even smaller scales of several tenths of nanometers, shear viscosity within each layer should be included. Finally, permeation through the membrane seems to be irrelevant on all length scales with the possible exception of membranes in the vicinity of a substrate [72].

Based on this hierarchy of dissipative mechanisms, for giant vesicles it is often permissible to ignore all but viscous dissipation in the surrounding liquid. Even though the permeability of the membrane for water is relatively large, the membrane can be modeled as impermeable for liquid. Any net flow of water through the membrane would lead to large osmotic pressure differences which cannot be sustained by the membrane. As a consequence, the normal velocity of the fluid at the membrane pushes along the membrane and leads to a shape change. Tangential motion of the fluid along the membrane induces lipid flow within the membrane. The exact boundary conditions between fluid and membrane are not entirely clear. Most work uses so-called nonslip boundary conditions for which the tangential velocity of liquid and the one of lipid are equal. However, one could allow for some slip with a phenomenological friction coefficient.

Based on these physical principles, it is possible to derive a general equation of motion for the shape change of a membrane or vesicle by eliminating the surrounding liquid [73–75]. The resulting equation of motion for the vesicle shape is nonlocal, which signifies the long-range interaction mediated by the hydrodynamics. Except for an expansion in simple geometries, no general or comprehensive solution of this equations exists. Therefore, most work deals with specific physical situations. Two of the most explored situations are discussed next.

## 4.3 Pearling instability

The discovery of the pearling instability of cylindrical vesicles induced by optical tweezers [64] generated enormous theoretical activity [76–80] in order to explain its various stages. Briefly, the experimental situation is as follows, (Figure 7.5). Initially, cylindrical vesicle tubes are prepared with a length of several hundred microns and radii in the one micron range. These cylinders terminate in somewhat undefined lipid globules. Visible thermal fluctuations show that the cylinders are not under tension initially. Then, a laser tweezer is focussed onto the tubes. After a short delay, a

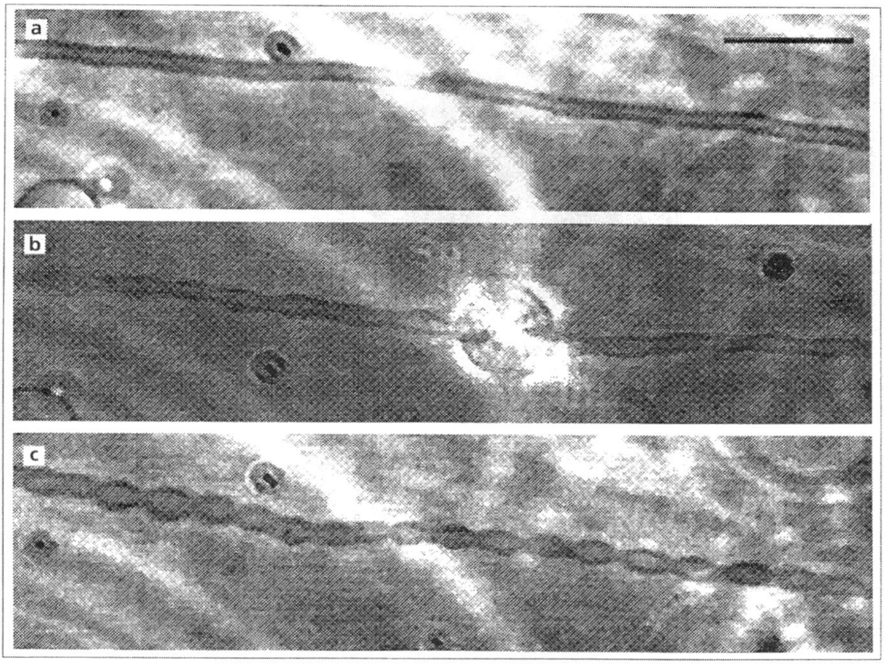


**Figure 7.5** Pearling instability of tubular vesicles made of DMPC: (a) the initially fluctuating tube after (b) being hit by a laser tweezer develops (c) an outwards propagating peristaltic shape mode. Bar corresponds to 10 μm. Courtesy of R. Bar-Ziv and E. Moses. Reproduced from *Phys. Rev. Lett.* **73**, 1392 (1994) by permission of the American Physical Society.

peristaltic shape modulation develops which propagates as a front outwards roughly with constant velocity. The remaining pattern has a well-defined wavelength. Upon longer action of the tweezer, the pattern develops into a string of pearls which self-aggregate finally in a necklace-like fashion even after the tweezer is shut off. Data have been collected for a whole set of tube radii and values for laser power [81]. The physical explanation of this phenomenon proceeds in four steps.

*(a) Spreading of tension*

The laser spot creates a sink for the membrane material. Dielectric energy is gained by bringing membrane into the spot and expelling liquid instead. As the energy gain is proportional to the area of lipid brought to the sink, this effect corresponds to a local external tension $\Sigma \simeq 10^{-3}$ erg cm$^{-2}$ pulling laterally on the membrane [76]. If the cylinder was perfectly straight and incompressible, such a local tension would spread infinitely fast along it. In reality, the cylinder initially fluctuates. Spreading the tension requires the straightening of these thermal fluctuations. It turns out that a thorough calculation of this effect is a surprisingly difficult problem. Progress can be made by exploiting the solution to the somewhat simpler problem of pulling at a semiflexible polymer [82,83] as follows.

When the tension has spread a length $\xi(t)$ along the cylinder, thermal fluctuations over this length have been pulled out. The area $A_\xi$ stored in these fluctuations scales as $A_\xi \sim \xi R k_B T/\kappa$. Thus, the membrane is pulled into the trap with a velocity $v \sim \mathrm{d}(A_\xi/R)/\mathrm{d}t \sim (k_B T/\kappa)\mathrm{d}\xi/\mathrm{d}t$. The power spent by the laser pulling the membranes is $P_\Sigma = 2\pi R\Sigma v$. Equating this power with the one dissipated in the fluid, $P_\eta \sim \eta(v/R)^2 R^2 \xi(t)$, one finds from

$$\xi(t) \sim (\kappa\Sigma R/k_B T\eta)^{1/2} t^{1/2} \tag{12}$$

a diffusive spreading of the tension.

### (b) Modified Rayleigh instability

Even though the above model has shown that the resulting tension obeys a diffusive profile, it is possible to simplify the problem and consider a membrane cylinder under homogeneous tension. By ignoring the bending energy, such a cylinder is unstable with respect to modulations for all wavelengths larger than $2\pi R$, according to the classical Rayleigh instability. With the bending energy included, the energy for a sinusoidal deformation of the local radius, $r(z) = R[1 + h_q \cos(qz/R)]$, becomes [76]

$$F_\kappa = (1/2)E_q|h_q|^2 \quad \text{with} \quad E_q \equiv \kappa[3/2 - \sigma + (\sigma - 1/2)q^2 + q^4] \tag{13}$$

where $\sigma \equiv \Sigma R^2/\kappa$ and $q$ is the dimensionless wavenumber. The bending energy thus sets a threshold for the instability. For $\sigma < 3/2$ nothing happens. Indeed, experiments have shown that a minimum laser power is required for the instability to set in. Transferring laser power into tension shows reasonable quantitative agreement for this threshold.

In order to select a wavelength, it is necessary to look at the dynamical problem. Solution of the hydrodynamic equations described qualitatively above, yields a dispersion relationship of the type [76]

$$\partial_t h_q = -\Gamma_q E_q h_q, \tag{14}$$

where $\Gamma_q$ is the wavenumber-dependent friction factor. The fastest growing mode occurs where $\Gamma_q E_q$ peaks which happens at about $q \simeq 0.65$, significantly smaller than the experimentally observed value ($q \gtrsim 0.8$). Indeed, selecting the fastest growing mode cannot be the full story because it is not clear how to obtain a front, let alone how to calculate its velocity, from this approach.

*(c) Front propagation*

Front propagation is obtained by applying the concept of marginal stability [77]. This theory enables calculation of the velocity of a front if the linear dispersion relationship eq. (14) is known. Physically, this criterion requires that the leading edge of the front is stable with respect to all possible deformations. Application to the pearling problem leads to good agreement for both the wavelength of the pattern and the front velocity [81].

*(d) Late stage dynamics*

The late stage dynamics of pearl formation and pearl motion is not yet fully understood even though some progress in modeling the hydrodynamics of this process has been made [80].

The interaction between lasers and membranes generates many other interesting phenomena, as described in an excellent review [81].

## 4.4 Vesicle in shear flow

The strong flexibility of membranes implies that equilibrium shapes of vesicles can deform substantially when placed in hydrodynamic flow fields. A paradigmatic hydrodynamic flow is simple shear. The behavior of an oblate vesicle under these conditions is shown in Figure 7.6. The shape gradually elongates and tilts after the flow is turned on. Finally, a stationary ellipsoidal shape is reached with an inclination angle that depends on the reduced volume but only slightly on the shear rate. As the membrane is fluid, it tank-treads around this stationary shape. These results are obtained by solving the equation of motion of the membrane subject to the appropriate boundary conditions on a dynamically triangulated surface [65].

The only shear-flow experiment on vesicles available so far operates in a slightly different regime [67]. Initially, these vesicles were quasi-spherical. Shear flow elongates them by pulling out area stored in the thermal fluctuations. The theory must therefore include fluctuations giving up the purely deterministic approach. For a general shape, this would be very hard but it becomes possible near the sphere [75]. The results show an inclination angle that depends on reduced volume and shear rate approaching the deterministic one in the limit of strong shear. Likewise, the ellipticity is linear in the applied shear and settles into a finite value for strong shear due to the geometrical constraints.

These studies are certainly only a first step towards understanding the behavior of vesicles in hydrodynamic flow. It will be interesting to investigate both mutual interactions and adhesion at a substrate under flow [84]. Finally, these results could become important in the rheology of dense vesicle suspensions.

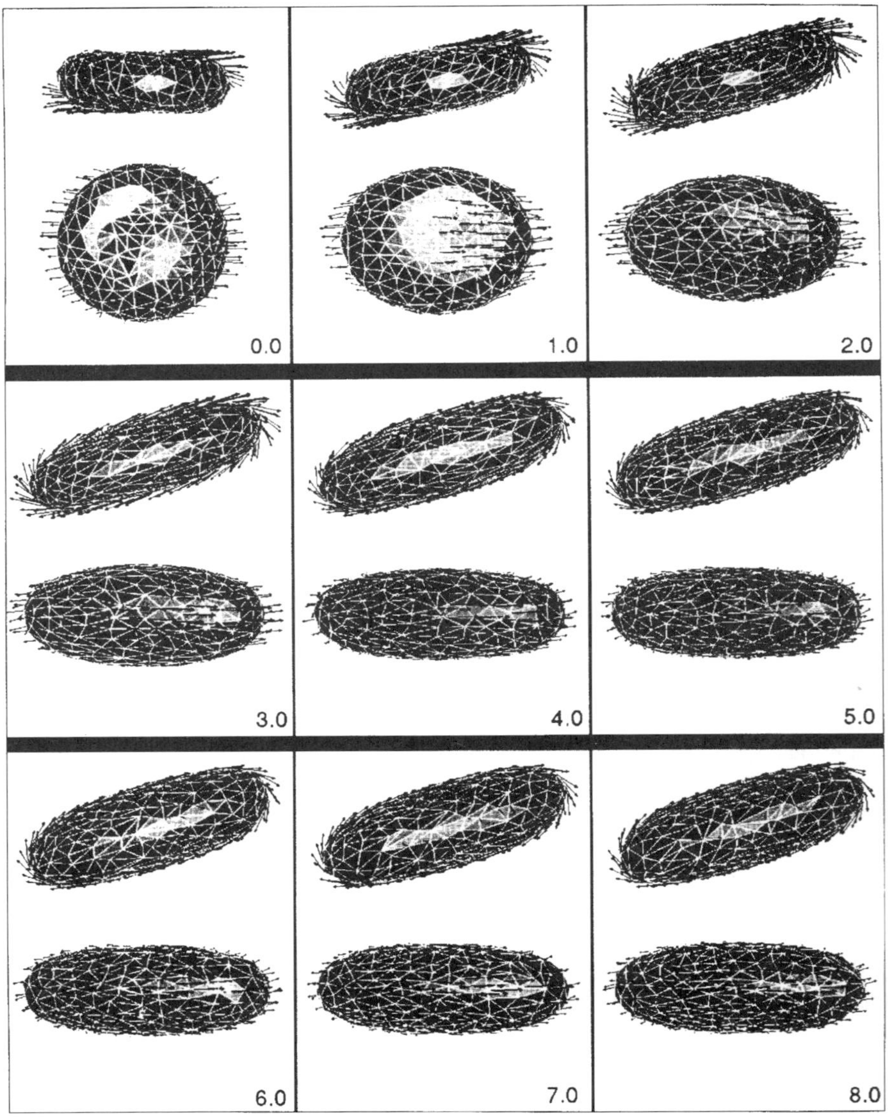


**Figure 7.6** Calculated shape transformation of an oblate vesicle in shear flow $v_x = \dot{\gamma} z$. The numbers in each frame refer to time in seconds after the shear flow is turned on. They hold for a vesicle with $v = 0.8$, shear rate $\dot{\gamma} = 1/\,s^{-1}$, bending rigidity $\kappa = 10^{-12}$ erg, and size $R = 10\,\mu m$. In each frame, the top picture shows the view onto the $(x, z)$ plane. The bottom one shows the view onto the $(x, y)$ plane. The arrows give the local velocity of the membrane.

## 5. CONCLUSIONS

The systematic theoretical study of fluid vesicles in close collaboration with experiments has lead to a comprehensive, quantitative understanding of the equilibrium properties of this paradigmatic soft matter system. They arise from a competition between curvature energy, area-difference elasticity, and geometrical constraints on area and volume. Adhesion and interactions involve a subtle interplay between fluctuations and constraints leading to two types of adhesion or unbinding transitions. Dynamics is ubiquitous for vesicle because of the slow time-scales. Case studies for both the pearling instability and vesicles in shear flow have identified basic principles acting in nonequilibrium.

Based on such solid ground, it will be possible to address more complex systems. Complexity necessarily introduces a larger parameter space for which a comprehensive analysis may not become possible or even desirable. It will then be crucial to focus on selected regions in this parameter space. The insight, gained with the simple model system described in this article, can help and guide such selection. Of course, experience also tells us that in this field the most interesting developments cannot be anticipated.

## 6. REFERENCES


1. U. Seifert, *Adv. Phys.*, **46**, 13 (1997).
2. H. Hotani, *J. Mol. Biol.*, **178**, 113 (1984).
3. E. Sackmann, H. P. Duwe and H. Engelhardt, *Faraday Discuss Chem. Soc.*, **81**, 281 (1986).
4. K. Berndl, J. Käs, R. Lipowsky, E. Sackmann and U. Seifert, *Europhys. Lett.*, **13**, 659 (1990).
5. J. Käs and E. Sackmann, *Biophys. J.*, **60**, 825 (1991).
6. E. Farge and P. F. Devaux, *Biophys. J.*, **61**, 347 (1992).
7. W. Wintz, H.-G. Döbereiner and U. Seifert, *Europhys. Lett.*, **33**, 403 (1996).
8. P. B. Canham, *J. Theoret. Biol.*, **26**, 61 (1970).
9. W. Helfrich, *Z. Naturforsch.*, **28c**, 693 (1973).
10. E. Evans, *Biophys. J.*, **14**, 923 (1974).
11. H. J. Deuling and W. Helfrich, *J. Physique*, **37**, 1335 (1976).
12. S. Svetina and B. Zeks, *Eur. Biophys. J.*, **17**, 101 (1989).
13. U. Seifert, K. Berndl, and R. Lipowsky, *Phys. Rev. A*, **44**, 1182 (1991).
14. L. Miao, B. Fourcade, M. Rao, M. Wortis and R. K. P. Zia, *Phys. Rev. A*, **43**, 6843 (1991).
15. L. Miao, U. Seifert, M. Wortis and H.-G. Döbereiner, *Phys. Rev. E*, **49**, 5389 (1994).
16. W. Helfrich, *Z. Naturforsch.*, **29c**, 510 (1974).
17. S. Svetina and B. Zeks, *Biochim. et Biophys. Acta*, **42**, 86 (1983).
18. M. Jaric, U. Seifert, W. Wintz and M. Wortis, *Phys. Rev. E*, **52**, 6623 (1995).
19. U. Seifert, *Z. Phys. B*, **97**, 299 (1995).
20. R. Homan and H. J. Pownall, *Biochim. Biophys. Acta*, **938**, 155 (1988).
21. F. Jülicher, U. Seifert and R. Lipowsky, *Phys. Rev. Lett.*, **71**, 452 (1993).
22. X. Michalet and D. Bensimon, *Science*, **269**, 666 (1995).
23. M. A. Peterson, H. Strey and E. Sackmann, *J. Phys. II France*, **2**, 1273 (1992).
24. M. Wortis, M. Jaric and U. Seifert, *J. Mol. Liq.*, **71**, 195 (1997).

25. M. B. Schneider, J. T. Jenkins, and W. W. Webb, *J. Phys. France*, **45**, 1457 (1984).
26. H. Engelhardt, H. P. Duwe and E. Sackmann, *J. Physique Lett.*, **46**, L395 (1985).
27. S. T. Milner and S. A. Safran, *Phys. Rev. A*, **36**, 4371 (1987).
28. J. F. Faucon, M. D. Mitov, P. Meleard, I. Bivas and P. Bothorel, *J. Phys. France*, **50**, 2389 (1989).
29. H. P. Duwe, J. Käs, and E. Sackmann, *J. Phys. France*, **51**, 945 (1990).
30. G. Niggemann, M. Kummrow and W. Helfrich, *J. Phys. II France*, **5**, 413 (1995).
31. W. Haeckl, U. Seifert and E. Sackmann, *J. Phys. II France*, **7**, 1141 (1997).
32. P. Meleard, C. Gerbeaud, T. Pott and L. Fernandez-Puente, *Biophys. J.*, **72**, 2616 (1997).
33. H.-G. Döbereiner, E. Evans, U. Seifert and M. Wortis, *Phys. Rev. Lett.*, **75**, 3360 (1995).
34. C. Gebhardt, H. Gruler and E. Sackmann, *Z. Naturforsch.*, **32c**, 581 (1977).
35. S. Leibler, *J. Physique*, **47**, 507 (1986).
36. U. Seifert, *Phys. Rev. Lett.*, **70**, 1335 (1993).
37. R. Lipowsky, *J. Phys. II France*, **2**, 1825 (1992).
38. F. Jülicher and R. Lipowsky, *Phys. Rev. Lett.*, **70**, 2964 (1993).
39. M. Goulian, R. Bruinsma and P. Pincus, *Europhys. Lett.*, **22**, 145 (1993).
40. N. Dan, A. Berman, P. Pincus and S. Safran, *J. Phys. II France*, **4**, 1713 (1994).
41. P. G. Dommersnes, J. B. Fournier and P. Galatola, *Europhys. Lett.*, **42**, 233 (1998).
42. R. Lipowsky, *Europhys. Lett.*, **30**, 197 (1995).
43. U. Seifert and R. Lipowsky, *Phys. Rev. A*, **42**, 4768 (1990).
44. A. Zilker, H. Engelhardt and E. Sackmann, *J. Physique*, **48**, 2139 (1987).
45. J. O. Rädler, T. J. Feder, H. H. Strey and E. Sackmann, *Phys. Rev. E*, **51**, 4526 (1995).
46. R. M. Servuss and W. Helfrich, *J. Phys. France*, **50**, 809 (1989).
47. E. Evans, *Biophys. J.*, **31**, 425 (1980).
48. E. Evans and W. Rawicz, *Phys. Rev. Lett.*, **64**, 2094 (1990).
49. R. Lipowsky and U. Seifert, *Mol. Cryst. Liq. Cryst.*, **202**, 17 (1991).
50. U. Seifert, *Phys. Rev. Lett.*, **74**, 5060 (1995).
51. S. M. Bailey, S. Chiruvolu, J. N. Israelachvili and J. A. N. Zasadzinski, *Langmuir*, **6**, 1326 (1990).
52. R. Lipowsky, in *Structure and dynamics of membranes – from cells to vesicles*, Vol. 1B of *Handbook of biological physics*, R. Lipowsky and E. Sackmann (eds), Elsevier, Amsterdam, 1995, pp. 521 – 602.
53. W. Helfrich, *Z. Naturforsch.*, **33a**, 305 (1978).
54. R. Lipowsky and B. Zielinska, *Phys. Rev. Lett.*, **62**, 1572 (1989).
55. W. Helfrich and R. M. Servuss, *Il Nuovo Cimento*, **3D**, 137 (1984).
56. R. Lipowsky and U. Seifert, unpublished results.
57. R. Lipowsky and S. Leibler, *Phys. Rev. Lett.*, **56**, 2541 (1986).
58. M. Mutz and W. Helfrich, *Phys. Rev. Lett.*, **62**, 2881 (1989).
59. W. Helfrich, in *Structure and Dynamics of Membranes – From Cells to Vesicles*, Vol. 1B of *Handbook of biological physics*, R. Lipowsky and E. Sackmann (eds), Elsevier, Amsterdam, 1995, pp. 691 – 722.
60. J. Happel and H. Brenner, *Low Reynolds number hydrodynamics*, Noordhoff International Publishing, 1973.
61. P. Meleard, J. F. Faucon, M. D. Mitov and P. Bothorel, *Europhys. Lett.*, **19**, 267 (1992).
62. H.-G. Döbereiner and U. Seifert, *Europhys. Lett.*, **36**, 325 (1996).
63. H. G. Döbereiner, E. Evans, M. Kraus, U. Seifert and M. Wortis, *Phys. Rev. E*, **55**, 4458 (1997).
64. R. Bar-Ziv and E. Moses, *Phys. Rev. Lett.*, **73**, 1392 (1994).
65. M. Kraus, W. Wintz, U. Seifert and R. Lipowsky, *Phys. Rev. Lett.*, **77**, 3685 (1996).
66. R. Bruinsma, *Physica A*, **234**, 249 (1996).

67. K. H. De Haas, C. Blom, D. van den Ende, M. H. G. Duits and J. Mellema, *Phys. Rev. E*, **56**, 7132 (1997).
68. L. Kramer, *J. Chem. Phys.*, **55**, 2097 (1971).
69. F. Brochard and J. F. Lennon, *J. Physique*, **36**, 1035 (1975).
70. E. Evans, A. Yeung, R. Waugh and J. Song, in *The Structure and Conformation of Amphiphilic Membranes*, Vol. 66 of *Springer Proceedings in Physics*, R. Lipowsky, D. Richter, and K. Kremer (eds), Springer, Berlin, 1992, pp. 148 – 153.
71. U. Seifert and S. A. Langer, *Europhys. Lett.*, **23**, 71 (1993).
72. J. Prost, J.-B. Manneville and R. Bruinsma, *Eur. Phys. J. B*, **1**, 465 (1998).
73. M. Doi and S. F. Edwards, *The Theory of Polymer Dynamics*, Clarendon Press, Oxford, 1986.
74. W. Cai and T. C. Lubensky, *Phys. Rev. E*, **52**, 4251 (1995).
75. U. Seifert, *Eur. Phys. J. B.*, **8**, 405 (1999).
76. P. Nelson, T. Powers and U. Seifert, *Phys. Rev. Lett.*, **74**, 3384 (1995).
77. R. E. Goldstein, P. Nelson, T. Powers and U. Seifert, *J. Phys. II France*, **6**, 767 (1996).
78. R. Granek and Z. Olami, *J. Phys. II France*, **5**, 1349 (1995).
79. P. D. Olmsted and F. C. MacIntosh, *J. Phys. II France*, **7**, 139 (1997).
80. J. L. Goveas, S. T. Milner and W. B. Russel, *J. Phys. II France*, **7**, 1185 (1997).
81. R. Bar-Ziv, E. Moses and P. Nelson, *Biophys. J.*, **75**, 294 (1998).
82. U. Seifert, W. Wintz and P. Nelson, *Phys. Rev. Lett.*, **77**, 5389 (1996).
83. A. Ajdari, F. Jülicher and A. Maggs, *J. Phys. I France*, **7**, 823 (1997).
84. U. Seifert, *Phys. Rev. Lett.*, **83**, 876 (1999).

# Chapter 8

# Free Energy of a Fluctuating Vesicle. Influence of the Fluctuations on the Laplace Law

*ISAK BIVAS*

*Bulgarian Academy of Sciences, Sofia, Bulgaria*

## 1. INTRODUCTION

It is known that if bending moments act on a deformed membrane, the Laplace law is modified [1]. The problem is even more complicated in the case of a fluctuating membrane. The membrane's out-of-plane fluctuations change its effective elasticity [2,3] and renormalize the tension of the membrane [4]. The aim of the present work is to find the relationship between the difference of the hydrostatic pressure $\Delta p$ inside and outside a fluctuating giant vesicle, its tension $\sigma$, and its elastic constants. For this purpose, the free energy of the vesicle is calculated.

## 2. DEFINITIONS

Let $k_s$ be the stretching elasticity of the membrane, and let $k_c$, $\bar{k}_c$, and $c_0$ be the bending elasticity, the saddle splay bending elasticity, and the spontaneous curvature of the membrane [5,6]. Later on, attention is restricted to a quasispherical flaccid fluctuating vesicle. Only the case of fluctuations with magnitudes much less than the radius of the vesicle are considered. Let the area of the tension-free flat membrane of the vesicle be $S_0$, and its volume (assumed constant in time) be $V$. Choose a laboratory frame of reference $XYZ$ with origin $O$ inside the vesicle. Let $\theta$, $\varphi$ be the

*Giant Vesicles* Edited by P. L. Luisi and P. Walde

polar angles measured in this system. Let $R'(\theta, \varphi, t)$ be the distance between $O$ and the point where the line passing through $O$ with direction $(\theta, \varphi)$ penetrates the membrane of the vesicle. Let $R$ be the radius of the sphere with volume $V$, $V = \frac{4}{3}\pi R^3$. Define a function $u(\theta, \varphi, t)$ in the following way:

$$R'(\theta, \varphi, t) = R[1 + u(\theta, \varphi, t)] \tag{1}$$

The function $u(\theta, \varphi, t)$ can be expanded in spherical harmonics

$$Y_n^m(\theta, \varphi) = \frac{1}{\sqrt{2\pi}} e^{im\varphi} \bar{P}_n^m(\theta)$$

where $\bar{P}_n^m(\theta)$ are the normalized Legendre polynomials:

$$u(\theta, \varphi, t) = \sum_{n=0}^{n_{\max}} \sum_{m=-n}^{n} u_n^m(t) Y_n^m(\theta, \varphi) \tag{2}$$

The quantity $n_{\max}$ is determined by the relationship $n_{\max} \sim N^{0.5}$ where $N$ is the number of lipid molecules in the membrane of the vesicle. Choose the origin $O$ of the laboratory frame of reference $XYZ$ in such a way that $u_1^m = 0$. By using the approaches of Helfrich [7] and Milner and Safran [8], the energy $H_e$ of the membrane of the fluctuating vesicle can be expressed in the following way:

$$H_e(t) = \sum_{n=2}^{n_{\max}} \sum_{m=-n}^{n} \frac{1}{2} k_c (n-1)n(n+1)(n+2)|u_n^m(t)|^2$$

$$-2k_c c_0 \frac{S(t)}{R} + \frac{1}{2} k_s \frac{(S(t) - S_0)^2}{S_0} + 8\pi k_c + 4\pi \bar{k}_c \tag{3}$$

where

$$S(t) = 4\pi R^2 + \delta S = 4\pi R^2 + \frac{R^2}{2} \sum_{n=2}^{n_{\max}} \sum_{m=-n}^{n} (n-1)(n+2)|u_n^m(t)|^2 \tag{4}$$

the excess area $\delta S$ being expressed as a function of the amplitudes $u_n^m(t)$. On the basis of these results, calculate the free energy $F(R, S_0, N_{\max})$ of the vesicle.

## 3. THEORY

The free energy $\boldsymbol{F}(R, S_0, n_{\max})$ of a vesicle with area $S_0$ of its tension free membrane, radius $R$ and $n_{\max} \sim N^{0.5}$ is

$$F(R, S_0, n_{\max}) = -kT \ln Z \tag{5}$$

where $k$ is the Boltzmann constant, $T$ is the temperature, and $Z$ is the statistical sum of the system. Later, this section presents a theoretical approach enabling calculation of the function $F(R, S_0, n_{\max})$.

Let $\Sigma_{\delta S}$ be a hypersurface in the space of amplitudes $|u_n^m|$ having the property, that in each of its points the area of the vesicle corresponding to this point is $\delta S$. Then,

$\delta S > 0$. Let $\Sigma_{\delta S + \mathrm{d}\delta S}$ have the same property, the corresponding area being $\delta S + \mathrm{d}\delta S$. Let $dW$ be the hypervolume in the space of amplitudes, enclosed between $\Sigma_{\delta S}$ and $\Sigma_{\delta S + \mathrm{d}\delta S}$. A normalized variable $s = k_c/kT \times \delta S/R^2$ is introduced; define the function $\phi(s, n_{\max})$ in the following way:

$$\mathrm{d}Z = Z_0' \int_{\mathrm{d}W} \exp\left[-\frac{1}{kT}(H_e(|u_2^{-2}|, \ldots, |u_{n_{\max}}^{n_{\max}}|, S_0, R))\right] d|u_2^{-2}| \cdots d|u_{n_{\max}}^{n_{\max}}|$$

$$= Z_0 \int_0^\infty \exp\left[-\frac{\Psi(s, R, S_0, n_{\max})}{kT}\right] \mathrm{d}s, \qquad (6)$$

where the function $\Psi(s, R, S_0, n_{\max})$ is:

$$\Psi(s, R, S_0, n_{\max}) = kT\phi(s, n_{\max}) - 2k_c c_0 \frac{4\pi R^2 + \frac{kT}{k_c} R^2 s}{R}$$

$$+ \frac{1}{2} k_s \frac{\left(4\pi R^2 + \frac{kT}{k_c} R^2 s - S_0\right)^2}{S_0} \qquad (7)$$

First, derive modifications to the Laplace law, assuming that the function $\phi(s, n_{\max})$ is known. For this aim, a standard procedure is applied. That is, the minimum and the second derivative of the function $\Psi(s, R, S_0, n_{\max})$ is found with respect to $s$, this function is then expanded in a Taylor series around its minimum, the terms containing the third and higher derivatives are truncated, and the integration in the right-hand side of eq. (6) will be explicitly done, giving the possibility to calculate $Z$ and $F(\mathrm{R}, S_0, n_{\max})$. The procedure is correct for high enough values of $k_s$, provided that the contribution of the term containing the second derivative in the Taylor series is much higher than the contributions of the terms containing higher order derivatives. This condition is fulfilled for typical values of the elastic constants of a lipid bilayer. The final result is as follows. Let $\bar{s}(R, S_0, n_{\max})$ be a solution with respect to $s$ of the equation

$$\left(\frac{\partial \Psi}{\partial s}\right)_{s=\bar{s}(R, S_0, n_{\max})} = 0 \qquad (8)$$

The free energy $F(R, S_0, n_{\max})$ is

$$F(R, S_0, n_{\max}) = \Psi[\bar{s}(R, S_0, n_{\max}), R, S_0, n_{\max}]$$

$$+ \frac{1}{2} kT\, ln\left[\frac{\left(\frac{\partial^2 \Psi}{\partial s^2}\right)_{s=\bar{s}(R, S_0, n_{\max})}}{kT}\right] + \text{constant} \qquad (9)$$

The hydrostatic pressure difference $\Delta p$ can be expressed via the derivatives of $F(R, S_0, n_{\max})$ in the following way [9]:

$$\Delta p = \frac{1}{4\pi R^2}\frac{\partial F}{\partial R} \tag{10}$$

Consequently, knowledge of the function $\phi(s, n_{\max})$ enables calculation of both $\bar{s}(R, S_0, n_{\max})$ and the modifications of the Laplace law due to the thermal fluctuations. In order to calculate $\phi(S, R, n_{\max})$ consider an imaginary vesicle, referred to from now on as the 'model vesicle', with zero spontaneous curvature, zero stretching elasticity, fixed tension $\sigma$, reduced tension $\bar{\sigma} = \sigma R^2/k_c$, and the same bending elasticity $k_c$, number of molecules $N$, and radius $R$ as the real vesicle. This is exactly the model proposed by Milner and Safran [8], and some of the results of their calculations are used below.

Let $Z_\sigma$ be the statistical sum of the model vesicle. Using the variables introduced above, this can be expressed as

$$Z_\sigma = Z_\sigma^0 \int_0^\infty \exp\{-[\phi(s, n_{\max}) + \bar{\sigma}s]\}\mathrm{d}s \tag{11}$$

The subscript 'mv' is used to distinguish terms relating to the model vesicle from those for the real vesicle. Using the Milner and Safran approach [8], it is possible to find the exact mean values $\bar{s}_{\mathrm{mv}}(\bar{\sigma}, n_{\max})$, $\overline{(s - s_{\mathrm{mv}})^2}$, $\overline{(s - s_{\mathrm{mv}})^3}$, etc., of the model vesicle. The result for the first three moments is:

$$\bar{s}_{\mathrm{mv}}(\bar{\sigma}, n_{\max}) = \frac{1}{2}\sum_{n=2}^{n_{\max}} \frac{2n+1}{n(n+1)+\sigma} \tag{12}$$

$$\overline{(s - \bar{s}_{\mathrm{mv}}(\bar{\sigma}, n_{\max}))^2} = \frac{1}{2}\sum_{n=2}^{n_{\max}} \frac{2n+1}{[n(n+1)+\bar{\sigma}]^2} \tag{13}$$

$$\overline{(s - \bar{s}_{\mathrm{mv}}(\bar{\sigma}, n_{\max}))^3} = \sum_{n=2}^{n_{\max}} \frac{2n+1}{[n(n+1)+\sigma]^3} \tag{14}$$

Let us expand the function $\phi(s, n_{\max}) + \bar{\sigma}s$ in a Taylor series around $s = \bar{s}_{\mathrm{mv}}(\bar{\sigma}, n_{\max})$:

$$\phi(s, n_{\max}) + \bar{\sigma}s = \phi(\bar{s}_{\mathrm{mv}}, n_{\max}) + \bar{\sigma}\bar{s}_{\mathrm{mv}} + \left[\left(\frac{\partial\phi}{\partial s}\right)_{s=\bar{s}_{\mathrm{mv}}} + \bar{\sigma}\right](s - \bar{s}_{\mathrm{mv}})$$

$$+\frac{1}{2}\left(\frac{\partial^2\phi}{\partial s^2}\right)_{s=\bar{s}_{\mathrm{mv}}}(s - \bar{s}_{\mathrm{mv}})^2 + \frac{1}{6}\left(\frac{\partial^3\phi}{\partial s^3}\right)_{s=\bar{s}_{\mathrm{mv}}}(s - \bar{s}_{\mathrm{mv}})^3 \cdots \tag{15}$$

Assuming that the contribution of the terms containing the second and third derivative to the statistical sum dominates the contributions of the higher order ones, truncate them in eq. (15). Using eqs. (12) and (13) this enables the expression of $\bar{s}_{mv}(\bar{\sigma}, n_{max})$ and $\overline{(s - \bar{s}_{mv})^2}$ via $\phi(s, n_{max})$:

$$\overline{(s - \bar{s}_{mv}(\bar{\sigma}, n_{max}))^2} = \frac{1}{\left(\frac{\partial^2 \phi}{\partial s^2}\right)_{s=\bar{s}_{mv}(\bar{\sigma}, n_{max})}} \tag{16}$$

$$\overline{(s - \bar{s}_{mv}(\bar{\sigma}, n_{max}))^3} = 2\frac{\bar{\sigma} + \left(\frac{\partial \phi}{\partial s}\right)_{s=\bar{s}_{mv}(\bar{\sigma}, n_{max})}}{\left[\left(\frac{\partial^2 \phi}{\partial s^2}\right)_{s=\bar{s}(\bar{\sigma}, n_{max})}\right]^2} \tag{17}$$

Equations (16) and (17) yield

$$\frac{\overline{(s - \bar{s}_{mv})^3}}{[\overline{(s - \bar{s}_{mv})^2}]^2} = 2\left[\bar{\sigma} + \left(\frac{\partial \phi}{\partial s}\right)_{s=\bar{s}_{mv}(\bar{\sigma}, n_{max})}\right] \tag{18}$$

It can be shown that the results (16), (17), and (18) are true when the ratio on the left-hand side of eq. (18) does not depend on $\bar{\sigma}$. The numerical calculations for $n_{max} = 30000$ (corresponding to a giant lipid vesicle with a radius $\sim 10\,\mu m$) show this ratio is practically equal to 2, when $\bar{\sigma} > 10$. Consequently, for model vesicles of such size, with $\bar{\sigma} > 10$, the following equation is fulfilled in addition to eq. (16):

$$\bar{\sigma} + \left(\frac{\partial \phi}{\partial s}\right)_{s=\bar{s}(\bar{\sigma}, n_{max})} = 1 \tag{19}$$

Then the function $\phi(s, n_{max})$ can be calculated from eqs (12) and (19).

The other limiting case is relatively easy to treat is when $\bar{\sigma}$ is only slightly greater than $-6$. Under this condition the excess area is mainly due to the elliptic modes ($n = 2$). Clearly, in this range of $\bar{\sigma}$ the fluctuations of the modes of the model vesicle with $n > 2$ are practically independent of $\bar{\sigma}$, whereas the modes with $n = 2$ diverge when $\sigma$ tends to $-6$. A simple model of a vesicle with only elliptic fluctuation modes ($n = 2$) enabled calculation of the function $\phi(s, n_{max})$:

$$\phi(s) = 6s - \tfrac{3}{2}\ln s + \text{constant} \tag{20}$$

In the crossover between the two regions the solutions should be sewn together. A numerical procedure for calculating $\phi(s, n_{max})$ was developed, with a known precision, in the crossover between the regions of high enough $\bar{\sigma}$ and $\bar{\sigma} \sim -6$.

The results obtained enabled calculation of the function $\phi(s, n_{max})$, and its related free energy $F(R, S_0, n_{max})$. Later on, some of consequences of these results are given.

## 4. RESULTS AND DISCUSSION

The area $S(t)$ of a real vesicle fluctuates, and therefore its tension $\sigma(t)$ fluctuates, too. Let $\langle\sigma(t)\rangle$ and $\langle S(t)\rangle$ be the time average value of $\sigma(t)$ and $S(t)$. They are related in the following way:

$$\langle\sigma(t)\rangle = k_s \frac{\langle S(t)\rangle - S_0}{S_0} = k_s \frac{\pi R^2 + \dfrac{kT}{k_c} R^2 \bar{s}(R, S_0, n_{\max}) - S_0}{S_0} \tag{21}$$

Let $\overline{\sigma'}$ be the reduced tension of the model vesicle introduced after Eq. (10), provided that the mean area of the real and the model vesicle have the same values. At high $\overline{\sigma}$ and from eqs (8), (19) and (21) it follows that:

$$\overline{\sigma'} = \frac{\langle\sigma(t)\rangle R^2}{k_c} - 2c_0 R + 1 \tag{22}$$

Consequently, the effective $\overline{\sigma'}$ is shifted by a factor of 1 in addition to the shift due to the spontaneous curvature.

In the case of $\bar{\sigma} \sim -6$, eqs (8) and (20) yield

$$\overline{\sigma'} = \frac{\langle\sigma(t)\rangle R^2}{k_c} - 2c_0 R \tag{23}$$

The last two equations show, that not depending on the spontaneous curvature $c_0$ shift of $\overline{\sigma'}$ is equal to 1 at high enough $\overline{\sigma'}$, tends to zero when $\overline{\sigma'}$ tends to $-6$, and smoothly decreases from 1 to zero in the crossover region. The spontaneous curvature can be either inherent to the membrane (arising from the differences of various origins of both sides of the membrane, different compositions of the two monolayers of the bilayer, etc.) or induced by the nonequilibrium number of molecules in the monolayers of the bilayer. The Laplace law in the case of a fluctuating vesicle can be obtained from eqs (8), (9) and (10). With $\Delta\tilde{p} = (\Delta p)R^3/k_c$, denote the normalized value of $\Delta p$. The classical Laplace law is $\Delta\tilde{p} = 2\overline{\sigma'}$. For fluctuating vesicles, in the case of high enough $\overline{\sigma'}$, the Laplace law, eq. (10), expressed in terms of $\Delta\tilde{p}$ and $\sigma'$ is:

$$\Delta\tilde{p} = 2(\overline{\sigma'} + c_0 R - 1)\left[1 + \frac{kT}{8\pi k_c}\sum_{n=2}^{n_{\max}} \frac{2n+1}{n(n+1) + \overline{\sigma'}}\right]$$

$$= 2(\overline{\sigma'} + c_0 R - 1)\left[1 + \frac{kT}{4\pi k_c}\bar{s}(R, S_0, n_{\max})\right] \tag{24}$$

In eq. (24) corrections due to the logarithmic term in eq. (9) are omitted because they are negligible.

In the case of $\bar{\sigma} \sim -6$, this law has the form:

$$\Delta\tilde{p} = 2(\overline{\sigma'} + c_0 R)\left[1 + \frac{kT}{8\pi k_c}\sum_{n=2}^{n_{\max}} \frac{2n+1}{n(n+1)+\overline{\sigma'}}\right]$$

$$= 2(\overline{\sigma'} + c_0 R)\left[1 + \frac{kT}{4\pi k_c}\bar{s}(R, S_0, n_{\max})\right] \tag{25}$$

The exact form of the corrections to the Laplace law in the crossover region can be determined after numerical calculation of the function $\phi(s, n_{\max})$ in this region. For high enough values of $\Delta p$ the Laplace law is valid with a high degree of precision for symmetrical membranes with $c_0 = 0$. This justifies its utilization in the analyses of the experiments where giant vesicles are sucked in micropipettes [3]. The domain of low pressures is more interesting for the analysis of the form fluctuations of the vesicles. Knowledge of the free energy of the fluctuating vesicle will enable estimation of deviations of the amplitude fluctuations of the elliptic modes ($n = 2$) with respect to the predictions of the Milner and Safran theory [8].

## 5. REFERENCES

1. E. A. Evans and R. Skalak, *CRC Grit. Rev. Bioeng.*, **3**, 181 (1979).
2. A. G. Petrov and I. Bivas, *Progr. Surf. Sci.*, **16**, 389 (1984).
3. E. Evans and W. Rawicz, *Phys. Rev. Lett.*, **64**, 2094 (1990).
4. D. Marsh, *Biophys. J.*, **73**, 865 (1997).
5. W. Helfrich, *Z. Naturforsh.*, **28c**, 693 (1973).
6. W. Helfrich, *Z. Naturforsh.*, **29c**, 510 (1974).
7. W. Helfrich, *J. Phys. France*, **47**, 321 (1986).
8. S. T. Milner and S. A. Safran, *Phys. Rev. A*, **36**, 4371 (1987).
9. L. D. Landau and E. M. Lifshitz, *Course of Theoretical Physics*, vol. 5: *Statistical Physics*, Pergamon Press, London-Paris, 1959.

# Part Four

# Physical Properties

# Chapter 9

# Use of Micropipet Manipulation Techniques to Measure the Properties of Giant Lipid Vesicles

*DAVID NEEDHAM AND DONCHO ZHELEV*

*Duke University, Durham, USA*

## 1. INTRODUCTION

The lipid bilayer membrane is a truly remarkable engineering material. It surrounds every cell on the planet providing a mechanical, chemical, and electrical barrier for the cell. It also acts as a two-dimensional solvent for the protein and other lipidic components of the cell membrane. Shown in Figure 9.1, in order of increasing complexity of composition and structure, are video micrograph images of three membrane-bound capsules: a lipid vesicle, an erythrocyte, and a neutrophil.

When viewed in real time, the membrane of the flaccid lipid vesicle is seen to undulate due to thermal motion. These thermally driven displacements have been used to characterize the mechanical stiffness in bending of otherwise nonstressed, flaccid bilayers by what might be called the zero mean tension methods [9,32,68,96,97]. The values for the bending modulus are only a few $k_B T$, revealing that the 5 nm thick membrane is extremely soft, and indeed is one of the thinnest and softest materials known. The membrane of the red blood cell is a composite of a lipid bilayer and attached superficial structures of carbohydrate polymers on the outside, and a spectrin network on the inside which is glued to the membrane via spot welds

*Giant Vesicles* Edited by P. L. Luisi and P. Walde

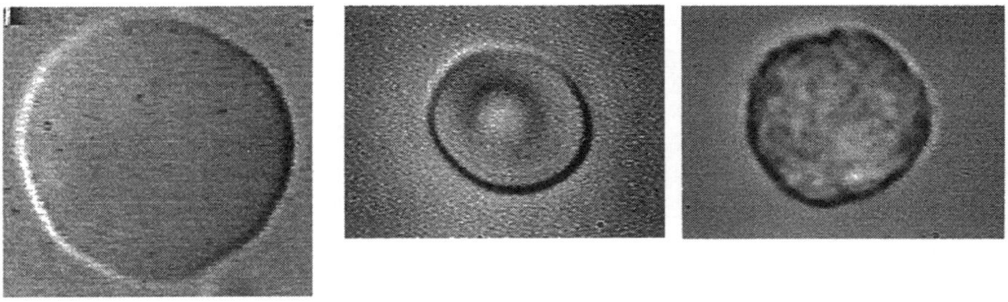

**Figure 9.1** Videomicrographs of a flaccid giant lipid vesicle (12 microns diameter), a red blood cell (7 microns diameter) and a neutrophil (8 microns diameter). All these capsules are bounded by the lipid bilayer membrane. Some information about the material properties, in particular bending stiffness and intrinsic curvature is obtained from microscopic observation of membrane undulations, but the full range of material behavior is obtained by aspirating individual vesicles or cells by micropipet.

composed of the transmembrane band-3 protein and other small globular proteins [3]. The red blood cell membrane undulates in much the same way as the lipid vesicle, which is somewhat surprising given that it contains an underlying superficial structure. It too is a soft material, yet is strong enough to survive 120 days in the blood stream, maintaining a discoid shape with excess area above that for a sphere of the same volume. This implies that the spectrin must add some shear rigidity to the composite structure because the lipid bilayer is essentially a two-dimensional liquid. The neutrophil jiggles in Brownian motion, but its composite membrane interface does not appear to move. Here, the same soft lipid bilayer covers a complex underlying cytoskeleton that is responsible for cell signalling and active crawling motion driven by actin polymerization and depolymerization, with the actin matrix coupled through the membrane to an array of extracellularly exposed adhesion receptors. Direct observation of these microscopic capsules bounded by just a 5 nm thick lipid membranes provides some, but limited, information about their material properties.

From a materials engineering perspective, what is needed in order to completely characterize these capsular structures, is a tool with which to probe their mechanical properties – an ability to manipulate individual giant lipid vesicles capsules and cells, that can not only apply well defined stresses for each of the three basic modes of deformation, (dilational, shear, and bending), but that can also measure the strain resulting from the applied stress, and therefore characterize the material behavior in terms of elastic moduli and viscous coefficients. The micropipet technique, initiated by Rand and Burton [92] and later perfected by Evans and Hochmuth [16], provides such an ability. It has been used extensively since the late 1970s to measure and characterize the material properties of red cells, white cells, and giant vesicles as reviewed in several recent publications [30,69,82].

Focusing on the giant lipid vesicles, these membranes are both fragile and inherently difficult to resolve optically as they do not contain any refractory internal structure or light-absorbing macromolecules. Direct measurements of the full range of material and interactive properties of lipid vesicle membranes have only been possible by the development of sensitive micropipet manipulation techniques and the creation of appropriate preparative procedures that provide the investigator with large (20–30 micron) single-walled lipid vesicles that can be seen in the optical microscope. New methods that have expanded the study of lipid vesicles can produce higher yields of giant structures ([2]; Chapters 3 and 4 of this book).

The glass micropipet, with precise control over applied hydrostatic pressures, has provided a unique way of applying well defined stresses to a single giant lipid vesicle while acting as a sensitive transducer of vesicle membrane area, shape and volume change. Using a suction pipet, a single lipid vesicle can be aspirated and manipulated, and several mechanochemical experiments can be performed ([82] and references therein) that characterize: membrane area expansion, tensile failure and bending for solid and liquid membranes; yield shear and shear viscosity for solid phase membranes; adsorption, uptake, and desorption of various membrane-binding and membrane-soluble components; membrane water permeability coefficient; transmembrane pore formation due to the action of an electric field or uptake of macromolecular structures; and thermal bilayer transitions. The ability to manipulate individual and pairs of vesicles has also allowed: the measurement of intermembrane adhesion energy that results from the cumulation of several attractive and repulsive colloidal potentials, including van der Waals, and polymer attraction, and hydration, electrostatic, undulation, and polymer–steric repulsions; the fusion between two vesicles to be observed as a result of electroporation and defect formation due to the inclusion of nonbilayer phospholipids; and adhesion mediated by specific receptor–ligand bonds to be quantified, even down to the level of single molecular bonds [67].

Much of this work using various micropipet methods to study the mechanochemical properties of lipid bilayer vesicles dates back to the early 1980s, and some of this has been previously reviewed [82]. Only a brief summary is given here of each of the micropipet experiments that have been used to characterize the mechanical and thermomechanical properties of vesicles as well as their colloidal and receptor-ligand medicated adhesion. This review focuses mainly on vesicle preparation and the micropipet technique itself, and its use in a series of recent micropipet experiments: water and urea permeability of lipid membranes, and the molecular exchange of surfactants and peptides, including the role of PEG in inhibiting molecular approach and binding to lipid vesicle interfaces. The information gained from these kinds of direct measurements made on single giant vesicles not only characterizes the membrane and its intermembrane interactions from a fundamental materials science perspective, it also provides essential materials property data that are required for the successful design and deployment of lipid vesicle capsules in applications such as drugs [45,76].

## 2. VESICLE PREPARATION AND THE MICROPIPET TECHNIQUE

### 2.1 Vesicle Preparation

Procedures for the preparation of giant (20–40 μm) lipid vesicles from stock organic lipid solutions are documented in several publications dating back to 1969 [70,72,74,79], but some new issues are addressed here. It is common practice to first make up lipids and lipid mixtures in an organic solution such as chloroform. The solution is then placed in contact with some surface upon which the lipid is to be deposited, such as glass (vial or flask), Teflon (disk), or metal (electrode). The organic solvent is then evaporated and traces of solvent are removed by vacuum drying [72]. When viewed under the microscope using interference contrast microscopy the phospholipid film displays a characteristic lamellar structure, presumably of dried lipid bilayer sheets, as the hygroscopic nature of such lamellae means they will take up some water from surrounding air and present the bilayer morphology, even in this relatively dry state. The layered structure is clearly visible as shown in the video micrograph of a dried lipid film in Figure 9.2(a). Rehydration of the lipid is simply achieved by adding an aqueous solution. It is usual to rehydrate with deionized water, or, if osmotic balance is required for later experimentation, (for example in physiological buffer), then a sucrose solution of equivalent osmolarity is used to swell the lipid. The vesicles then contain this sucrose solution. Thus, the lipid is found to spontaneously hydrate from the dried lipid film to form lipid vesicles. However, as pointed in the late 1960s by Reeves and Dowben [93], the giant single-walled capsules that are essential for micromechanical measurements are only formed by rehydrating lipid in nonelectrolytes such as de-ionized water or sucrose solutions. If physiological salt solutions are used, the lipid does not form giant unilamellar vesicles. Hydration of the lipid in salt solution does clearly occur (such media are used routinely in X-ray diffraction measurements on bilayer structures [64]), but what is observed are extremely multilamellar and optically refractory structures (good scatterers for X-ray), and any apparently thin-walled vesicles are still multilamellar, and unusable in micropipet experiments. Typical examples of dried lipid that is then hydrated by sucrose are shown in the video micrographs presented in Figures 9.2(a) and (b), along with images taken of the subsequent addition of equiosmotic salt solution to the sucrose vesicles (Figures 9.3(a) and (b)). Also shown in Figures 9.4(a) and (b), is the rehydration of a dried lipid film with salt solution.

These micrograph pictures clearly show that hydrating a dried lipid film (Figure 9.2(a)), with sucrose solution (or water) (Figures 9.2(b) and (c)) immediately produces giant vesicles, some single walled, that are independent of each other, and not adherent. Single bilayers in paucilamellar stacks are clearly separated by water gaps. The addition of salt solution to these preformed vesicles produces massive aggregation of the vesicles throughout the chamber (Figure 9.3). In direct contrast to the hydration in nonelectrolytes, initial hydration with salt solution does not produce vesicles at all, but multilayered aggregates that are only partially

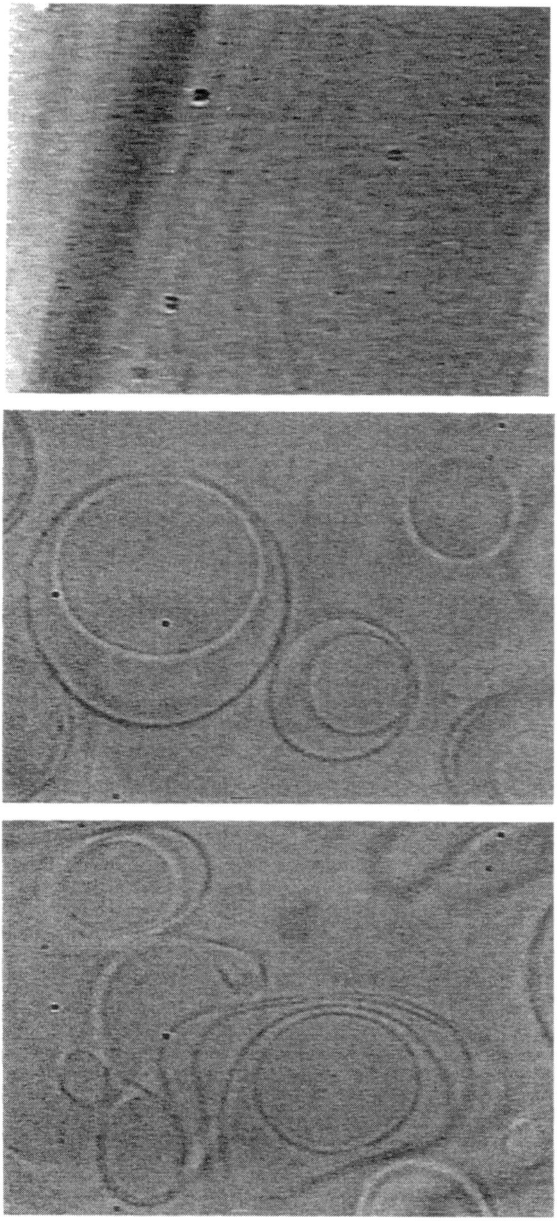

(a) (b) (c)

**Figure 9.2** (a) Dried lipid film showing typical edge defect in a dried lipid film of SOPC. Note the layers of lipid lamellae stepping down at the edge of the defect. (b) The same lipid film as in (a), rehydrated with a nonelectrolyte solution of 200 mM sucrose. The same result is obtained if hydration is carried out in water. (c) The same lipid film as in (a), rehydrated with a nonelectrolyte solution of 200 mM sucrose.

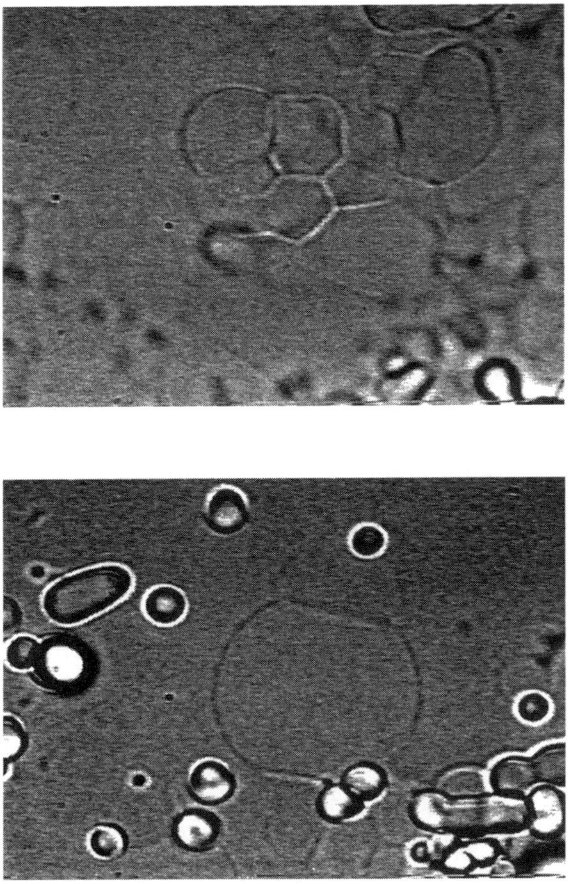

(a)

(b)

**Figure 9.3** Salt solution added to sucrose solution vesicles. (a) The same lipid vesicle sample that was rehydrated with a nonelectrolyte solution of 200 mM sucrose (Figure 9.2(b), to which was subsequently added just less than an equal, amount of 290 mM phosphate buffered saline, pH 7.4, making the salt concentration approximately 100 mM. As the two solutions mixed, vesicles were seen to adhere to each other and form aggregates, with the lipid vesicle bilayers forming characteristic adhesion angles. (b) The same sample as in (a) showing three vesicles in an adhesive aggregate. Some partially hydrated refractory lipid structures are seen in the background. Once the salt was added, all undulation of vesicle bilayers ceased and no further rehydration of the lipid film occurred. Aggregation was widespread.

hydrated and optically refractory (Figure 9.4). Thus, free-floating single-walled giant vesicles of the kind that are required for micropipet manipulation experiments do not form when lipid is simply rehydrated in salt solution.

Why do vesicles form so well in nonelectrolyte solutions but not in high ionic strength buffers? It appears that neutral bilayers are in fact not so neutral! Ostensibly,

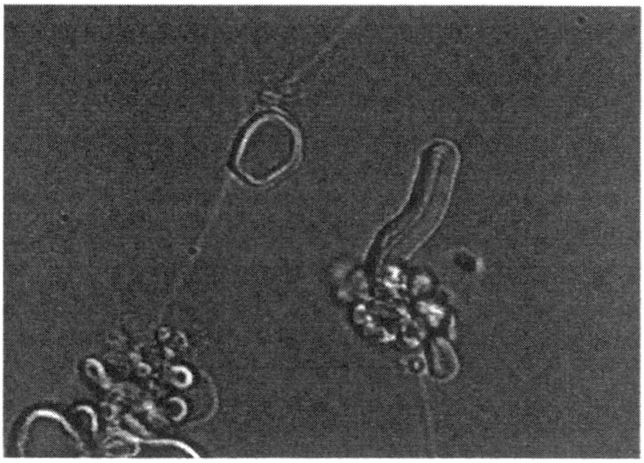

(a)

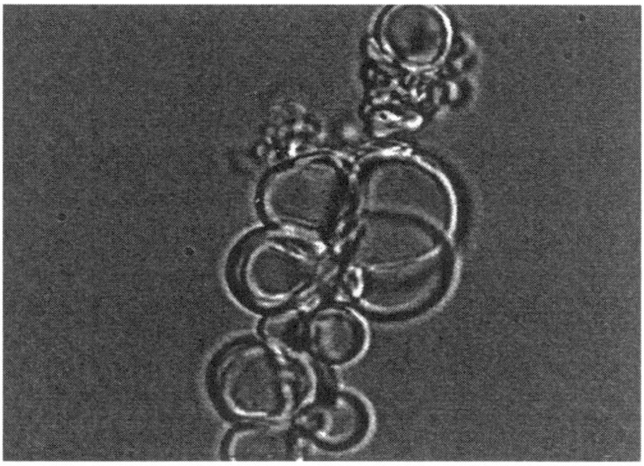

(b)

**Figure 9.4** Lipid rehydrated in 90 mM salt solution. (a) Typical partially rehydrated myelin-like structures that are obtained when 90 mM salt solution is added to a dried lipid film. The film does rehydrate, but only to the extent of forming the partially hydrated refractory lipid structures. (b) The only vesicle-like structures that could be seen were very multilamellar, aggregated, and clearly not suitable for micropipet manipulation experiments.

neutral vesicles can have a slight net negative charge due to the presence of trace amounts of negatively charged impurities. Thus, the difference between hydrating in sucrose solution versus sodium chloride salt solution originates from the fact that in salt solution the electrostatic repulsion that normally allows the vesicles to form in nonelectrolytes (where Debye lengths can be of the order of microns) is quenched. Under electrostatically quenched conditions, van der Waals attraction between bilayers can then dominate to keep the hydrating lamellae in adherent contact, as we and others have shown in vesicle–vesicle adhesion experiments carried out in 100 mM salt solution [13,20–26,71].

The perception that neutral membranes can possess a slight negative charge has existed anecdotally since the early 1980s (Dennis Haydon, 1980, personal communication). However, few papers have attempted to prove this by, for example, reporting measurements of the electrophoretic mobilities for neutral vesicles. Most tend to focus on purposefully charged bilayers [66]. However, Janzen *et al.* have recently studied vesicles composed of egg phosphatidylcholine (EPC) and EPC : cholesterol [46]. They showed (and were concerned by the fact) that vesicles made from neutral EPC and EPC : cholesterol, did have a net negative electrophoretic mobility. For EPC it was found to be $-0.21\ \mu m\ cm\ V^{-1}\ s^{-1}$ in 2 mM NaCl. This electrophoretic mobility is equivalent to a zeta potential (using the Huckel equation) of −4.2 mV. The mobility was higher for liposomes containing 40 mol % cholesterol in 100 mM NaCl, measured at $-0.5\ \mu m\ cm\ V^{-1}\ s^{-1}$ (equivalent zeta potential is 10.6 mV) which indicated that lipid and, especially, cholesterol oxidation products could add a significant charge to the bilayers.

For the purpose of this review these results were checked with other synthetic lipids (Avanti Polar Lipids, Pelham). Experiments were conducted to measure the zeta potential of neutral bilayers made from dioleoylphosphatidylcholine (DOPC) and a charged series with dioleoylphosphatidylglycerol (DOPG) as 100 nm extruded liposomes in 0.1 mM NaCl. The results are shown in Figure 9.5.

Starting with a 100 % DOPG bilayer, the zeta potential was measured at ∼ −75 mV. As the amount of DOPG was reduced by mixing with neutral DOPC,

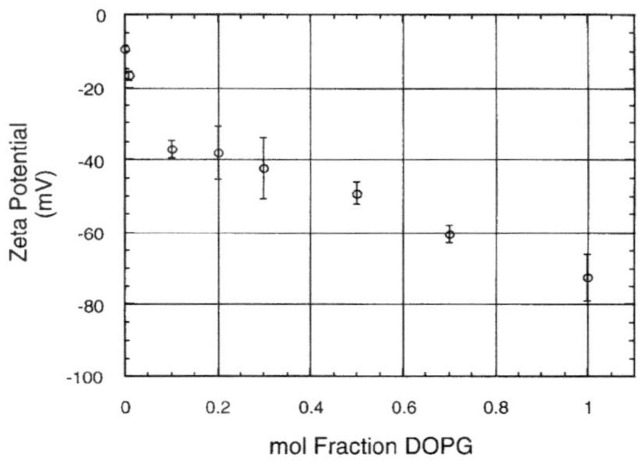


**Figure 9.5** Zeta potential versus concentration of the charged lipid DOPG in DOPC vesicles. Eight samples, each containing 10 mg of lipid (of different DOPC : DOPG ratios) were dried down, rehydrated with 1 ml of 0.1 mM NaCl and then extruded through 200 nm filters in a Lipex Biomembranes extruder 11 times. The samples were then diluted with 1 ml of 0.1 mM NaCl. 1.3 ml of each sample was placed in a different cuvette. The zeta potential of the extruded vesicles was measured 10 times for each sample using a Brookhaven Zeta PlusPCS machine.

the zeta potential decreased in a nonlinear manner, consistent with Gouy–Chapman–Stern theory [66]. The observation was that for ostensibly neutral DOPC vesicles the zeta potential in 0.1 mM NaCl was not zero, but $\sim -9$ mV. It seems that oxidation products, formed perhaps just on storage and perhaps during rehydration, can create a negative charge on liposomes and so could be a source of inconsistency in lipid composition. Devaux has also shown that small amounts of degradation products are produced in DMPC vesicles when a procedure is employed where vesicles are swelled overnight at only slightly elevated temperature (40°C) [31].

These procedures involve only simple rehydration and yet appear to produce lipid degradation products, so one might be concerned that electrode effects during the more chemically active process of electroformation could possibly make this situation worse. It would appear prudent to take extra care when preparing vesicles by electroformation, especially when using vesicles in situations where small amounts of spurious charge are not fully screened, i.e. in low electrolyte solutions. Experimentalists using this technique are encouraged to check if such electroformation vesicles are charged or not by, for example, carrying out electrophoretic mobility measurements on the vesicle samples.

In any event, for the simple swelling of dried lipid off a glass or Teflon surface, the presence of a small amount of charge impurity, when immersed in nonelectrolytes, appears to be responsible for allowing vesicles to rehydrate from the dried lipid film. When placed in salt solutions this charge is well screened and the van der Waals interaction is allowed to take over, producing intervesicle adhesion. If it were not for this small amount of charge, the lipid bilayers, that are being rehydrated from tightly stacked lamellae, would strongly adhere even in nonelectrolyte. Additionally, there may also be a role for thermal undulations, but these should be the same in pure water as in salt buffer and so do not represent all the story.

Once hydrated, only the lowest level of imposed agitation on the swelling lipid promotes the formation of large lipid vesicles that can still vary in size and degree of multilamellarity. This agitation can be provided by convection in the suspending medium by recycling into and out of a low temperature oven set at, say, 40°C. Even so, the single-walled vesicles necessary for micropipet experiments are few and far between, especially for neutral lipids, and have to be selected from the largely multilamellar population. Other techniques that complement micromechanical measurements, such as X-ray methods, require multiwalled liposomes that provide detectable repeat periods; these methods yield information about the bilayer geometry, phase behavior and colloidal interaction between membranes [65]. If vesicles are made from 100 % charged lipids (such as DOPG) or if PEG–lipids are included in the composition, then every bilayer is separated from its neighbor and a large number of giant vesicles are unilamellar. Therefore it is routine to purposefully include a small amount of charged lipid or PEG–lipid in vesicle preparations that are being used in mechanical or permeability studies. This is obviously not an option for vesicles where intermembrane, vesicle–particle, or vesicle–macromolecule interactions are being evaluated, unless the surface charge or steric polymer is the variable under test.

In preparation for an experiment, the lipid vesicle suspension is diluted into an equi-osmotic glucose solution and placed in a simple glass microchamber on a microscope stage into which the micropipet(s) is subsequently inserted. In order to prevent electrostatic effects that tend to charge the pipet tip, all glass surfaces are preequilibrated with 1 mg ml$^{-1}$ albumin solution. This solution is then removed prior to experiments in the desired solution or buffer, even though the presence of albumin does not affect, for example, vesicle–vesicle adhesion measurements, [26] (and unpublished results). If single vesicle experiments are to be carried out involving transfer of the vesicle to a second chamber, for example to measure water permeability, then additional vesicles can be added to this second chamber to also help passivate the glass chamber surfaces [84].

### 2.2 Micropipet Experimentation

The micromanipulation system has been described in detail [12,30,70,71]. Briefly, it is centered around an inverted microscope that has the capacity for up to four micromanipulators to be mounted directly on the microscope stage plate. Control over micropipet suction pressure is in the range of microatmospheres to tenths of atmospheres (0.1 N m$^{-2}$ to 100 000 N m$^{-2}$) and is achieved by a water-filled manometer (Figure 9.6) equipped with a sensitive micrometer-driven displacement and coarser syringe control; positive and negative pressures are recorded by in-line pressure transducers (Validyne, Northridge, CA).

A glass micropipet of desired internal diameter and flat tip (usually in the range of 1–15 μm depending on the experiment to be performed) is used to both apply the force to the aspirated vesicle during the vesicle deformation tests and to measure the resulting vesicle deformation. Micropipets are made from 0.75 mm internal diameter glass capillary tubing (A-M Systems, Inc., Everett, WA), formed into microneedles by a heated pipet puller. The micropipet tips of desired diameter are broken by quick fracture on a microforge. The pipets are filled with a NaCl or sugar solution that matches the osmolarity of the solution in which the vesicle properties are to be measured. The measuring micropipet is mounted in a micromanipulator (Research Instruments Inc., Durham, NC), via a wet chuck that serves to connect the pipet to the water-filled manometer system that controls the pipet pressure, as shown in Figure 9.7.

The De Fonbrune micromanipulator, shown in Figure 9.8 (Research Instruments Inc., Durham, NC), allows the pipet to be held absolutely stationary and to be moved in the axial, lateral and vertical directions by the transduction of three separate and variable air pressures provided by a joystick via flexible transmission tubes. The whole assembly is mounted on the microscope stage so that the pipet enters the microchamber horizontally, as depicted in Figure 9.9.

Experiments are recorded on videotape, and information, such as time, pipet suction pressure and chamber temperature, is displayed directly onto the videotape using video multiplexing (Vista Electronics, La Mesa, CA; Colorado Video, Inc., Boulder Colorado). A series of experiments can thus be recorded such that geo-

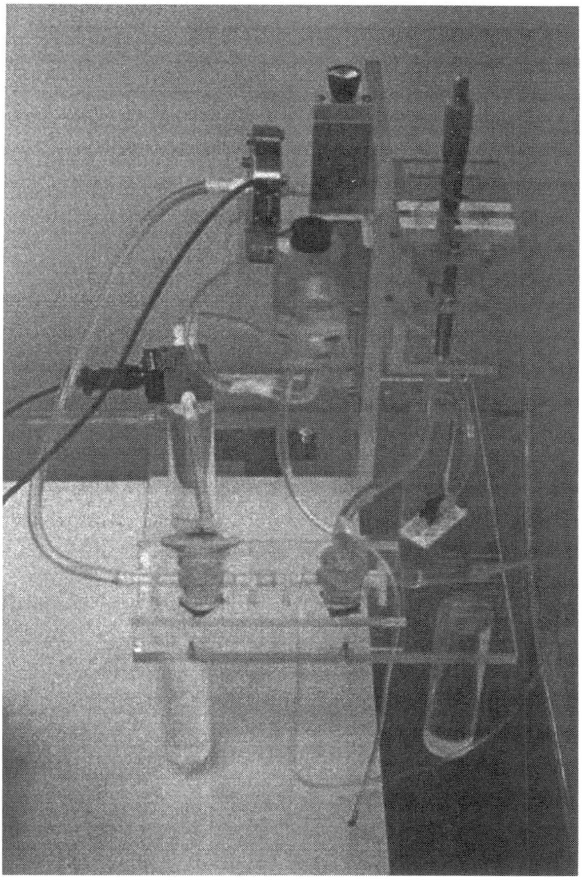

**Figure 9.6** Manometer system for applying negative or positive pressures to the micropipet. Coarse control is achieved by use of a simple syringe to displace volume above the front reservoir (on the right). Fine control, microatmospheres, is achieved by adjusting the position of the front reservoir using the sprung micrometer drive. The whole manometer system can be leveled with the pipet tip in order to achieve zero pressure at the pipet tip, using the central coarse screw. In-line pressure transducers measure small pressures (0–1000 dyne $cm^{-2}$) and large pressures (1000–100 000 dyne $cm^{-2}$) [84].

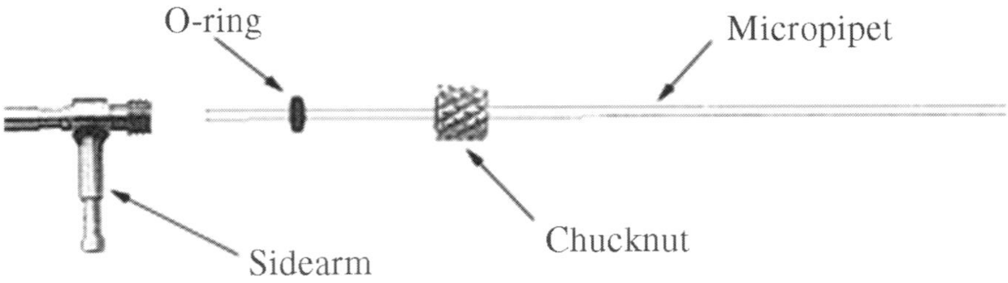


**Figure 9.7** Wet chuck and micropipet assembly for holding pipet. The chuck is inserted and locked into the nose of the micromanipulator. (Courtesy of Research Instruments Inc., Durham, NC).

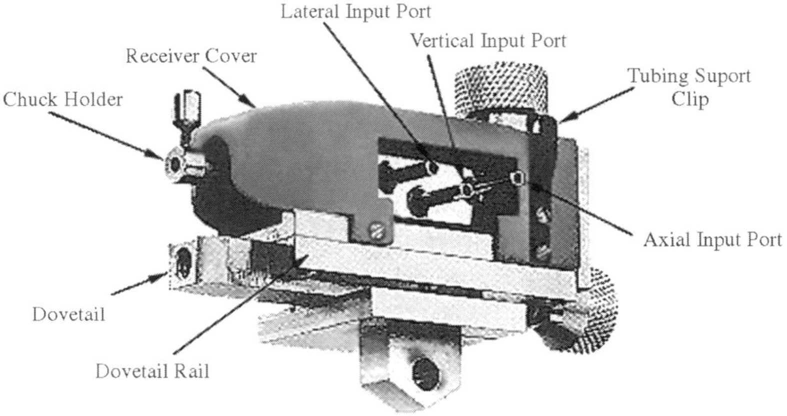


**Figure 9.8** Micromanipulator for holding the micropipet via the chuck holder. $X$, $Y$, and $Z$ motion is achieved using the coarse screw-controlled pins on the top, back side, and rail. Fine control of position is achieved by connecting a pneumatic joystick to the three axial input ports that drive steel bellows in the manipulator (Courtesy of Research Instruments Inc., Durham, NC).

metrical analyses (vesicle and pipet dimensions) can be made sub-sequent to the experiment using a video caliper system (Vista Electronics, La Mesa, CA).

## 3. VESICLE MEMBRANE MATERIAL PROPERTIES AND INTERVESICLE INTERACTIONS

### 3.1 Modes of Deformation

As discussed at length by Evans [29], and reviewed previously [5,82], three independent shape changes are usually considered in analyses of vesicle membrane

**Figure 9.9** Temperature-controlled chamber on the microscope stage with two micropipets inserted into each side of the chamber [84].

deformations. These are area expansion, in-plane shear, and bending. Together they characterize the deformation and rate of deformation of the lipid bilayer. These independent shape changes are produced by the application of external forces to the membrane elements. For example, a hydrostatic pressure, such as that provided by the micropipet aspiration of a single lipid vesicle, acts normal to the membrane and is balanced by membrane tension components multiplied by the curvature of the membrane vesicle (a form of the law of Laplace).

Proportionalities between intensive forces and static deformations describe the three modes of deformation that are used to characterize the membrane's material properties and give rise to first-order constitutive relationships. For a liquid lipid bilayer, which does not support shear stress, the two important elastic relationships are area dilation and bending:

*(a) Area dilation*

This is characterized by an isothermal area expansion (or compressibility); compressibility and expansibility are used interchangeably when referring to the elastic area modulus $K_a$, given by the equation

$$\Delta\tau = K_a\alpha \qquad (1)$$

where $\alpha$ is the fractional change in membrane area ($\Delta A/A_0$) produced by isotropic membrane stress $\Delta\tau$, and $A_0$ is the initial vesicle area at a tension of $\sim$0.5 mN m$^{-1}$. Below this threshold level of tension, area change is largely due to extending membrane undulations and so does not represent true expansion of the bilayer itself [27].

*(b) Membrane bending*

Membrane bending is characterized by the bending rigidity $k_c$, which is the ratio of the change in membrane bending moment $\Delta M$ to changes in total membrane curvature $\Delta c$,

$$k_c = \Delta M/\Delta c \tag{2}$$

At a given position in the membrane, the curvature change is the change in the principal radii of curvature $R_1$ and $R_2$, i.e.

$$\Delta c = \Delta(1/R_1 + 1/R_2) \tag{3}$$

$K_a$ and $k_c$ are related when a certain distribution of stress is assumed across the bilayer. In both modes of deformation the area per molecule at the membrane interface changes: for expansion, there is an increase in area per molecule in both monolayers of the bilayer; for bending, the area per molecule increases in one side of the bilayer and decreases in the other.

Gel-phase lipid bilayers as solids do support shear stress elastically and so, for this solid bilayer state, shear deformation is characterized as follows. The surface shear rigidity for a membrane element $\mu$ is the ratio between surface shear stress $\tau_s$ and shear deformation $e_s$,

$$\tau_s = 2\mu e_s \tag{4}$$

where the shear deformation $e_s$ is given by the in-plane extension $\lambda_e$ as,

$$e_s = (\lambda_e^2 - \lambda_e^{-2})/2 \tag{5}$$

*(c) Viscous coefficients*

These characterize the liquid behavior (viscosity) for each of these three modes and are given by proportionalities between forces (including moments) and rates of deformation [5,29]. For the liquid lipid bilayer though, time constants for dilational and bending deformation are on a molecular time-scale ($10^{-5}$ s to $10^{-10}$ s [5]), associated with acyl chain conformational changes. These time-scales are not measurable in micropipet experiments where observation is limited to times around $10^{-2}$ s because of the 1/60th second speed of video signals. Viscous resistance to shear for liquid bilayers has been measured by pulling bilayer tethers from a vesicle held under known tension [106]. For solid bilayers a shear viscosity has been measured using the micropipet technique [22].

## 3.2 Mechanical and Thermomechanical Properties

As mentioned earlier, many of the basic mechanical and thermomechanical experiments have been reviewed previously [82]. For completeness, some of the more important conclusions of the work are now given.

### *(a) Mechanical*

The simple application of a suction pressure to a giant vesicle held in a micropipet induces a well defined tension in the vesicle membrane and produces an expansion of the area of the vesicle, measured from the length change of the membrane projection in the pipet (see Figure 9.13 later). When carried out for a series of increasing then decreasing suction pressures, a plot of tension versus area change gives directly four parameters that characterize the cohesion of the giant vesicle bilayer, namely, the area expansion modulus, tensile strength, critical area strain, and strain energy [78]. The strongest bilayers have been shown to be the least compressible. The strength and compressibility of bilayers made from single or even mixtures of common PC and PE lipids are limited in range to around 200–300 $mN\,m^{-1}$ [22]. The presence of multiple double bonds, as in diarachidonylphosphatidylcholine (DAPC), reduces the modulus to $\sim 100\,mN\,m^{-1}$ [27,79], which is of the order of the minimum modulus expected based on hydrophobic energy arguments, i.e. twice the interfacial tension of an oil–water interface. In a series of micropipet tests carried out on $C_{18}$ dichain lipids with varying degrees of unsaturation from di18 : 0/1 to di18 : 2 (where 18 refers to the hydrocarbon chain length, and the number after the colon refers to the number of double bonds for the first or the second chain, thus di18 : 01 is SOPC), the elastic modulus changed very little, all data clustering within 10 % of a mean value of 198 $mN\,m^{-1}$ [84]. Thus, more than two double bonds per chain are required to reduce the modulus significantly from the monounsaturated SOPC value.

Going the other way, the addition of cholesterol is the single most effective way to increase the strength and area elastic and bending moduli of lipid bilayers. The highest area dilation modulus and highest strength so far measured is achieved for lipid bilayers composed of lipids with saturated acyl chains and 50 mol % cholesterol. Interestingly, the strength limit is around 42 $mN\,m^{-1}$, which, when converted to an equivalent bulk strength by dividing by the bilayer thickness of $\sim 4$ nm (giving $10^7\,N\,m^{-2}$) is equal to the tensile strength of the hydrocarbon polymer polyethylene. Similarly, the highest area modulus of $\sim 4000\,mN\,m^{-1}$ converts to a bulk modulus of $10^9\,N\,m^{-2}$, again equivalent to the Young's modulus of polyethylene. This indicates that, for these cholesterol-rich bilayers, strength and elasticity over and above any hydrophobic effect is derived from actual van der Waals bonding between largely all-*trans* hydrocarbon chains and the planar cholesterol ring structure. For these bonded bilayers, the hydrophobic effect then makes only a small contribution to the overall strength and elastic modulus.

When comparisons are made with the mechanical properties of gel phase bilayers [26,82], an underlying lesson from these data is that, while both a reduction in temperature to form gel phase bilayers and the addition of cholesterol decrease the area per lipid molecule and produce more condensed and ordered bilayer structures, the addition of cholesterol to a given bilayer lipid, such as dimyristoylphosphatidylcholine (DMPC) [74], is a more effective way to maximize the bilayer strength than freezing the bilayer into a gel phase. Moreover, although ordered, the liquid state of the bilayer is maintained. This is an important feature for cell membranes where the presence of cholesterol means that the bilayer's essential role as a liquid solvent for transmembrane proteins is preserved, while enhancing the membrane's mechanical properties. This is not to say that the presence of cholesterol will not change the partitioning or solubility of a variety of molecules in the bilayer. Membrane cohesion is an underlying determining factor as to whether drug molecules, such as paclitaxel [95], and protonatable polymers, such as polyethylacrylicacid (PEAA) [77] can enter into the cholesterol-rich bilayer structure.

With respect to membrane bending, four different micropipet experiments have been devised that, together with analyses of thermal fluctuations of vesicle contours, and theoretical predictions [9,27,32,96,97,106,116], have shown that the bending modulus for typical lipid bilayers such as SOPC is around $1 \times 10^{-19}$ J. These studies have also shown that cholesterol increases the bending modulus in parallel with its increase in area expansion modulus.

Finally there follows a brief discussion concerning the work for membrane breakdown for all the lipid systems studied to date. The work of membrane breakdown $w_0$ has been compared to the critical membrane tensions for breakdown. The parameter $w_o$ is calculated from

$$w_o = \frac{\tau_c^2}{2KN} \tag{6}$$

where $\tau_c$ is the critical membrane tension, $K$ is the membrane area expansion modulus, and $N$ is the number of molecules per unit membrane area [115].

The work of membrane breakdown for membranes made from different lipids and lipid–cholesterol mixtures is shown in Figure 9.10. It is seen that, for membranes with small critical membrane tensions, i.e. largely single-component liquid-phase membranes, the work of membrane breakdown increases almost linearly with the critical membrane tension. For higher-strength membranes, i.e. saturated lipids with maximum cholesterol content, the work of membrane breakdown reaches a value on the order of 0.025 kT and remains almost constant regardless of the increase of the critical membrane tension. Thus, the question arises, as to whether tension creates new pores or acts on existing defects.

To answer this question, pores have been created in giant lipid membranes and the far-field tensions sufficient to balance micrometer radius pores have been measured

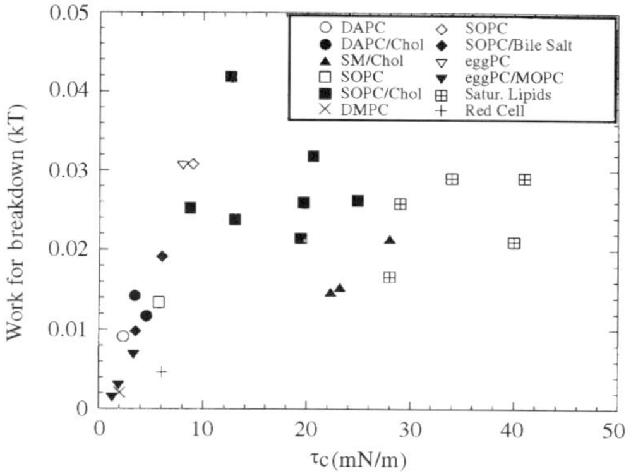


**Figure 9.10** Work of membrane breakdown versus the critical membrane tension for different lipids and lipid–cholesterol mixtures. (Data from references [82], [28] and [113]).

[111,115]. The line tensions, are of the order of 0.01 mN m$^{-1}$. This value is three orders of magnitude smaller than the measured critical membrane tension for tensile failure which is of the order of 6 mN m$^{-1}$ for the SOPC membrane. Thus, even if the pore radius is of the order of 5 nm (which is the bilayer thickness), the values of the balancing far-field tension will be of the order of 2 mN m$^{-1}$, which is smaller than the measured critical membrane tension for breakdown. Also, critical membrane tensions on the order of 1.9 mN m$^{-1}$ were measured by Zhelev and Needham [111] for SOPC membranes just after their electroporation. The above calculation and the measured critical membrane tension of electroporated membranes show that bilayer membranes in their resting state are most probably free of pores, and that the applied stress creates new defects that result in failure. Therefore, the work of membrane breakdown shown in Figure 9.10 is a measure of the work for pore formation. Surprisingly, this work is much smaller than kT (the same observation was made for lipid–cholesterol membranes by Needham and Nunn [78]) which suggest that pore formation is not a single molecular event but involves many molecules. Also, the observation that the maximum values of the work of membrane breakdown for several lipid systems do not depend on the membrane composition, suggests that the work of pore formation depends mainly on the packing of the membrane hydrocarbon region [61,62,79].

### *(b) Thermomechanical*

It is well established that lamellar phases of lipids are either anisotropic liquids with fluid acyl chains (the $L_\alpha$ liquid crystalline state) or solids with crystalline chains (the gel state) [43,44,54,55,90]. By using the micropipet manipulation technique, it is possible to support all the excess membrane area that is created and removed at these transitions. The thermal membrane area changes can therefore be measured directly for single bilayer vesicles as a result of temperature change [22,72,74]. Isothermal area transitions can also be produced in response to the level of stress applied to frozen bilayers in the so-called, rippled phase [22,72,74]. These micropipet experiments then have measured the area changes associated with temperature change in nontransition regions of the lipid phase diagram (i.e. thermal area expansivity of the bilayer) as well as thermal transitions for single lipids and lipid mixtures (including the effects of cholesterol). The thermal area expansivity of the liquid phase DMPC bilayer, was found to be $4 \times 10^{-3}\,\mathrm{C}^{-1}$, and for the planar $L^*_{\beta'}$ phase was $3.5 \times 10^{-3}\,\mathrm{C}^{-1}$. Thus, although obviously in a more condensed state, the pretransition bilayer is still relatively soft, containing gauche bonds that still provides for area changes due to thermal contraction. In the transition regions, the pre- and main transitions (between $L_{\beta'}$ and $P_{\beta'}$, and, $P_{\beta'}$ and $L_\alpha$, phases respectively) have a quite complicated stress history. The expected 25 % decrease in projected vesicle area that accompanies the molecular condensation at the main $L_\alpha$ to $P_{\beta'}$ transition ($T_c$) is seen only when the membrane is under zero or low stress [19,72]. The relative magnitudes of the membrane area changes at the $L_\alpha$ to $P_{\beta'}$ (22 %) and $P_{\beta'}$ to $L_{\beta'}$ (4 %) transitions are in the same proportion as the excess specific heats measured by calorimetry [52, 60].

The rippled super lattice of the $P_{\beta'}$ phase introduces additional complexity to the temperature and stress history of the gel phase. By being able to apply tension to the vesicle as it cools through the main transition, micropipet experiments have demonstrated that the formation of the rippled superlattice phase ($P_{\beta'}$) for saturated PCs is sensitive to applied membrane tension. Once formed, the removal of the $P_{\beta'}$ to give an $L_{\beta'^*}$ phase by applying tension of 1–3 dyne cm$^{-1}$ to the rippled structure resulted in an initial elastic response followed by a yield failure and plastic flow producing large (7–9%) expansion of the vesicle projected area. This experiment allowed quite accurate measurement to be made of the lipid molecule tilt angles as a function of temperature in the gel phases of DMPC and extended the approximate value of 30° obtained by X-ray diffraction for the whole phase region [44].

As in liquid-phase experiments, our micropipet experiments gave direct measurements of the area and bending moduli for gel-phase bilayers showing that, despite a major transitional condensation for DMPC from $L_\alpha$ to $P_{\beta'}$, the membrane underwent only a modest decrease in membrane compressibility. This change in the mechanical property is in good agreement with structural studies using X-ray [6], Raman [88,101,104], and $^2$H NMR [7,56], that show that a significant fraction of gauche bonds and chain disorder still exists. It is not until after the pretransition, and the formation of the $L_{\beta'}$ phase that the membrane shows significant solid character and a much higher modulus of area elasticity of 855 mN m$^{-1}$. The same kind of thermo-mechanical experiments have also been performed with lipid mixtures, specifically, mixtures of DMPC and cholesterol [74], and SOPC and palmitoylphosphatidylethanolamine (POPE) [21,26].

Finally, in this section on mechanical properties, membrane yield shear and shear viscosity, that are peculiar to gel phase bilayers, are obtained from an analysis of the threshold suction pressure for initial deformation into the mouth of the pipet and from the rate of flow of the vesicle into the pipet due to the application of the excess pressure [22]. The yield shear and shear viscosity increase for the $P_{\beta'}$ gel-phase bilayer with decreasing temperatures below the main transition at 24°C. Here, the surface viscosity is $\sim 10^4$ times greater than that expected for the $L_\alpha$ phase based on the lateral mobility of membrane fluorescent probes [17]. Then, as the temperature is reduced to below the pretransition, there is an abrupt augmentation of these material properties as the bilayer forms the more highly condensed $L_{\beta'}$ phase. The yield shear and shear viscosity principally reflect the density and mobility of crystal defects, such as grain boundaries and intragrain dislocations, in the solid bilayer membranes. When abstracted to three dimensions, these values for the yield shear ($5 \times 10^3\ \mathrm{N\,m^{-2}}$) and shear viscosity ($2 \times 10^6\ \mathrm{Ns\,m^{-2}}$) are comparable to the same properties of polyethylene at room temperature, again indicating the secondary (van der Waals) bonding nature of this 5 nm thick material.

### 3.3 Inter-vesicle interactions

Giant lipid bilayer vesicles provide unique and versatile systems for the study of a range of colloidal and surface interactions; the membrane interfaces are essentially molecularly smooth with homogeneity of vesicle composition at the molecular scale. Through judicial choice of lipids and other membrane-compatible components, the lipid surface can be made to be neutral or charged (positive or negative), poorly or highly hydrated, and bare or polymer covered. Appropriate choice of lipid acyl chain composition allows the surface structures to be mobile in a fluid substrate or immobile in gel state lipids. Finally, specific surface groups of special relevance to cell adhesion and surface recognition, such as receptors, antigens or antibodies, can be co-incorporated with lipid-grafted polyethyleneglycol (PEG–lipid) into the lipid bilayer for controlled reconstitution studies in which background attraction is eliminated [83].

As conceived of by Evans, the interfacial free energy density for the adhesion of unsupported lipid bilayers as giant vesicles can be measured using micropipet manipulation [29]. Vesicle manipulation techniques have been developed and a series of experiments have been carried out that measure mutual adhesion energies for bilayers that exhibit a range of colloidal interaction potentials [15,20,21,23–26,29]. These experiments, have investigated the full compliment of interactions between lipid bilayer membranes in aqueous media. These interactions include: electrodynamic, electrostatic, and solvation forces, commonly recognized as van der Waals' attraction, electric double layer repulsion and hydration repulsion [42,51,64,85,86,91,103], as well as other steric and structural interactions including the weak opposition to formation of adhesive contact due to thermally driven bending undulations of the ultrathin membrane [14,18,21,27]; enhanced adhesion due to depletion flocculation for nonadsorbing polymers such as plasma proteins, dextrans and PEGs [23,26,49]; and the resistance to close approach afforded by the steric repulsion due to lipid-grafted water-soluble polymer chains [39,40,47,75]

### *(a) Micropipet Measurement*

In the simple vesicle–vesicle adhesion experiment, two vesicles are aligned and maneuvered into close proximity; the pipets are maintained in these fixed positions so that no axial force is exerted during the adhesion test [20,22,24,29]. The extent of adhesion is then controlled via the tension in the left-hand adherent vesicle membrane, which in turn is controlled by pipet suction pressure.

The reversible work of adhesion $w_a$ to assemble two bilayers [27,29] is the cumulation of the action of densely distributed interbilayer forces from large separation (infinity) to intimate contact. Mechanical equilibrium at stable contact is the balance between the free energy required to increase the contact area ($w_a$) and the mechanical work to deform the vesicle contour as represented by the Young–Dupre equation,

$$w_a = \tau_m(1 - \cos\theta_c) \tag{7}$$

where $\tau_m$ is the membrane tension in the adherent vesicle and $\theta_c$ is the included angle between the adherent and test surface membranes exterior to the contact zone.

As summarized in the next sections, the micropipet technique has been used to measure adhesive interbilayer interactions based on these attractions that are enhanced by depletion flocculation produced by nonadsorbent polymer, and are attenuated by short-range hydration repulsion, thermal undulations and electrostatic double-layer repulsion energies [14,15,22,25].

### *(b) Van der Waals versus Hydration*

The energy of attraction between neutral bilayers has been measured by this method to be ~0.015 mJ m$^{-2}$ [20,22,26]. This attraction energy is well within the range for this technique that can measure adhesion energies up to 40 mJ m$^{-2}$, i.e. limited only by the tensile strength of the adherent bilayers. The results are well modeled by the van der Waals power law attraction limited by a steep, close-in, exponential hydration repulsion, establishing adherent contact at an interbilayer separation of ~25 Å, as measured by X-ray diffraction, [64].

### *(c) Van der Waals versus Electrostatic Repulsion*

As expected from DLVO theory [103], the presence of negative charges in the bilayer surface opposes the van der Waals attraction with double layer forces at large distances and attenuates the measured adhesion energy. Negative charges are simply introduced into lipid vesicles by incorporating negatively charged lipids such as phosphatidylserine (PS) or phosphatidylglycerol (PG), both having one net negative charge per molecule. For example, a series of experiments were carried out that studied the lipid mixtures: SOPC/POPS in 100 mM salt solution, in which the bilayer concentration of the negatively charged lipid was varied from 0 to 6 mol % [26]. It was found that the adhesion energy was attenuated with increasing POPS concentration in agreement with Guoy–Chapman theory, and when ~5 mol % negative lipids were present in the bilayers the vesicles simply did not adhere.

### *(d) Depletion Flocculation*

The addition of large polymers and protein macromolecules to aqueous suspensions of lipid vesicles has been found to greatly augment the adhesion between lipid bilayers over and above that provided by the relatively weak van der Waals attraction. This enhanced adhesion originates from the exclusion of polymer in the vicinity of the membrane surface and a corresponding decrease of lipid-associated water. Essentially, attraction between surfaces is caused by interaction of depleted concentration profiles associated with each membrane surface, which leads to a depreciated polymer segment concentration at the center of the gap. The concentration reduction in the gap relative to the exterior bulk solution gives rise to an excess osmotic pressure that acts to draw the surfaces together PEGs [23,26].

Such polymer-induced adhesion has been tested by micropipet manipulation using PC vesicles in solutions of dextran and polyethylene glycol (PEG or PEO) of molecular weights ranging from 11 000 to 150 000 g mol$^{-1}$ [12,23,24,25]. Measured adhesion energies increased rapidly without attenuation (well beyond the threshold level of adhesion due to van der Waals attraction at 0.015 mJ m$^{-2}$), with increasing concentration of both polymers. The nonadsorbing polymer-induced attraction has also been shown to overcome other long-range repulsive potentials such as electrostatic stabilization. For example, the adhesion energy induced by a dextran solution of 0.057 volume fraction, is attenuated as a function of surface charge provided by inclusion of POPS in SOPC vesicles, but is not abolished completely even at 33 mol % POPS, especially for the higher-molecular-weight polymers.

### *(e) Grafted Polymer Surfaces*

With grafted polymers included in the lipid bilayer interfaces, the micropipet manipulation of giant vesicles has shown that adhesion is prevented [75]. However, when experiments were carried out in aqueous solutions of nonadsorbing polymers such as dextrans and PEGs, otherwise polymer-stabilized vesicle bilayers were in fact made to be adherent by the action of nonadsorbing polymers in solution. This depletion flocculation interaction is well modeled by simply placing the interacting surfaces at the end of the grafted polymer [49].

### *(f) Specific Ligand Receptor Interactions*

Finally, in a departure from long-range and nonspecific interactions, experiments were carried out involving vesicle membranes made from neutral lipids, co-expressing a polymer–lipid layer and biotinylated lipids at low concentration [83]. Avidin, bound to one vesicle surface then provided the linker between two biotinylated vesicles. The adhesion experiment involved the presentation of an avidin-coated vesicle to a biotinylated vesicle and measurement of the adhesion energy with time as the biotin receptors became bound in the contact zone. Unlike the spontaneous adhesion of vesicles due to the continuum action of forces, adhesion mediated by biotin–avidin–biotin cross-bridges is slow, with the size of the adhesion zone increasing over the order of minutes. As the spreading proceeds, the contact angle between the vesicles increases which, for a constant-suction pressure, provides a measure of the increase in membrane tension that balances the spreading energy.

The results indicate that diffusion and accumulation of receptors in the contact zone is the rate-determining step and a diffusion coefficient was calculated for the receptor accumulation of between 1 and $2 \times 10^{-12}\ m^2\ s^{-1}$, in close agreement with that for the lipid translational diffusion coefficient measured in vesicle membranes [10], thereby supporting earlier theoretical predictions [11].

## 4. WATER PERMEABILITY, MOLECULAR EXCHANGE AND PEG BARRIER

The micromechanical experiments discussed so far have evaluated the properties of membranes of constant composition (mainly lipids and cholesterol), i.e. in these systems the aqueous solubility for membrane lipid components is so low (less than $10^{-10}$ M) that bilayer mass can essentially be considered constant, at least during the time of an experiment. However, other membrane-compatible materials, such as the lysolecithin monooleoylphosphatidylcholine (MOPC, 1-oleoyl-*sn*-glycero-3-phosphocholine), can have a significant solubility in the aqueous phase as monomers and, at higher concentrations in the micromolar range can exist as micellar phases. Thus in solutions of this [81] and other surfactants, such as bile salts [28], or other molecules such as peptides, polymers, polyphenolics, and drugs, molecules in solution can exist in a dynamic and thermodynamic equilibrium with lipid bilayers, and bilayer composition can be changed in seconds. Following earlier work concerning pore formation in a vesicle membrane by the application of electric fields and how surfactants might influence the poration and resealing [73,111, 112,114,115], new micropipet techniques have been developed in order to measure the exchange of a variety of molecules with giant vesicle membranes, (uptake into and desorption from the membranes) and to study their role in formation and evolution of porous defects in the membranes [80,81,100,113,117].

The remaining sections focus on these new experiments involving the transport of water and other small molecules, such as globular macromolecules, surfactants, polyphenolics and peptides to, into, and across bilayers. Once again, the micropipet brings the unique ability of not only being able to measure area and volume changes for single vesicles and of controlling the level of membrane stress, but also to manipulate giant vesicles (and other particles [48]), rapidly changing their solution environment. Because of the nature of the interaction between lipid bilayers and these different kinds of molecules, it is conceivable that they can trigger the spontaneous disruption of liposome membranes when configured into drug delivery systems, thereby releasing the encapsulated drug at site. Studies like these will be important to the design of the more sophisticated liposome and other carriers currently envisaged by the industry.

## 4.1 Micropipet Experimentation for Molecular Exchange

### *(a) Microchamber Transfer*

Before being able to understand the influence of such biologically active molecules and surfactants on cellular and technological processes, it is fundamental to first characterize the kinetic and equilibrium characteristics of the giant vesicle bilayer–aqueous solution system. What is required are direct measurements of the uptake into and desorption from the bilayer of these molecules. Uptake includes both adsorption to the aqueous bilayer interface, involving a binding to a specific grafted moiety or an attractive interaction between the molecule or micelle with the lipid headgroup region, and the possible intercalation of the molecule in between the lipid molecules of each of the lipid monolayers. Desorption is simply the re-entry of the molecule into the aqueous phase as a monomer, due to wash out. The ideal would be a measure of the actual number of molecules that are taken up by the membrane for a single giant vesicle as a function of concentration in the aqueous phase and tension in the membrane [28]. As the micropipet provides an accurate measurement of area change, changes in membrane area can give a direct measure of the number of molecules taken up by knowing the area per molecule of the host lipid and adsorbing molecule. If this is not known a priori, (from X-ray diffraction for example) then it can in principle be measured from uptake experiments carried out as a function of membrane tension [28]. Also, the rate of molecular uptake is likely to be influenced by the fluid dynamics of solution flow around the vesicle, and so experimental methods have been developed that allow solutions of test molecules to be delivered by simple diffusion and by convective transport.

In the simplest micropipet experiment, a single vesicle supported with a constant pipet suction pressure is transferred from a chamber without the test molecule into one with the test molecule at a desired concentration. The test molecule can be a molecule that might bind to the bilayer, intercalate into it, or pass through it, including water, (i.e. a solution of different concentration). In the following description, an experiment using lysolipid, the lysophosphatidylcholine (LPC) monooleoylphosphatidylcholine (MOPC), is featured as an example. Upon transfer of the giant vesicle into, say, a 1 μm solution of MOPC (a concentration that is less than the critical micelle concentration which is at 3 μm), a gradual increase in the projection length $L_p$ inside the pipet is observed over a period of 300–400 s, as plotted in Figure 9.11.

Transfer of the vesicle back into the LPC-free solution results in a decrease in the projection length almost back to the original position in ~100 s. A subsequent transfer back to the LPC solution shows an approach to a similar plateau and is followed by the same desorption. This uptake–desorption cycle can be repeated several times and clearly demonstrates the reversibility of the LPC exchange dependent on bathing solution concentrations and time of exposure to test and washout solutions.

The maximum change in projection length is ~6 μm, and is readily measurable by the micropipet technique. If the vesicle maintains constant volume, this length change is proportional to the relative change in vesicle membrane area $\Delta A$ [50,79]. For this particular size of vesicle and pipet, the $L_p$ change of 6 μm represents only a 2 % increase in vesicle area up to the apparent saturation of the membrane. The area change is then converted to a concentration of LPC in the bilayer if it is assumed that

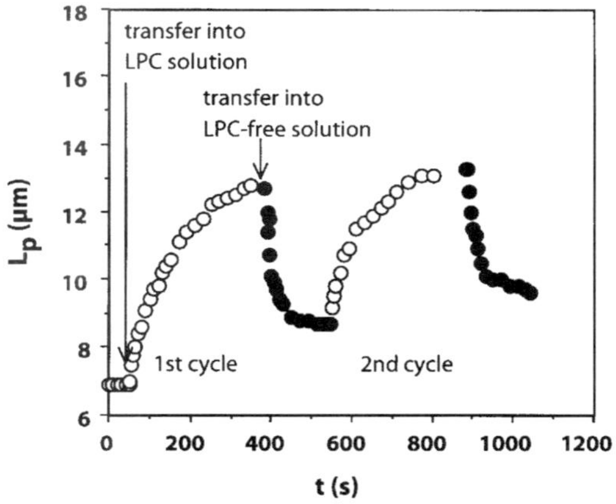


**Figure 9.11** Uptake for lysolecithin exchange with egg PC vesicle bilayer measured by a change in the vesicle membrane projection length $L_p$ of the vesicle membrane in the micropipet versus time. Concentration of MPOC in the bathing solution is 1 μm.

the area per molecule of egg PC is 0.64 $nm^2$ [63] and that of LPC is 0.44 $nm^2$ [41,59]. Extrapolating the experimental points to the asymptote according to a model developed by Zhelev [113] gives the amount of LPC in the membrane, at the apparent saturation, as ~3.3 mol %.

When LPC desorbs, the driving force is the concentration gradient from bilayer to solution. The final vesicle area returns to a value that is close to the area before adsorption. An important feature of the experimental technique is that during this whole process the membrane tension is controlled via the pipet suction pressure. In the present example, the suction pressure in the pipet is kept constant at 800 N $m^{-2}$ which, for this particular vesicle and pipet size, converts to a constant membrane tension of 1.5 mN $m^{-1}$. Tension-dependent uptake in this manner can give the area per molecule of the adsorbing species. For example, using this method, it was determined that bile salts have an area in the bilayer of 29 Å$^2$ [28].

Thus, this simple transfer experiment of a giant lipid vesicle into a lipid surfactant solution provides a measure of the extent and rate of uptake of lysolecithin upon exposure to the surfactant and, similarly, upon transfer back to a surfactant-free solution measures the rate and extent of desorption for the same vesicle. The method gives the whole time history of surfactant exchange for a single bilayer vesicle with controlled membrane tension for a range of concentrations of the surfactant in the solution around the vesicle. Further details of these experiments involving lysolipids are discussed in later section, featuring the effects of incorporated polyethyleneglycol-grafted lipids in the bilayer in limiting macromolecule and surfactant exchange.

*(b) Hydrodynamic Flow*

In order to establish a constant boundary condition for the concentration of the test molecule at the bilayer surface, a micropipet flow method was used to deliver solution directly to a vesicle at known rates of flow [80]. As shown in Figure 9.12, the vesicle in the micropipet is held stationary and two opposing pipets can deliver either bathing (control solution) or test solution at predetermined flow rates.

The flow pipets are connected to a standard water-filled manometer system and so the flow of solution from the pipet can be accurately controlled by displacement of the water reservoir. For example, a pressure difference of 20 N $m^{-2}$ produces a flow rate of 240 μm $s^{-1}$ at the tip of a 40 micron flow pipet. With flow of solution over the vesicle, a much more rapid exchange is achieved for both uptake and wash out.

How these techniques have been used in several exchange experiments involving water, nonelectrolytes, polyphenolics, peptides, globular macromolecules, and surfactants will now be described.

## 4.2 Water Permeability

The permeability of lipid bilayers to water and other bilayer-soluble components depends strongly on the state of lateral cohesion that the bilayer exhibits and the presence of defects [5]. Permeability data have been obtained using micropipet

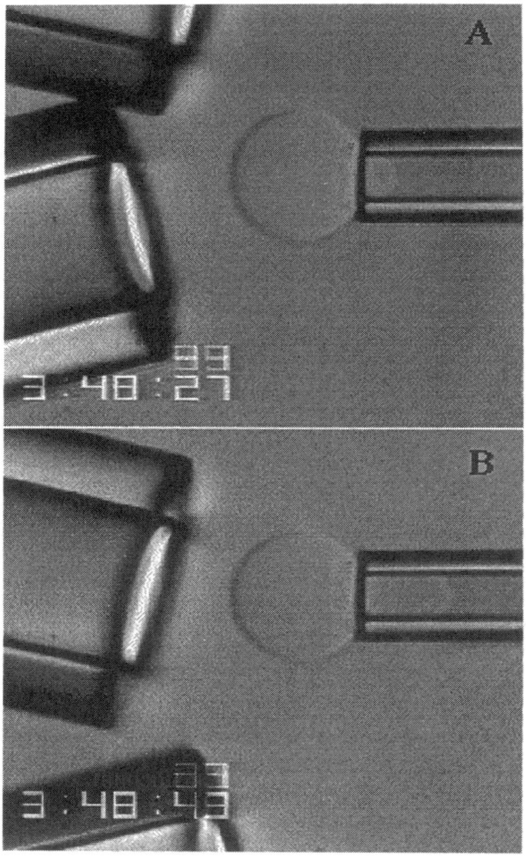


**Figure 9.12** Video micrograph showing arrangement of the holding pipet and the two flow pipets that deliver the reference solution (bottom pipet) and the solution containing the exchangeable molecule (top pipet). (A) Initially, the vesicle is exposed to the reference solution and the vesicle projection length in the holding pipet establishes its reference position. Then the vesicle is exposed to the test solution containing the exchangeable molecule (B) which is delivered from the top pipet. Immediately after the exposure to the second solution the projection length of the vesicle membrane in the pipet ($L_p$) starts to increase, indicating the intercalation of the exchangeable molecule in the vesicle membrane. Finally, when the vesicle is exposed to the reference solution again, i.e. back to (A), its projection length decreases and eventually reaches its initial position as shown in (A). The holding pipet is 6 microns in diameter.

manipulation for several lipids and lipid compositions for which compressibility data are also available [5,82]. Water permeability through bilayers progressively decreases as the compressibility of the bilayer decreases, showing the role of fluctuations in surface density which determine both compressibility and permeability to water. Again, cholesterol makes enormous changes in permeability. One of the least compressible bilayers, sphingomyelin–cholesterol, shows permeabilities about 100 times less than single-component phospholipids such as DMPC.

Apart from these studies on the effects of cholesterol-induced bilayer condensation on water permeability, relationships between lipid structure and water permeability for pure bilayers are not well understood. Knowledge of how structural features such as unsaturated bonds of the lipid interact with structure/size of the solute will shed more light on how a range of molecules partition into and cross lipid membranes.

A new series of studies [84] has focused on bilayers composed of unsaturated lipids. This was motivated in part by the fact that: (1) bilayer permeability to glycerol was previously found to increase with lipid unsaturation [8] and (2) the solubility of water in alkenes is slightly greater than in alkanes. Transport of various molecules across the bilayer occurs by diffusion down a concentration gradient. The permeability, that characterizes this mass transfer is a property of the membrane composition, the solute, and the temperature,

$$\mathrm{d}m_i/\mathrm{d}t = -P_i A \Delta c_i \tag{8}$$

where $m_i$ is the mass of solute $i$ inside the vesicle, $P_i$ is the permeability coefficient for that solute, $A$ is the membrane area, and $\Delta c_i$ is the concentration gradient of the solute across the membrane. In the experiment, a single giant vesicle is held in the micropipet (Figure 9.13), which allows all the geometric features (membrane area, vesicle volume, pipet dimensions) and applied stresses (membrane tension) to be measured and controlled [82].

After a prestress, to take up all excess membrane area, a small tension is applied ($\sim 0.3\,\mathrm{mN\,m^{-1}}$) and the vesicle is transferred into an adjacent microchamber where the solution is at $\sim 10\,\%$ higher osmolarity (hyperosmotic) than that in the vesicle interior. Water then leaves the vesicle by osmosis. The subsequent change and rate of change of vesicle volume due to water efflux, at constant vesicle membrane area, is thus measured from the increase in the projection length of the membrane in the pipet. The result is a plot of relative volume change $V/V_o$ versus time (Figure 9.14) from which the permeability coefficient is derived.

As shown in Figure 9.14 the permeability of water through the membrane of a single giant vesicle was measured by following the vesicle volume $V$ with time $t$ in response to an imposed osmotic gradient. The time course of the volume change was fit to the following equation:

$$(V/V_{\mathrm{inf}} - 1)\exp(V/V_{\mathrm{inf}}) = (V_o/V_{\mathrm{inf}} - 1)\ \exp(kt + V_o/V_{\mathrm{inf}}) \tag{9}$$

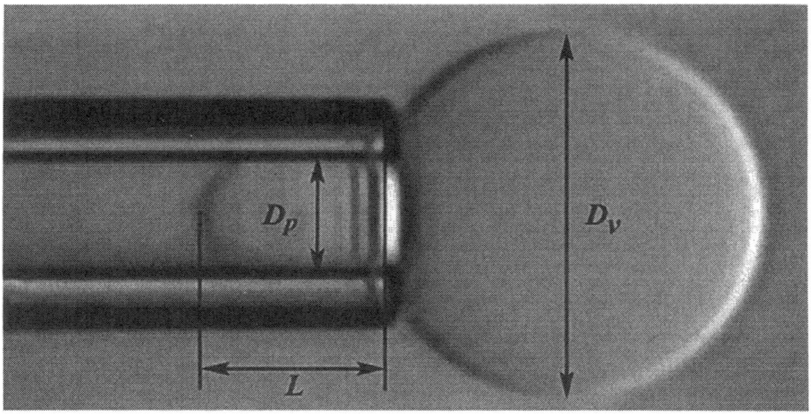


**Figure 9.13** Videomicrograph of lipid vesicle held by suction micropipet. The single-walled vesicle contains sucrose solution and the bathing solution is glucose, giving an index difference for better visualization in the interference contrast microscope. Dimensions used in calculations of area, volume and tension are shown [84].

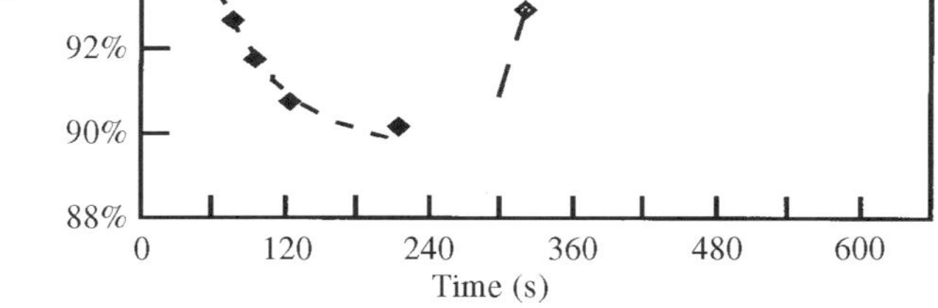


**Figure 9.14** Graph of relative volume versus time for a typical water permeability experiment showing volume change in response to a transfer from a 200 mOsm glucose solution into a hyperosmotic (220 mOsm) glucose solution. The vesicle was returned to the original solution at about 300 s. Note that the volume change is completely reversible. The lipid used in this experiment was SOPC at 15°C. The area of the vesicle was 1937 μm$^2$, and the volume was 11 183 μm$^3$.

Where $V_{inf}$ is the final equilibrium volume at $t \rightarrow$ infinity, $V_o$ is the initial volume, and $k$ is the rate constant for the decay ($s^{-1}$). Note that this is not a simple exponential decay because the efflux of water changes the concentration gradient driving the water flow. The permeability coefficient is determined from the rate constant by the following relationship:

$$P_f = \frac{kV_\infty}{AC_\infty \bar{v}_w} \tag{10}$$

Where $\bar{v}_w$ is the specific volume of water [84].

Using this method on single giant lipid vesicles has shown that the compressibility of the lipid bilayer is directly correlated with the water permeability coefficient for membranes that were monounsaturated, or saturated and mixed with cholesterol. However, for the unsaturated lipids studied, an unexpected result was that, although the area expansion modulus of bilayers made from lipids that contained multiple unsaturation did not change much as the unsaturation was changed from one to six double bonds per molecule, the water permeability coefficient increased from $20\,\mu m\, s^{-1}$ for 18:0/1 bilayers to $100\,\mu m\, s^{-1}$ for di18:2 bilayers, as graphed in Figure 9.15.

This observation indicates that although membrane interface cohesion is important in restricting water transport for the more condensed membranes, such as those containing cholesterol, it appears to be the absolute solubility of water in the membrane that dictates the rates of transport for the softer membranes.

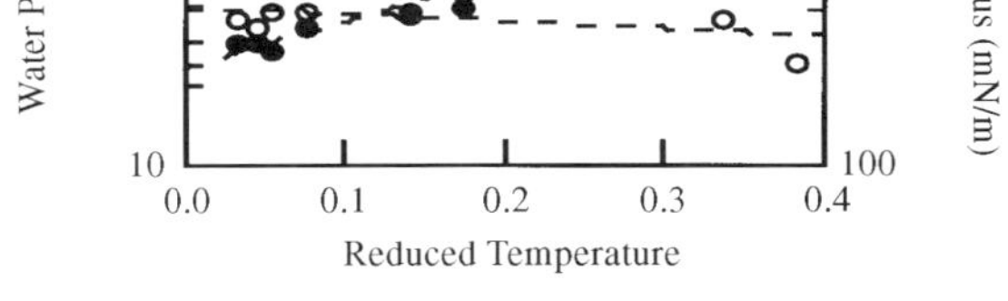


**Figure 9.15** (●) Water permeability and (○) area expansion modulus versus reduced temperature ($T_r = (T/T_m) - 1$) of unsaturated lipids at 21°C. Highly unsaturated lipids have low transition temperatures ($T_m$), and thus have higher reduced temperatures.

### 4.3 Transport of Nonelectrolytes

In order to characterize lipid vesicle membrane transport properties further, we have begun to study the transport of other molecules and test whether correlations might also exist with mechanical compressibility and water permeability. Here, although some data exists for black lipid membranes [33], little attention has so far been paid to the transport of small molecules and ions for giant vesicular membranes (i.e. free of solvent and boundary-support effects) [87] and none using the micropipet technique on single vesicles. The new series of transport studies considers first urea, which is both a water-soluble and membrane-soluble molecule.

As mentioned above, one of the main techniques for characterizing the transport of molecules uses BLM systems to track transport using radiolabeled solutes or other tracers [33,105]. However, the micropipet method offers some distinct advantages over existing methods for determining solute permeability. As changes in vesicle volume are used to report on the concentration of the solute, the micropipet technique does not require any special labeling or chemical analysis for measuring the solute concentration. The vesicles used for micromanipulation also have well-defined areas and volumes, and do not contain solvents, which may affect the measured permeability. Having established the rate of water transport across vesicle bilayers, it becomes possible to determine the permeability of other permeant solutes, such as urea or glycerol.

The permeability of small molecules other than water can be determined by simply transferring vesicles into a hyperosmotic solution containing the test solute. Preliminary studies transferred a vesicle (200 mOsm sucrose inside) from a glucose solution (200 mOsm) into a urea solution (240 mOsm). As shown in Figure 9.16, the vesicle volume first reduced then increased.

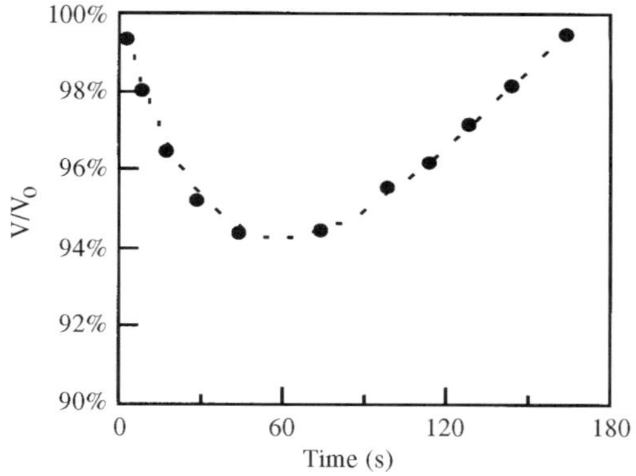


**Figure 9.16** Fractional volume change showing volume versus time for a urea permeability experiment. The vesicle was composed of SOPC, with an area of 5981 μm$^2$ and a volume of 42 951 μm$^2$. The experiment was conducted at 25°C.

Upon exposure of the vesicle (200 mOsm sucrose inside) to the urea solution (240 mOsm), there is an initial volume loss from the vesicle followed by a subsequent inflation of the vesicle, all over a period of around 3 min. The initial decrease in vesicle volume is due to the rapid efflux of water down the osmotic gradient into the more concentrated urea solution. This efflux of water slows and reverses as urea permeates the vesicle membrane bringing water with it, which overtakes the original efflux of water as the osmotic gradient of 40 mOsm collapses. The vesicle swells and can eventually burst if the gradient is large enough and the initial vesicle excess area is small, eventually being reduced to zero as the concomitant flow of urea and water continues. The permeability of the membrane to the urea solute can then be determined from a numerical simulation of the time course of the volume change as the permeability of the membrane to water is already known.

The volume change can therefore be used as a reporter of urea concentration in the vesicle and so determine its permeability coefficient for lipid vesicle membranes of varying composition. Preliminary experiments indicate that urea has a permeability coefficient of $6.8 \times 10^{-7}$ cm s$^{-1}$ in SOPC, and $1.4 \times 10^{-6}$ cm s$^{-1}$ in DOPC at 25°C. These values compare extremely well with the earlier reported values of urea permeability ($4.0 \times 10^{-6}$ cm s$^{-1}$ at 25°C in eggPC [33], or $1.4 \times 10^{-6}$ cm s$^{-1}$ at 25°C in DOPC [87]). These experiments now lay the founding protocols for experiments aimed at quantifying drug transport across giant vesicle bilayers.

## 4.4 Exchange of Tannic Acid

Molecular exchange can involve intercalation into the outer exposed monolayer of the vesicle bilayer, as discussed above for lysolipid. Following outer monolayer uptake the molecule can then pass through the bilayer limited by flip-flop across the bilayer midplane [81]. In contrast, other less membrane-soluble molecules can bind into the interface but not cross the bilayer. One such molecule is tannic acid, a polyphenolic compound composed of a central ring derivatized with five digallic acid residues as shown in Figure 9.17 [99].

Upon exposure to this molecule, the giant vesicle membrane undergoes a small expansion in area of only a few per cent, before breaking. This area change is readily detected by holding the giant vesicle in the micropipet and measuring the small length change of the membrane projection in the pipet [99]. As shown in Figure 9.18, the rate of uptake, as measured by the increase in relative vesicle area, is dependent on the concentration of tannic acid in solution and the rate determining step is therefore simply diffusion limited.

Membranes are stable up to concentrations of 1 μM, but for $5 \times 10^{-5}$ M and $10^{-4}$ M they fail in seconds. The concentration for failure coincides with the precipitous collapse of the interbilayer space in multilamellar vesicles measured by X-ray diffraction, representing a dehydration of the bilayer surface and the

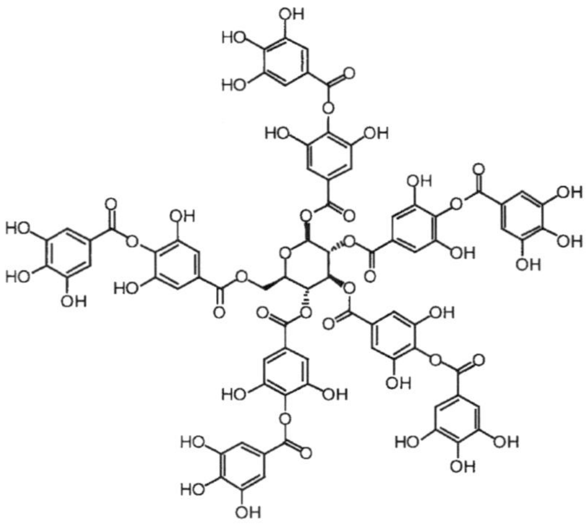


**Figure 9.17** Chemical structure of tannic acid.

possible bridging of the molecule between adjacent bilayers [100]. It appears that, because the molecule cannot cross the bilayer, its adsorption to the outer exposed monolayer induces a concomitant expanding tension in the inner monolayer. The micropipet assay therefore gives a measure of the adsorption of tannic acid to the bilayer interface and demonstrates that this adsorption must involve intercalation into the headgroup region but that it does not cross the membrane. It is expected that this effect will be measurable with other molecules, for example the PEGylated lipids that are relatively soluble in aqueous media as monomers, but because of the large hydrophilic polymer should not cross the bilayer midplane.

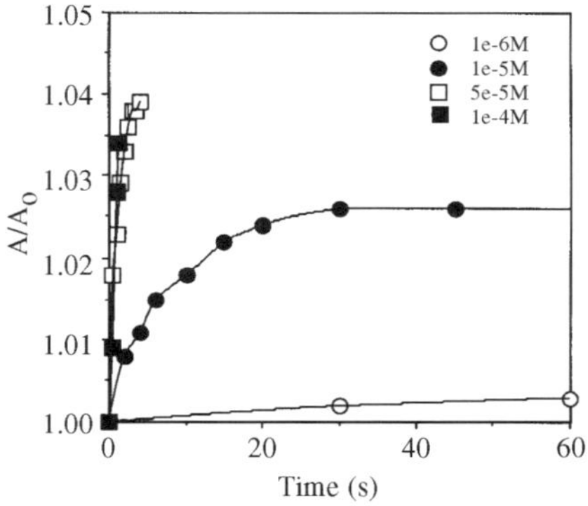


**Figure 9.18** Relative change in vesicle area upon transfer into equiosmolar tannic acid solutions of $10^{-6}$ M (open circles), $10^{-5}$ M (filled circles), $5 \times 10^{-5}$ M (open squares), and $10^{-4}$ M (closed squares).

## 4.5 Exchange of Peptides

The micropipet technique has also been used to study the interactions of water-soluble peptides and proteins with membranes [117]. These interactions are important for a variety of biological processes such as entry of viruses and microorganisms into cells, the cytosolic action of toxins of microorganisms, defense toxins of the immune system, and the action of antibiotics. More specifically, these initial studies are focusing on characterizing the interaction of viral fusion peptides and pore-forming bacterial toxins with membranes. Applications of the studies are in the design of potentially new drugs and drug delivery systems for gene therapy where vesicle fusion with the plasma membrane or endosome membrane is desirable in order to deliver the gene directly into the cytoplasm of the cell.

Fusion reactions of viruses are mediated by short peptides that fall into two main groups: one dependent on a shift to low pH and the other pH-independent [108]. Initial interests here have focused on peptides involved in pH-dependent fusion because of their potential for application in environmentally sensitive, and therefore controllable, model systems. Of the pH-dependent viruses (influenza virus, the vesicular stomatitis virus, the Semliki forest virus and mouse mammary tumor virus [108]), influenza hemagglutinin (HA) is probably the most studied [4]. HA alone is enough to facilitate the fusion of the influenza virus with the host membrane [102,107]. Through the involvement of at least three subunits, it mediates both the initial attachment of the virus to receptors that contain sialic acid [94] and its fusion with the endosomal membrane [57,58].

This discussion narrows to the HA2 subunit because it is involved in fusion and therefore is expected to insert into lipid bilayers. The fusogenic part of HA2 is hidden inside HA and is exposed only at low pH [34,98,109]. Within seconds after exposure to peptide solution, the peptide is inserted into the host membrane [35,37]. This insertion is independent of membrane composition [110] which suggests that the insertion occurs in the lipid part of the membrane. Other fusion proteins require a specific ligand on the host cell membrane (e.g. HIV). In this case, the insertion of the fusion peptide occurs in close proximity to the ligand, and so cannot be considered to occur in the lipid alone. The inserted portion is composed of about 20 amino acids and is called a fusion peptide [37,38]. The importance of this amino acid residue for HA fusion is revealed by the fact that site-specific mutations severely affect HA fusion activity [35].

The initial measurements have studied the insertion of the fusion peptide into the membrane of giant SOPC vesicles. Similar studies have also been carried out [53]. The mutant HA2-G4E synthesized by Wharton *et al.* [118] was used, which has the sequence GLFEAIAGFIENGWEGMIDGK. This mutant is similar to the wild-type peptide, except that the glycine at the fourth position is replaced by glutamic acid (underlined). This peptide has similar lytic activity [118] and occupies a similar area in monolayers [89] as its wild-type counterpart. Therefore, the insertion of this peptide in SOPC membranes is considered a good model for the insertion of HA2.

Because the lytic activity of HA2 is pH dependent, it is expected that its insertion in the membrane will also be pH dependent. Figure 9.19 shows the measured change of the apparent vesicle area for different pH values.

It is seen from the data that the area of the vesicle increases immediately after its exposure to the peptide buffer, indicating peptide insertion in the membrane. Also the insertion is pH dependent and increases when the pH is decreased from 7.5 to 5. With time, the increase of the apparent vesicle area reaches a semistationary value. For pH below 4.8, the membrane breaks down due to the action of the peptide. Data from experiments like this carried out over the whole pH range is summarized in Figure 9.20 which shows how the apparent vesicle area increases continuously with decreasing pH from 7.5 to 5.5 and then membranes become unstable below pH 4.8.

There are then two regions in Figure 9.20, one above pH 4.8 where the membrane is stable, and another below pH 4.8 where the membrane breaks down. The tendency of the peptide to insert into the membrane when the pH decreases suggests that the peptide has increased its own tendency for self-aggregation. This hypothesis was tested using light scattering and concentrated peptide solutions (1.4 mM). The light-scattering experiment showed that in the range of pH from 7.5 to 5 the peptide macromonomer is in the singlet form. However, as the pH was reduced in the range pH 6 to pH 5, peptide aggregates appeared which scattered a measurable amount of light. The conclusion is that even though the vast majority of peptide molecules are in monomer form, the peptide showed a tendency for aggregation as the pH decreased. When the pH becomes close to 4.5 the peptide precipitated. This is close to the pH (pH 4.8) for membrane breakdown and the p*K* (4.7) of glutamic acid. Therefore, membrane breakdown below pH 4.8 and the peptide precipitation are

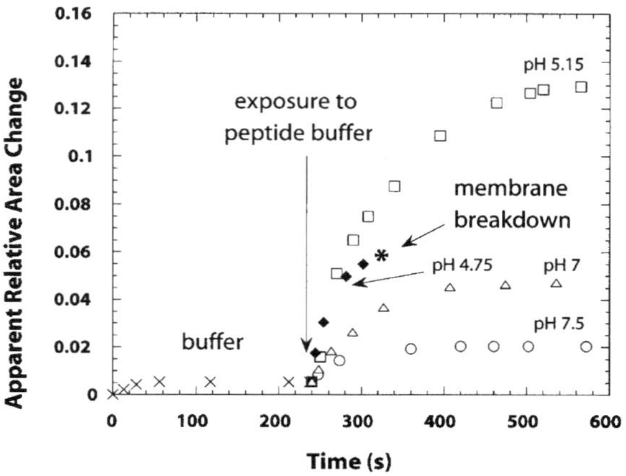


**Figure 9.19** Increase of the apparent vesicle area versus time after exposure of the vesicle to a buffer containing 10 μM HA2-G4E and at different pH values. The apparent rate of area increase as well as the magnitude of this increase are larger for lower pH. The exception is the area increase at pH 4.75, which ends abruptly because of vesicle breakdown.

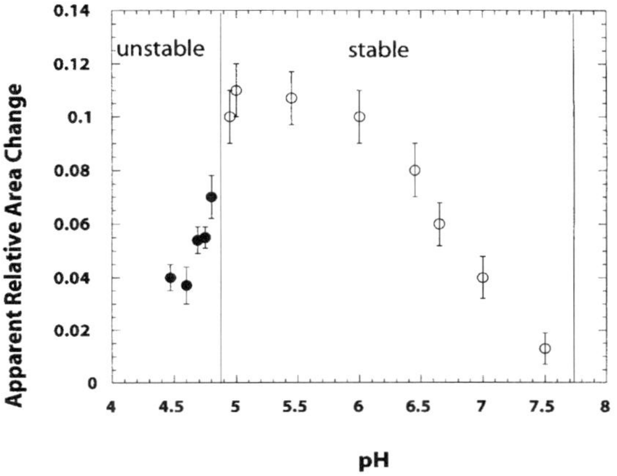


**Figure 9.20** Dependence of the apparent vesicle area increase on the pH of the buffer solution. The amount of peptide is 10 μM.

most probably related to the protonation of the glutamic acid resulting in a decreased solubility for the peptide in water and an increased solubility of the peptide in the giant vesicle membrane.

The reversibility of peptide binding is simply tested by washing the peptide-loaded vesicle membrane with peptide-free buffer. In these experiments, the apparent vesicle area did not recover to its initial value. To study this effect in more detail two sets of experiments were performed (Figure 9.21). In one of them, the vesicle was continuously exposed to peptide solution for 20 min and then it was washed with peptide-free buffer. In the other set of experiments, the vesicle was exposed to peptide buffer three times and after each exposure it was washed with peptide-free buffer. The total time of exposure of the second vesicle was the same as that for the first one. The result was that in all cases the change of the apparent vesicle area during peptide desorption was the same. This suggested that the change of the apparent vesicle area during desorption represents the true area occupied by the desorbing peptide in the outside membrane monolayer. (A model relating the resultant membrane change, and the change of number of molecules in the two membrane monolayers, is presented by Zhelev [117]. This model predicts that the resultant membrane area weakly dependends on the unequal lipid distribution between the monolayers). The issue of the incomplete recovery of the apparent vesicle area is not resolved. It may be due to the incomplete peptide desorption (possibly because of peptide partitioning in the internal membrane monolayer) or to volume change because of pores.

Other studies have measured conditions for pore formation and pore reversibility [117]. Here again, being able to work with a single lipid vesicle, and to distinguish between area and volume changes as the peptide solution or washing solution is

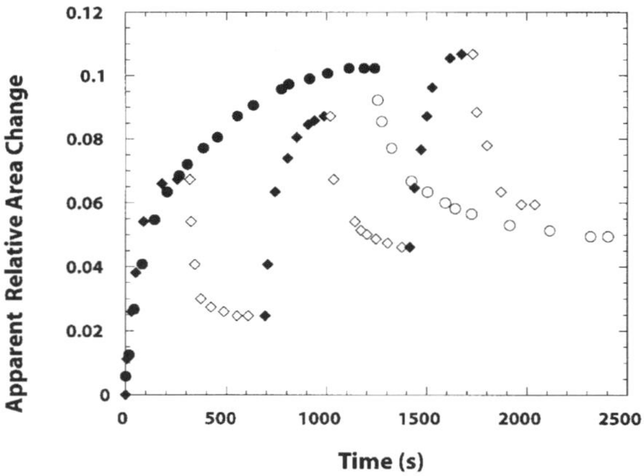


**Figure 9.21** Peptide insertion and desorption at pH 5.4 (10 μM peptide). One vesicle is exposed to peptide buffer for 20 min and then it is washed with peptide-free buffer. The other vesicle is exposed to peptide solution three times (each time for 7 min) and each time the vesicle was washed with peptide-free buffer. The apparent area change during peptide desorption is similar in all washes.

blown directly onto the vesicle, has proved invaluable and essential. These experiments have shown that pre-inserted peptide did not produce lysis of the vesicle when the peptide-loaded bilayer was exposed to low pH, but that lysis was only brought about by interaction between the vesicle bilayer and aggregated peptide from solution. Also, when the vesicle was exposed to low concentrations of peptide solution at pH 4.7 (that did not cause complete breakdown), selective nanopores were formed which initiated vesicle swelling. During swelling the vesicle was washed with peptide-free buffer, which almost immediately arrested the swelling. The conclusion from this experiment is that peptide-induced pores are reversible. The fact that different-sized pores are formed under different experimental conditions suggests that HA2-G4E-formed pores contain a different number of monomers.

## 4.6 Role of PEG in Preventing Membrane–Particle Interactions

Finally, the micropipet technique has been used to test how lipid-grafted polyethyleneglycol (PEG) affects the binding and exchange of macromolecules and association-colloid particles with the lipid bilayer [76,80,83].

### *(a) Avidin Binding to Biotinylated Lipid*

As a model system for ligand–receptor interactions the micropipet technique has been used to measure the rate and amount of binding of avidin to its biotinylated lipid receptor in a single giant vesicle membrane [83]. Here the assay for binding was the build-up of fluorescence intensity at the vesicle surface due to the binding of fluorescent avidin to the biotinylated lipid. The ~5 nm sized avidin has two binding pockets for biotin on each side of the molecule. The biotin was linked to the head group of the DPPE lipid and so could readily be incorporated into a lipid bilayer at defined concentrations. The biotin group was attached to the lipid via a 12 Å extension arm that made it more accessible to bind up and into the relatively deep binding $\beta$-barrel pocket of the avidin molecule [36]. 750 MW PEG polymer was incorporated at the surface via PEG–lipids (PEG750–lipid) at surface densities of 0–10 mol % PEG750–lipid.

In this experiment, the transfer technique was used to move a vesicle with a given amount of incorporated biotin and PEG–lipid from the control chamber into an adjacent microchamber containing 0.1 mg ml$^{-1}$ avidin. After the desired time, the vesicle was then transferred back to the avidin-free chamber and the image of the fluorescent vesicle was analyzed by image processing to determine the relative fluorescence intensity. The maximum value at the circumference of the vesicle was taken to be directly proportional to the amount of bound avidin. This experiment was performed for a fixed biotin concentration of 5 mol % with and without PEG750–lipid as a function of time, and for a 2 min incubation for vesicles that had 0–10 mol % PEG750–lipid in their membranes. The kinetics of this adsorption showed that only 2 mol % PEG750–lipid was required to significantly reduce the rate of avidin binding, slowing down the diffusion of avidin to the vesicle surface by a factor of about five.

Increasing amounts of surface-grafted PEG750–lipid significantly decreased the binding of avidin to a bilayer containing 5 mol % biotin, as shown in Figure 9.22.

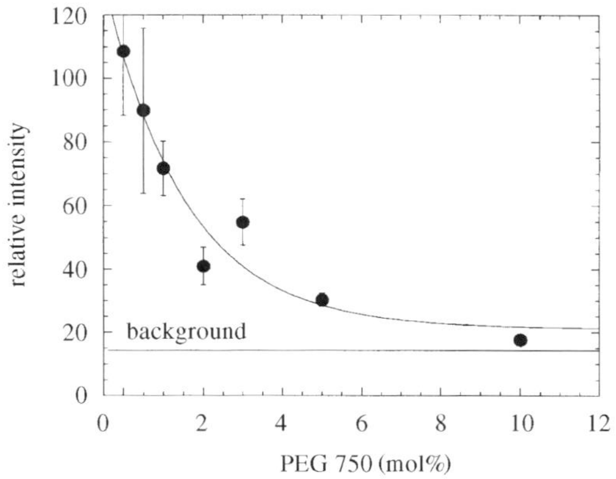


**Figure 9.22** Dependence of avidin binding on molar PEG750 concentration. 5 mol % biotin was incorporated in the vesicle membranes and the incubation time in 0.1 mg ml$^{-1}$ avidin was 2 min. An exponential decay according to $\exp(-A_{\mathrm{Av}})\Pi_{\mathrm{p}}/kT$ [1] was fitted to the data.

For a 2 min incubation (i.e. under diffusion conditions), it was possible to block binding almost completely by incorporating PEG750–lipid in the membrane at concentrations greater than ~6 mol %. At this concentration, the polymer layer is still in the mushroom regime [47] and the surface is not completely covered with PEG750. Based on X-ray diffraction data for the same PEG-grafted bilayers [47], the zone over which this retardation occurred was only within 25 Å of the lipid surface.

These experiments have shown that small PEGs can retard the access of relatively small macromolecules (40 Å × 50 Å) to a bilayer surface. As discussed in subsection (b) below, in connection with the exchange of lysolipid micelles, and presented in more detail in Needham *et al.* [76], this retardation of binding can be explained in terms of two parameters: an additional energy barrier that the surface pressure $\Pi_p$ of PEG mushrooms provides against diffusion to the surface, and the cross sectional area $A_{Av}$ of the adsorbing avidin molecule. Conceptually, the avidin has to push the polymer aside in order to create free area at the interface for binding to biotin, and so the rate of adsorption, and therefore the absolute number of adsorbed molecules in a given time, is decreased by a factor that is proportional to $\exp(-A_{Av}\Pi_p/kT)$ [1,76]. Thus, even at the low density of 2 mol % PEG750–lipid, the binding of avidin to a biotinylated vesicle is hindered but not prevented (reduced by ~50%) by the surface pressure of the lipid-linked PEG750.

### *(b) Uptake of Lysolipids as Monomer and Micelle*

Following the earlier studies on lysolipid exchange with giant vesicle bilayers composed of egg PC or SOPC [81,113], the same exchange was studied for bilayers that contained varying amounts of PEG–lipids. The interest was to determine whether the same 750 molecular weight PEG-lipid (PEG750–lipid), that inhibited binding of avidin to biotinylated lipid vesicle bilayers, would also inhibit surfactant uptake and micelle fusion in a similar fashion.

PEG–lipid was incorporated into the lipid vesicle bilayers at membrane concentrations from 0 to 20 mol %. A diagram of the micelle in relation to the PEG-lipid membrane is shown in Figure 9.23.

Results of the micropipet experiment, using the flow technique to deliver the lysolipid solution to the vesicle, showed that access of micelles, and to a lesser extent monomers, to the surface of the lipid bilayer did depend strongly on the surface density of the polymeric barrier at the bilayer interface; micelles could in fact be inhibited from fusing with the bilayer. In developing a model to interpret these results it was important to realize that the binding of avidin and the fusion of micelles have unique features that make the lysolipid exchange processes very different to those for avidin binding from a kinetic standpoint. Unlike the adsorption of avidin, where the macromolecule binds to its biotin receptor and does not desorb during the course of an experiment (its off rate (desorption rate) is many months$^{-1}$!), MOPC monomer uptake and micelle–membrane fusion are coupled with the rapid

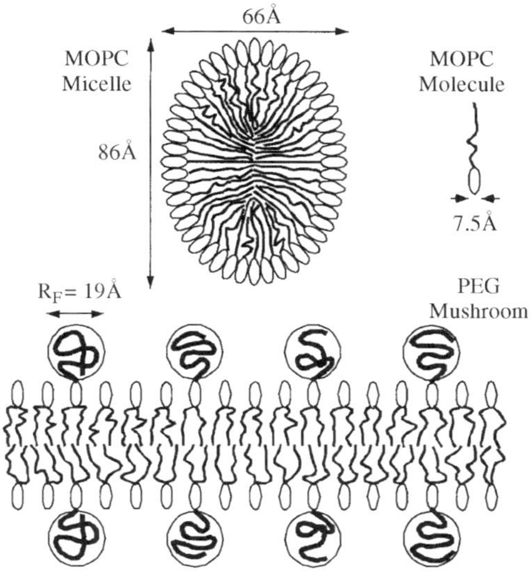


**Figure 9.23** Scale drawing of a MOPC micelle at the surface of a PEG-grafted bilayer. The PEG-lipid density shown is approximately 5 mol %.

desorption of MOPC monomers from the bilayer back to the bulk solution [81]. This coupling between uptake and desorption establishes a stationary equilibrium concentration of MOPC in the membrane. This stationary MOPC concentration is again uniquely measured in the micropipet experiment by a corresponding vesicle area change. Assuming that, at least initially, MOPC only entered the outer monolayer, the relative fractional area change $\Delta A/A_o$ is converted to a molar percentage for exchanged MOPC molecules relative to the total number of lipids in the bilayer, using the approximation,

$$\mathrm{MOPC}_{\mathrm{bilayer}}(\mathrm{mol}\,\%) = \frac{\Delta \mathrm{A}}{\mathrm{A_o}}\frac{\mathrm{A_{SOPC}}}{\mathrm{A_{MOPC}}} \times 100 \tag{11}$$

where $\Delta A$ is the change in vesicle area due to MOPC uptake relative to the starting area $A_0$, $A_{\mathrm{SOPC}}$ is the area per lipid molecule of 67 Å$^2$, and $A_{\mathrm{MOPC}}$ is the area per MOPC molecule (projected area of the acyl chain) of 44 Å$^2$ [113].

The protective effect of the grafted PEG750 in preventing micelle fusion with the membrane is shown in Figure 9.24. Without PEG–lipid in the bilayer, exposure to 100 μM MOPC causes rapid expansion and rupture of the membrane. With 20 mol % PEG–Lipid in the bilayer, uptake in 100 μM MOPC is essentially the same as that observed at the CMC (3 μM) and bilayers are stable.

Additional experiments were then carried out to determine the rate and magnitude of MOPC uptake from 100 μM solutions as the concentration of PEG750–lipid in

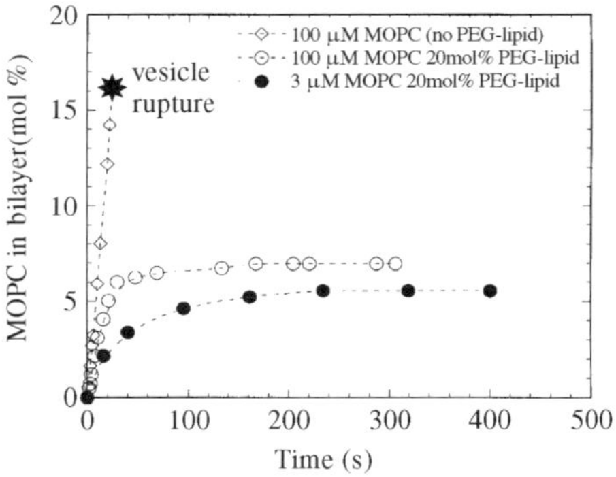


**Figure 9.24** Exposure of single SOPC vesicles to a flow of MOPC solution: (◇) SOPC (no PEG–lipid) exposed to 100 μM MOPC; (○) 20 mol % PEG–lipid exposed to 100 μM MOPC; (●) 20 mol % PEG-lipid exposed to 2 μM MOPC. Dotted lines are simply curve fits through the data.

the membrane was increased from 0.5 mol % to 20 mol % PEG–lipid, and as a function of MOPC concentration in solution. For the unmodified vesicles and vesicles with 0.5 mol % PEG–lipid, the presence of micelles in the bathing solution produced rapid uptake of MOPC resulting in vesicle breakage. However, for PEG750–lipid concentrations between 1 mol % and 20 mol %, the amount of lysolipid partitioning into the membrane from 100 mM MOPC solution reached a stationary equilibrium, and the value of this equilibrium partitioning decreased monotonically with increasing PEG750–lipid concentration and bilayers remained stable [80].

The desorption, or wash-out, of the lysolipid from MOPC-rich membranes involves only monomers and occurred very rapidly upon exposure of the MOPC-loaded bilayers to the MOPC-free solution. The kinetics of MOPC desorption from loaded bilayers containing 1, 2.5, 10 and 20 mol % PEG750-lipid, was not significantly affected by the presence of PEG–lipid over the whole range of PEG surface densities, i.e. high densities of PEG750–lipid did not decrease the off rate of the monomer from the bilayer.

The uptake and desorption results, both as a function of lysolipid concentration in solution and PEG–lipid concentration in the bilayer, were analyzed in terms of true thermodynamic (for uptake of monomer below the CMC) and by a stationary equilibrium model for monomer, and micelle exchange [80]. This model was extended by Needham and Zhelev [76] to predict the on rates of macromolecules and other particles as a function of PEG–lipid molecular weight, graft density and macromolecule and particle size.

A compilation of the membrane uptake data for fixed bulk solution concentrations of MOPC (3 μM and 100 mM) as a function of PEG–Lipid (0–20 mol %) are shown in Figure 9.25 along with the predictions of the stationary equilibrium model.

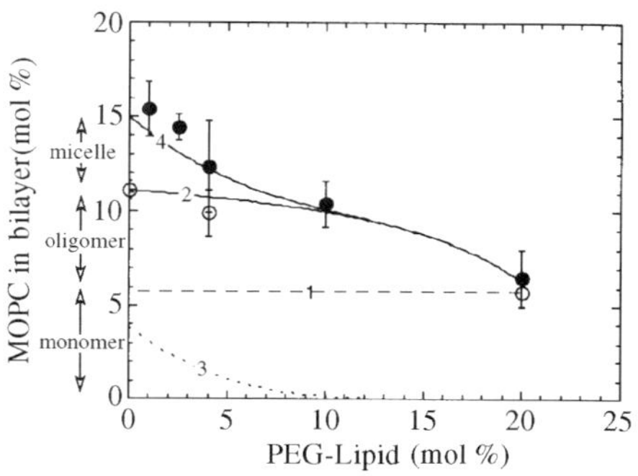


**Figure 9.25** Theoretical model fitted to the data for MOPC uptake by membranes containing different amounts of PEG–lipid from MOPC solutions of 3 μM (curve 2 and open circles) and 100 μM (curve 4 and closed circles). True monomer uptake is shown by the dashed line (curve 1). The amount of MOPC taken up as micelles according to the model is shown by the dashed curve (curve 3).

This plot is fairly complex, but it allows the breakdown in the contributions of all species, true monomer, premicellar aggregates (trimers) and micelles, to be seen. The figure consists of the following: experimental data for 3 μM and 100 μM MOPC, and the theoretical fits for MOPC solution concentrations of 3 μM (curve 2, monomer and trimer) and 100 μM (curve 4, all species), and the uptake due to true monomer alone (horizontal dashed line (1)) and for micelles alone (dotted curve 3) for the 100 μM case. At 20 mol % PEG–lipid for both 3 μM and 100 μM MOPC, all but true monomer is excluded and so the horizontal dashed line (1) represents the level of true monomer uptake for all PEG-lipid concentrations, i.e. the membrane–monomer equilibrium accounts for about 5 mol % of MOPC in the membrane. For MOPC solution concentrations of 3 μM (curve 2), there are no micelles and the measured uptake is fitted by considering only true monomer and trimer grouped as monomer. For the higher 100 μM MOPC solution concentration, as the PEG–lipid concentration in the membrane is reduced from 20 mol % to zero (curve 4), the data are coincident with uptake of monomer (curve 2) down to ~10 mol % PEG–lipid. Below 10 mol % PEG–lipid, uptake of micelles (curve 3) accounts for the rest. Thus, although monomer passes freely through the PEG mushroom region at all PEG–lipid concentrations, trimers are excluded at 20 mol % PEG–lipid but start to find enough free spaces between mushrooms as the PEG–lipid surface density is reduced, and account for an additional 5 mol % of uptake at the CMC for the bare membrane. Similarly, micelles only begin to find enough free space to reach the membrane below 10 mol % PEG–lipid. Ultimately, at PEG–lipid concentrations below 1 mol %, micelles gain complete access to the surface and micelle–membrane exchange results in bilayer disruption. There is, therefore, a molecular size cut-off for MOPC transport through the 25 Å thick PEG–lipid layer somewhere between the dimensions of the monomer and the micelle that can, in principle, be controlled by the surface density of PEG. Thus, as measured using the micropipet techniques, the incorporation of PEG as PEG–lipids in liposomes is a very effective way to control the way in which macromolecules, particles and larger surfaces gain access to the lipid vesicle surface.

## 5. CONCLUSIONS

It is hoped to have shown that micropipet manipulation is an incredibly versatile technique. It enables the study of a wide range of giant vesicle behavior and properties, including the material properties of bilayers, their colloidal interactions, and the uptake and desorption of a variety of macromolecules. These direct measurements on single giant vesicles then help to characterize the properties of natural cell membranes and also provide the essential information needed for the design and ultimate application of lipid vesicle and lipid-coated systems in biotechnology and medicine.

## 6. ACKNOWLEDGMENTS

This work was supported in part by grant GM 40162 from the National Institutes of Health. Acknowledgment is also made to those who have helped to establish and use the micropipet technique for the experiments described above, including Evan Evans and Wieslawa Rawicz at the University of British Columbia, and Natalia Stoicheva, Doris Noppl-Simson, and Kevin Olbrich at Duke University.

## 7. REFERENCES

1. J. D. Andrade (ed.), *Surface and Interfacial Aspects of Biomedical Polymers*, Plenum Press, New York, 1985.
2. M. I. Angelova, S. Soleau, P. Méléard, J. F. Faucon and P. Bothorel, *Prog. Colloid Polym. Sci.*, **89**, 127–131 (1992).
3. V. Bennet, *Annu. Rev. Biochem.*, **54**, 273–304 (1985).
4. J. Bentz, *Viral Fusion Mechanisms*, CRC Press, Boca Raton, 1993.
5. M. Bloom, E. Evans and O. G. Mouritsen, *Quart. Rev. Biophys.*, **24**, 293–397 (1991).
6. G. W. Brady and D. B. Fein, *Biochim. Biophys. Acta.*, **464**, 249–259 (1977).
7. J. H. Davis, *Biophys. J.*, **27**, 339–358 (1979).
8. J. de Gier, J. G. Mandersloot and L. L. M. van Deenen, *Biochim. Biophys. Acta.*, **150**, 666–675 (1968).
9. H. P. Duwe, H. Engelhardt, A. Zilker and E. Sackmann, *Mol. Cryst. Liq. Cryst.*, **91**, 1–7 (1987).
10. M. Edidin, *Ann. Rev. Biophys. Bioeng.*, **3**, 179–201 (1974).
11. E. Evans, *Biophys. J.*, **48**, 175–183 (1985).
12. E. Evans, in *Physical Basis of Cell-Cell Adhesion*, (ed.) P. Bongrand, CRC Press, Boca Raton, 1988.

13. E. Evans, *J. Coll. Int. Sci.*, **43**, 327–347 (1989).
14. E. Evans, *Langmuir*, **7**, 1900–1908 (1991).
15. E. Evans, *Adv. in Colloid and Interface Sci.*, **39**, 103–128 (1992).
16. E. Evans and R. M. Hochmuth, *J. Membr. Biol.*, **30**, 351–362 (1977).
17. E. Evans and R. M. Hochmuth, *Current Topics in Membranes and Transport*, (eds) F. Bonner and A. Kleinzeller, Academic Press, New York, 1978, p. 1
18. E. Evans and J. Ipsen, *Electrochimica Acta.*, **36**, 1735–1741 (1991).
19. E. Evans and R. Kwok, *Biochemistry*, **21**, 4874–4879 (1982).
20. E. Evans and M. Metcalfe, *Biophys. J.*, **46**, 423–426 (1984).
21. E. Evans and D. Needham, *Faraday Discuss. Chem. Soc.*, **81**, 267–280 (1986).
22. E. Evans and D. Needham, *J. Phys. Chem.*, **91**, 4219–4228 (1987).
23. E. Evans and D. Needham, *Macromolecules*, **21**, 1822–1831 (1988).
24. E. Evans and D. Needham, *Molecular Mechanisms of Membrance Fusion*, (eds) S. Ohki, D. Doyle, T. Flanagan, S. W. Hui and E. Mayhew, Plenum Press, New York and London, 1988, pp. 83–99.
25. E. Evans and D. Needham, *Physics of Amphiphilic Layers*, (eds) J. Meunier, D. Langerin and N. Boccara, Springer Proceedings in Physics, Springer Verlag, 1988, pp. 178–198.
26. E. Evans, D. Needham and J. Janzen, *Proteins at interfaces. Physicochemical and biochemical studies* (eds) Brash and Horbett, ACS Symposium Series 343, 1987, pp. 88–103.
27. E. Evans and W. Rawicz, *Phys. Rev. Lett.*, **64**, 2094–2097 (1990).
28. E. Evans, W. Rawicz and A. F. Hofmann, *XIII International Bile Salt Meeting*. Bile Acids in Gastroenterology: Basic & Clinical Advances, Falk Symposium. San Diego, CA. 1994.
29. E. Evans and R. Skalak, *Mechanics and Thermodynamics of Biomembranes*, CRC Press , Boca Raton, 1980.
30. E. A. Evans, *Methods Enzymol.*, **173**, 3–35 (1989).
31. E. Farge and P. D. Devaux, *Biophys. J.*, **61**, 347–357 (1992).
32. J. F. Faucon, M. D. Mitov, P. Meleard, I. Bivas and P. Bothorel, *J. Phys. France*, **50**, 2389–2414 (1989).
33. A. Finkelstein, *J. General Physiology*, **68**, 127–135 (1976).
34. Y. Gaudin, R. Ruigrok and J. Brunner, *J. General Virology*, **76**, 1541–1556 (1995).
35. M.-J. Gething, R. W. Doms, D. York and J. M. White, *J. Cell Biol.*, **107**, 11–23 (1986).
36. M. Green, *Adv. Protein Chem.*, **29**, 85–133 (1975).
37. C. Harter, T. Bachi, G. Semenza and J. Brunner, *Biochemistry*, **27**, 1856–1864 (1988).
38. D. Hoekstra, *J. Bioenergetics & Biomembranes*, **22**, 121–155 (1990).
39. K. Hristova and D. Needham, *Liquid Crystals*, **18**, 423–430 (1994).
40. K. Hristova and D. Needham, *J. Coll. Int. Sci.*, **168**, 302–314. (1994).
41. S. W. Hui and C. Huang, *Biochemistry*, **25**, 1330–1335 (1986).
42. J. Israelachvili, *Quart. Rev. Biophys.*, **6**, 341–387 (1974).
43. M. J. Janiak, D. M. Small and G. G. Shipley, *J. Biol. Chem.*, **254**, 6068–6078 (1975).
44. M. J. Janiak, D. M. Small and G. G. Shipley, *Biochemistry*, **15**, 4575–4580 (1976).
45. A. S. Janoff (ed.), *Liposomes. Rational Design*, Marcel Dekker, New York 1999.
46. J. Janzen, X. Song and D. E. Brooks, *Biophys. J.*, **70**, 313–320 (1996).
47. A. K. Kenworthy, K. Hristova, T. J. McIntosh and D. Needham, *Biophys. J.*, **68**, 1921–1936 (1995).
48. P. F. Kiser, G. Wilson and D. Needham, *Nature*, **394**, 459–462 (1998).
49. E. Evans, D. J. Klingenberg, W. Rawicz and F. Szoka, *Langmuir*, **12**, 3031–3037 (1996).
50. R. Kwok and E. Evans, *Biophys. J.*, **35**, 637–652 (1981).
51. D. M. LeNeveu, P. R. Rand, V. A. Parsegian and D. Gingell, *Biophys. J.*, **18**, 209–230 (1977).

52. B. R. Lentz, E. Friere and R. L. Biltonen, *Biochemistry*, **17**, 4475–4480 (1978).
53. M. L. Longo, A. J. Waring and D. A. Hammer, *Biophys. J.*, **73**, 1430–1439 (1997).
54. V. Luzzatti, in *Biological Membranes*, (ed.) D. Chapman, Academic Press, New York, 1968, p. 71.
55. V. Luzzatti and A. Tardieu, *Annu. Rev. Phys. Chem.*, **25**, 79 (1974).
56. A. L. MacKay, *Biophys. J.*, **35**, 301–313 (1981).
57. T. Maeda and S. Ohnishi, *FEBS Lett.*, **122**, 283–287 (1980).
58. K. Martin, H. Reggio, A. Helenius and K. Simons, *J. Cell Biol.*, **91**, 601–613 (1981).
59. J. Matti, N. M. Witzke, R. Bitman and G. G. Shipley, *Biochemistry*, **26**, 623–633 (1987).
60. S. Maybrey and J. M. Sturtevant, *Proc. Natl. Acad. Sci. USA*, **73**, 3862–3866 (1976).
61. T. McIntosh, *Biochim. Biophys. Acta*, **513**, 43–58 (1978).
62. T. McIntosh, S. Advani, R. Burton, D. Zhelev, D. Needham and S. Simon, *Biochemistry*, **34**, 8520–8532 (1995).
63. T. J. McIntosh, A. D. Magid and S. A. Simon, *Biochemistry*, **28**, 7904–7912 (1989).
64. T. J. McIntosh and S. A. Simon, *Biochemistry*, **25**, 4058–4066 (1986).
65. T. J. McIntosh and S. A. Simon, *Annu. Rev. Biophys. Biomol. Struct.*, **23**, 27–51 (1994).
66. S. McLaughlin, *Ann. Rev. Biophys. Biophys. Chem.*, **18**, 113–136 (1989).
67. R. Merkel, P. Nassoy, A. Leung, K. Ritchie and E. Evans, *Nature*, **397**, 50–53 (1999).
68. S. T. Milner and S. A. Saffran, *Phys. Rev. A*, **36**, 4371–4379 (1987).
69. V. C. Mow, F. Guilak, R. Tran-Son-Tay and R. Hochmuth (eds) *Cell Mechanics and Cellular Engineering*, Springer-Verlag, New York 1994.
70. D. Needham, *J. Mat. Educ.*, **14**, 217–238 (1992).
71. D. Needham, *Membrane Fusion Techniques*, (ed.) N. Düzgünes, Academic Press, 1993, pp. 111–129.
72. D. Needham and E. Evans, *Biochemistry*, **27**, 8261–8269 (1988).
73. D. Needham and R. M. Hochmuth, *Biophys. J.*, **55**, 1001–1009 (1989).
74. D. Needham, T. J. McIntosh and E. Evans, *Biochemistry*, **27**, 4668–4673 (1988).
75. D. Needham, T. J. McIntosh and D. D. Lasic, *Biochim. Biophys. Acta*, **1108**, 40–48 (1992).
76. D. Needham, T. J. McIntosh and D. V. Zhelev, *Liposomes: Rational Design*, (ed.) A. Janoff, Marcel Dekker, New York, 1999, pp. 13–62.
77. D. Needham, J. Mills and G. M. Eichenbaum. *J. Chem. Soc., Faraday Trans.*, **111**, 103–110 (1999).
78. D. Needham and R. S. Nunn, *Biophys. J.*, **58**, 997–1009 (1990).
79. D. Needham and R. S. Nunn, *Biophys. J.*, **58**, 997–1009 (1990).
80. D. Needham, N. Stoicheva and D. V. Zhelev, *Biophys. J.*, **73**, 2615–2629 (1997).
81. D. Needham and D. V. Zhelev, *Annals of Biomedical Engineering*, **23**, 287–298 (1995).
82. D. Needham and D. V. Zhelev, *Vesicles*, (ed.) M. Rosoff, Marcel Dekker, New York, 1996.
83. D. Noppl-Simson and D. Needham, *Biophys. J.*, **70**, 1391–401 (1996).
84. K. Olbrich, *Water Permeability and Mechanical Properties of Unsaturated Lipid Membranes and Sarcolemma Vesicles*, Ph. D. Thesis Duke University, 1997.
85. V. A. Parsegian, N. Fuller and R. P. Rand, *Proc. Natl. Acad. Sci. USA*, **76**, 2750–2754 (1979).
86. V. A. Parsegian and B. W. Ninham, *J. Theor. Biol.*, **38**, 101–109 (1973).
87. S. Paula, A. G. Volkov, A. N. Van Hoek, T. H. Haines and D. W. Deamer, *Biophys. J.*, **70**, 339–348 (1996).
88. D. A. Pink, T. J. Green and D. Chapman, *Biochemistry*, **19**, 349–356 (1980).
89. M. Rafalski, A. Ortiz, A. Rockwell, L. C. van Ginkel, J. D. Lear, W. F. DeGrado and J. Wilshut, *Biochemistry*, **30**, 10211–10220 (1991).
90. P. R. Rand, D. Chapman and K. Larsson, *Biophys. J.*, **15**, 1117–1124 (1975).
91. R. P. Rand, *Ann. Rev. Biophys. Bioeng.*, **10**, 277–314 (1981).

92. R. P. Rand and A. C. Burton, *Biophys. J.*, **4**, 115 (1964).
93. J. P. Reeves and R. M. Dowben, *J. Cell Physiol.*, **73**, 49–60 (1969).
94. G. N. Rogers, J. C. Paulson, R. S. Daniels, J. J. Skehel, I. A. Wilson and D. C. Wiley, *Nature*, **304**, 76–78 (1983).
95. R. Sarpal and D. Needham, *J. Liposome Research.*, **8**, 147–163 (1998).
96. M. B. Schneider, J. T. Jenkins and W. W. Webb, *Biophys. J.*, **45**, 891–899 (1984).
97. R. M. Servuss, W. Harbich and W. Helfrich, *Biochim. Biophys. Acta*, **436**, 900 (1976).
98. T. Shanguan, D. P. Siegel, J. D. Lear, P. H. Axelsen, D. Alford and J. Bentz, *Biophys. J.*, **74**, 54–62 (1998).
99. S. A. Simon, E. A. Disalvo, K. Gawrisch, V. Borovyagin, E. Toon, S. S. Schiffman, D. Needham and T. J. McIntosh, *Biophys. J.*, **66**, 1943–1958 (1994).
100. S. A. Simon, T. J. McIntosh, A. D. Magid and D. Needham, *Biophys. J.*, **61**, 786–799 (1992).
101. R. G. Snyder, D. G. Cameron, H. L. Casal, D. Compton and H. H. Mantsch, *Biochim. Biophys. Acta*, **684**, 111–116 (1982).
102. T. Stegmann and A. Helenius, *Viral fusion mechanisms*, (ed.) J. Bentz, CRC Press, Boca Raton, FL, 1993, pp. 89–111.
103. E. J. W. Verwey and J. T. G. Overbeek, *Theory of the stability of lymphobic colloids*, Elsevier, Amsterdam, 1948.
104. H. Vogel and F. Jahnig, *Chem. Phys. Lipids*, **29**, 83–101 (1981).
105. A. Walter and J. Gutknecht, *J. Membrane Biology*, **90**, 207–217 (1986).
106. R. E. Waugh, J. Song, S. Svetina and B. Zeks, *Biophys. J.*, **61**, 974–982 (1992).
107. J. M. White, A. Helenius and M.-J. Gething, *Nature*, **300**, 658–659 (1982).
108. J. M. White, M. Kielian and A. Helenius, *Q. Rev. Biophys.*, **16**, 151–195 (1983).
109. J. M. White and I. A. Wilson, *J. Cell Biol.*, **105**, 2887–2896 (1987).
110. J. Wilshut, J. Scholma and T. Stegmann, in *Biotechnological Applications of Lipid Microstructures*, (eds) B. Gaber, J. Schnur and D. Chapman, Plenum Press, New York, 1988, p. 105.
111. D. Zhelev and D. Needham, *Biochim. Biophys. Acta*, **1147**, 89–104 (1993).
112. D. Zhelev and D. Needham, in *Biological Effects of Electric and Magnetic Fields*, (ed.) D. O. Carpenter, Academic Press, Orlando, FL, 1994, pp. 105–142.
113. D. V. Zhelev, *Biophys. J.*, **71**, 257–273 (1996).
114. D. V. Zhelev, D. S. Dimitrov and I. Tsoneva, *Bioelectrochem. Bioenerg.*, **19**, 217–225 (1988).
115. D. V. Zhelev and D. Needham, *ASME. Advances in Heat and Mass Transfer in Biotechnology*, **34**, 47–49 (1996).
116. D. V. Zhelev, D. Needham and R. M. Hochmuth, *Biophys. J.*, **67**, 720–727 (1994).
117. D. V. Zhelev, N. Stoicheva and D. Needham, unpublished results.
118. S. A. Wharton, S. R. Martin, R. W. Riugrok, J. J. Skehel and D. C. Wiley, *J. Gen. Virol.*, **69**, 1847–1857 (1988).

# Chapter 10

# Fluctuating Vesicle Shapes

*HANS-GÜNTHER DÖBEREINER*

*Max-Planck-Institut für Kolloid- und Grenzflächenforschung, Golm, Germany*

## 1. INTRODUCTION

The preferred curvature of a fluid bilayer membrane is determined by the physical chemistry and statistical physics of the amphiphilic interface [1–3]. This review summarizes the current experimental and theoretical picture on the shapes and shape transitions of lipid vesicles [4,5]. Figure 10.1 shows an experimental example of the so-called budding transition where a small satellite is expelled from a larger vesicle. Further, various mechanisms leading to a spontaneous curvature of membranes are discussed, see Figure 10.2. Indeed, a thorough understanding of vesicle shapes allows their use as morphological probes for general membrane–solvent–solute interactions affecting elasticity. In turn, controlling the curvature of amphiphilic interfaces via these interactions is of great technological importance. For example, the various phases of microemulsions depend on the preferred curvature of the oil–water interface [6,7].

Vesicle shapes are not static entities but show quite pronounced thermal fluctuations due to the extreme softness of fluid membranes [8–13]. Thus, it is necessary to understand the interplay of equilibrium thermal fluctuations with the mean shape determined by the membrane material parameters and vesicle geometry [14,15]. The signatures of several shape transitions in the fluctuation spectrum of a prolate vesicle are described.

Experiments and theoretical models discussed in this review are restricted to single, isolated vesicles in thermodynamic equilibrium with respect to their morphological degrees of freedom. Dynamical aspects and the physics of vesicle adhesion are covered in Chapter 7. Giant vesicles are generally not in thermo-

*Giant Vesicles* Edited by P. L. Luisi and P. Walde

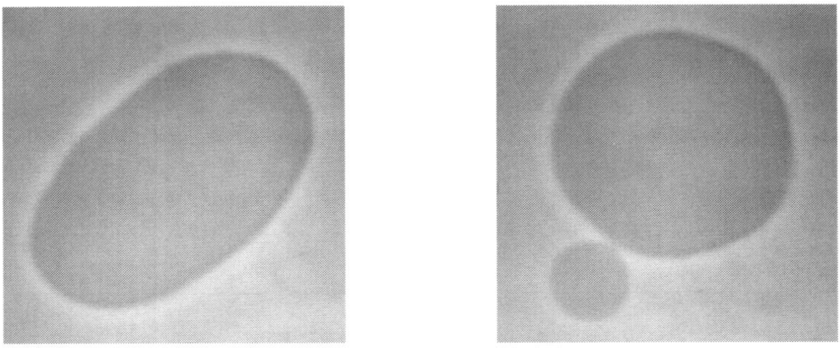

**Figure 10.1** Example of a vesicle shape transition. Budding of a prolate vesicle was induced by raising the effective spontaneous curvature $\bar{c}_0$ of the membrane. In this case, this was accomplished simply by increasing temperature [36]. The budding transition is relevant for vesicular transport in the cell. The pictures were taken by video phase contrast microscopy [22]. Generally, vesicle shape transitions are largely independent of the particular phospholipid under investigation. All experimental vesicles shown in this chapter were swollen from 1-stearoyl-2-oleoyl-*sn*-glycero-3-phosphocholine (SOPC) in sucrose–glucose solution and have a typical size of the order of 10 μm.

dynamic equilibrium with respect to the exchange of molecules with the bulk, on experimentally relevant time scales. The reason for this is the often low water solubility of the amphiphilic species studied. Thus, vesicles have to be considered as single objects. The discussion concentrates on giant vesicles with a typical size of the order of 10 μm, but the ideas and concepts developed to describe membrane

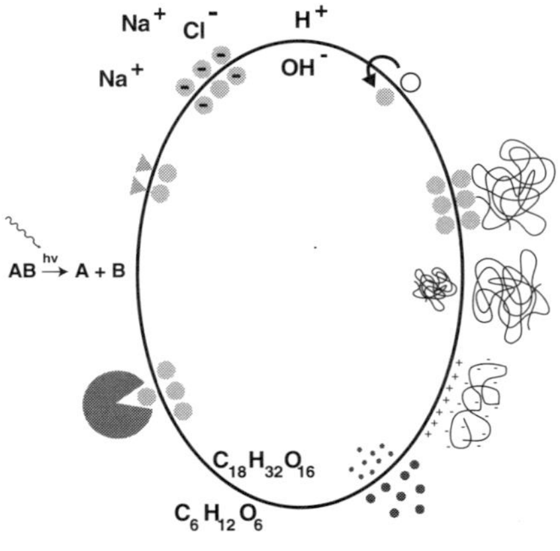


**Figure 10.2** Diagram showing several mechanisms for membrane asymmetry, as described in the text. Generally, any type of asymmetry across the bilayer induces curvature of the membrane. Asymmetry may be created by surface fields, as well as by boundary values of bulk fields at the interface.

curvature are applicable on length scales from 10 to $10^5$ nm. For instance, morphological transitions in the cell take place at the level of the intricately structured organelles, as well as on the length scale of the more extended plasma membrane [16,17]. In addition to pure bending of the fluid lipid bilayer, shape transitions of biomembranes involve shear elastic deformations of the cytoskeleton, which have to be included in a full analysis. Recent progress has been made in this respect [18–21]. This review excludes shear elastic contributions from the discussion.

This chapter starts with an exposition of the bending theory of vesicle shapes. Then, the physical and chemical origin of spontaneous curvature is discussed. After a brief description of vesicle image analysis, general and particular features of the phase diagram of vesicle shapes are recalled. It concludes with a few remarks on future directions and perspectives of the field.

## 2. AREA DIFFERENCE ELASTICITY (ADE) MODEL

Fluid lipid bilayers respond elastically to mechanical deformations which include bending, and stretching or compression of the membrane [23–25]. There is no resistance to shear. The bilayer may be idealized as a quasi two-dimensional sheet consisting of two monolayers with separately fixed number of molecules. (This implies we consider only membranes prepared from molecules which exhibit no flip-flop between the monolayers and are insoluble in water.) Measurements of the elastic moduli show that it is much harder to stretch a membrane than to bend it [26]. In building a theory for the morphology of membranes [27–32], one can thus assume to a good approximation that the overall surface area of the membrane is fixed. However, it turns out that the energetic contributions of the relative stretching and compression of the two monolayers, which arise when the membrane is bent, are of the same order of magnitude as the pure bending energy, which originates from splaying the molecules in the two monolayers. Therefore, there are two contributions to the bending energy of a closed bilayer: the bending of the two monolayers at fixed bilayer area $A$, and the relative stretching and compression of the monolayers. This implies a fixed mean area $A = (A^{\text{out}} + A^{\text{in}})/2$, but allows for changes in the differential area $\Delta A = A^{\text{out}} - A^{\text{in}}$. Thus, we have identified the geometrical quantities needed to formulate the expression for the bending energy: (i) local membrane curvature $C_i(r)$ along the two principal directions, and (ii) global area difference between the monolayers $\Delta A$. It can be shown that the latter quantity is, to a good approximation, proportional to the integrated mean curvature of the vesicle:

$$\Delta A = D \int \mathrm{d}A(C_1 + C_2), \tag{1}$$

where $D$ is the bilayer thickness. This equation makes the connection between shape and differential area explicit. One finds for the total energy [32]

$$E_{\rm B} = \frac{\kappa}{2}\int dA(C_1 + C_2 - C_0)^2 + \frac{\bar{\kappa}}{2}\frac{\pi}{AD^2}(\Delta A - \Delta A_0)^2, \tag{2}$$

where $C_0$ is the spontaneous curvature of the membrane, which accounts for a possible non-flat relaxed state of the bilayer, $\Delta A_0 = A_0^{\rm out} - A_0^{\rm in}$ is the difference of the unstressed monolayer areas, and $\kappa$ and $\bar{\kappa}$ are the local and nonlocal bending moduli of the membrane. In the first term, one simply integrates the local deviations in mean curvature from the relaxed state over the whole vesicle area. The second term accounts for the difference between the actual geometrical area difference and the relaxed one, which would be preferred by the monolayers. This preferred differential area is proportional to the difference in the number of molecules: $\Delta A_0 = a_{\rm lipid}^{\rm out} N^{\rm out} - a_{\rm lipid}^{\rm in} N^{\rm in} \simeq \overline{a_{\rm lipid}} \Delta N$. Both terms are quadratic and correspond to a generalization of Hooke's law for the spring to bending and differential area elasticity. The relative importance of the two contributions to the bending energy is measured by the ratio of the elastic moduli $\alpha = \bar{\kappa}/\kappa$. For $\alpha = 0$, there is no resistance to bending from differential area elasticity and one recovers the spontaneous curvature model [23]. For $\alpha = \infty$, the geometrical area difference $\Delta A$ does not deviate from the preferred area difference $\Delta A_0$ of the monolayers. The latter statement is the essence of the bilayer-couple hypothesis [33] which was formulated independently in the context of elasticity theory [24,25,34]. However, one estimates $\alpha$ to be of the order of one for all phospholipids [32]. This estimate is corroborated by measurements of this quantity [35]. Both terms are therefore important. Indeed, experiments on shapes and shape transitions in giant vesicles have shown that the general model is required to account quantitatively for the observations [22].

The relevant geometrical parameters controlled by the experiment [22,36–40] are the area and volume of the vesicles. Given a mean area per molecule, the membrane area can be considered fixed at constant temperature and a constant number of molecules. The volume is controlled by the osmolarities of the inner and outer solutions due to the high water permeability of the membrane. Any osmolarity differences would lead to large osmotic forces inducing water flow through the membrane. It is thus the *permeability* of the membrane to water which keeps the enclosed water volume constant at given external osmolarity.

For a vesicle of radius $R_A = (A/4\pi)^{1/2}$, the area available for shape changes is measured by the dimensionless volume-to-area ratio

$$v = \frac{V}{(4\pi/3)R_A^3} \tag{3}$$

This reduced volume $v$ is scaled so that a sphere is characterized by unity, the maximal possible volume-to-area ratio for a geometrical object. Deflated vesicles have values smaller than one, and zero would correspond to an empty membrane bag without any aqueous content. The reduced volume $v$ represents a convenient measure of the enclosed volume of a vesicle independent of size. Apart from this purely geometrical quantity, another parameter is needed to describe the overall tendency of the membrane to curve. It is provided by a dimensionless combination of the preferred area difference and the spontaneous curvature;

$$\overline{\Delta a_0} = \Delta a_0 + \frac{1}{2\pi\alpha} c_0 \tag{4}$$

where $\Delta a_0 = \Delta A_0/(8\pi D R_A)$ and $c_0 = C_0 R_A$ are scaled quantities. Again this scaling singles out the sphere. A spherical vesicle with radius $R_A$, a scaled spontaneous curvature $c_0 = 1$, and a scaled area difference $\Delta a_0 = 1$, is in a totally relaxed state with zero energy. This vesicle would have no differential area stresses and no bending stresses in its membrane. It is clear that this is a rather special situation. In general, a vesicle will be in a stressed state, its morphology being determined by the most relaxed shape consistent with the constraints imposed. The effective differential area $\overline{\Delta a_0}$ reflects the different sources of membrane curvature. In addition to the ratio of the elastic moduli $\alpha$, and the reduced volume $v$, it is the third relevant quantity for the classification of shapes. Vesicle shapes are obtained by minimizing eq. (2) under the constraints of fixed area $A$ and volume $V$ [32] In analogy to thermodynamics, minimal energy shapes can be placed in a two-dimensional phase diagram which is described in detail below. Strictly speaking, the notion of a phase diagram for experimental vesicle shapes makes no sense. At a finite temperature, a vesicle fluctuates around its mean shape which is characterized by a minimum in elastic energy. Statically, this mean shape is assumed only at zero temperature. In general, the vesicle will explore all available configurations, and the notion of well-defined shape classes separated by transition lines looses its meaning. In practice, however, when activation energies are large compared to thermal energies, an experimental vesicle in the laboratory is, indeed, found to fluctuate around a mean shape corresponding to a local minimum in elastic energy. The globally minimal shapes constitute the shape phase diagram.

The ratio of the elastic moduli $\alpha$ is given by the lipid system studied and is essentially fixed for each set of experiments. Although $\alpha$ is important for the topology of the phase diagram, existing shapes and shape classes, in fact, do not depend on it. Thus, it is convenient to define the effective spontaneous curvature [22,32]

$$\bar{c}_0 = c_0 + 2\pi\alpha(\Delta a_0 - \Delta a) \tag{5}$$

This parameter fully characterizes the shape of a vesicle with reduced volume $v$. Note that eq. (5) captures the physics of the two sources of preferred membrane curvature. Whereas $c_0$ is only a property of the vesicle membrane, the result of changing $\Delta a_0$ depends also on the reference shape via its integrated mean curvature, which is proportional to $\Delta a$. A large preferred differential area $\Delta a_0$ greater than $\Delta a$ tends to curve the membrane further outward, whereas a smaller value would lead to more invaginated shapes. Let us now discuss a few mechanisms for membrane curvature.

## 3. PHYSICAL AND CHEMICAL ORIGIN OF SPONTANEOUS CURVATURE

The previous section shows that there are two different elastic parameters that control the curvature of a vesicle membrane. These parameters, the spontaneous curvature $c_0$ and the preferred differential area $\Delta a_0$, depend on the properties of the membrane molecules and the surrounding aqueous medium. Any asymmetry across the membrane will lead to a curved bilayer state characterized by an effective spontaneous curvature. A technique for measurement of this quantity has recently been developed [22,41]. Figure 10.2 shows several situations leading to membrane curvature. The following discussion gives a brief overview of these mechanisms and the techniques to characterize them.

One factor is the effective molecular shape [42]. Molecules may be cone-shaped and, thereby, induce monolayer bending. (Note however, that the concept of a molecular shape should be taken with some caution. In fact, molecular configurations are a function of the local environment and, strictly speaking, cannot be viewed as an independent property of a molecule.) Thus, different monolayer densities on both sites of the membrane result in a spontaneous curvature of the bilayer. Similarly, an asymmetric change in the number of molecules or mean molecular area on the two sides of the membrane will lead to a change in the differential area. The mean molecular area depends, for example, on temperature and surface charge. The difference in the number of molecules may be free [43,44] or driven flip-flop [39,45]. Further, surface charge and the equilibrium distribution of molecules across the membrane depend on the electrolyte strength and pH of solution [46]. However, even for symmetric surface charge, a gradient in electrolyte concentration across the membrane leads to an electrostatic spontaneous curvature [47–51]. Interestingly, even a pH gradient across zwitterionic membranes induces curvature of the interface [52]. Note that the differential area is not an intrinsic quantity of the system, but depends on the particular (preparation) history of the vesicle studied. The distribution of molecules between the inner and outer monolayer is fixed at the time of membrane closure and, generally, varies widely from vesicle to vesicle. This explains why one usually observes a great variety of different vesicle shapes within one sample. However, taking these individual offsets into account relative changes of the effective spontaneous curvature are well defined and can be measured quantitatively [41,53].

A particular interesting class of effects can be observed monitoring the interactions of polymers [53–57] or colloids [58–61] with membranes. The molecules or particles may be adsorbed or depleted from the interface. In either case, the membrane reacts by adjusting its curvature. Water-soluble polymers can be grafted to one or both sides of the membrane with hydrophobic anchors. This leads to an increase of the differential area due to the insertion of the anchor molecule. In addition, the grafted mushroom-shaped polymer induces spontaneous curvature by balancing its gain in configurational entropy in the curved state against the bending elastic energy of the bare membrane [61]. For polymers not associated with the membrane, there is a depletion layer with a thickness on the order of the radius of gyration. In this situation, the membrane bends towards the side with the larger molecule with a corresponding gain in translational entropy. This depletion effect is not restricted to polymers but can also be observed for colloidal particles [61] and, possibly, even for small molecules such as simple sugars [53].

In polyelectrolytes, the interplay of configurational entropy and electrostatic energies leads to quite a number of phenomena. Especially interesting are DNA and cationic–lipid interactions [62]. Recently, the complexes formed by these molecules, which are relevant for gene therapy [63] have been characterized [64]. DNA-induced membrane curvature may be important in the initial process of formation of these complexes. Endocytosis of single or multiple DNA strands into cells can be understood as a morphological transition which is triggered by a change in the effective spontaneous curvature as the DNA associates with the plasma membrane.

Another important mechanism for membrane curvature are chemical reactions at or near the interface. During the course of the reaction, the chemical environment changes. The increase (respectively decrease) of products (and educts) in the bulk alters the various interactions of the membrane with the aqueous medium. The amphiphilic molecules building the membrane itself may also be modified. This can happen via reactions involving small molecules or molecular machines. Enzymatic cleavage of lipids by, for example, phospholipase molecules are biological relevant examples [65]. Any asymmetric change in molecular properties in the interface and/or the bulk will induce membrane curvature by the effects discussed above or via a specific mechanism. The resulting change in the stress profile within the membrane may in turn alter the functional state of membrane spanning protein machinery, such as channels and pumps [66,67]. Recently, a light-induced chemical reaction has been observed to couple strongly to vesicle morphology, see Chapter 25. The observed shape changes could be explained by pH-driven adsorption [52]. The sensitivity of membrane morphology to changes in pH is one possibility with which to couple vesicle shapes to oscillating chemical reactions. The mechanism of membrane curvature discussed in this review have been observed mainly in micrometer-size vesicles. However, the ideas and concepts developed for these systems carry over into the nanometer regime. Controlling the curvature of soft interfaces on the nanometer scale is an important subject of materials research.

## 4. IMAGE ANALYSIS OF FLUCTUATING VESICLES

The introduction emphasizes that vesicles are thermally fluctuating objects. Thus, single snapshots of a vesicle shape, in general, do not allow deduction of the specific properties of the vesicle under investigation. Instead, it is necessary to perform a careful statistical analysis of the full shape ensemble. In this respect, experimental conditions have to be stabilized and maintained for a few hours. Experimental mean shapes can then be compared to theory in order to deduce the elastic parameters of the vesicle membrane. Such a procedure has been performed for the spontaneous curvature of membranes [22,41]. Experimental vesicle shapes are analyzed on line at video frequency (25 frames per second) and mapped into the theoretical phase diagram which is described in the next section. Vesicle shapes are characterized by their contours in the focal plane of a phase-contrast microscope. These contours can be obtained with a relative spatial resolution of a few tens of nanometers. Note that it is the membrane position, *not* its structure, that can be obtained with such a high precision; that is, the shift of the diffraction pattern, not the fine details of the pattern itself, are monitored. For optical point resolution, the usual limits given by the wavelength of light and numerical aperture apply. On-line analysis includes the actual image analysis and subsequent Fourier transformation of the vesicle contours. The availability of these data during experiment allows adjustment of the specific experimental protocol in order to optimize the measurement of a particular feature. Image analysis of fluctuating vesicles has been done before but was restricted mainly to quasi-spherical vesicles or the analysis of single snapshots or sequences of vesicle shapes [36–38,68]. From the fluctuation spectrum of quasispherical vesicles, bending elastic moduli for several systems have been obtained [9,11,12,69].

## 5. THE ADE PHASE DIAGRAM

### 5.1 Overview

The shape of an individual fluid bilayer vesicle is obtained by minimizing the bending energy (eq. (2)) given the geometrical parameters area $A$ and volume $V$, as well as the material parameters, spontaneous curvature $C_0$, and preferred differential area $\Delta A_0$. When these parameters are tuned externally, the shape changes. Different regions of the parameter space correspond to different classes of shapes. Figure 10.3 shows experimental examples of a stomatocyte and a discocyte vesicle. Theoretically, these classes result from the existence of multiple solutions to the shape equations, derived by minimizing eq. (2). At the boundary between those regions, shape transitions occur, which are manifested, for instance, in a change of symmetry or another shape attribute. The bending modulus $\kappa$ sets the overall energy scale. Its ratio $\alpha$ with the second elastic modulus governs the relative energies between different shape classes, and, thus, the exact location and quality of the shape boundaries. However, the shapes themselves do not depend on $\alpha$. In Figure 10.4, the ADE phase diagram [32,70–72] is shown for the typical value $\alpha = 1.4$. On the horizontal axis, the reduced volume $v$ is plotted. The vertical axis shows the effective differential area $\overline{\Delta a_0}$. For medium values of $\overline{\Delta a_0}$, and not too small $v$, the stable shapes are prolate vesicles. The prolate phase extends into the upper left of the phase diagram in the form of long tubular vesicles (Figure 10.5). For large effective

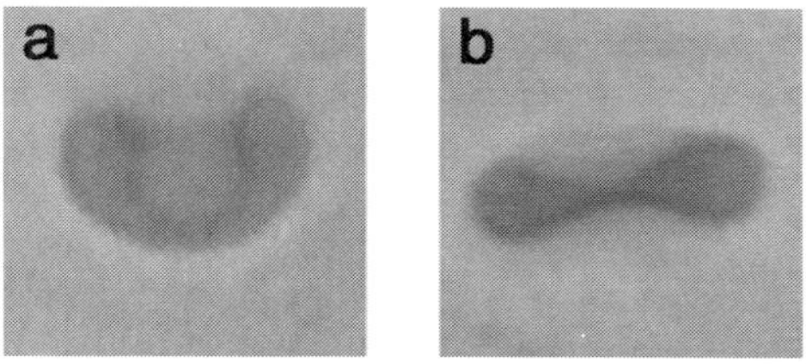


**Figure 10.3** Example of (a) a stomatocyte and (b) discocyte vesicle. The discocyte was obtained from the stomatocyte by raising the effective spontaneous curvature via grafting of an amphiphilic polymer onto the outer monolayer from solution [53].

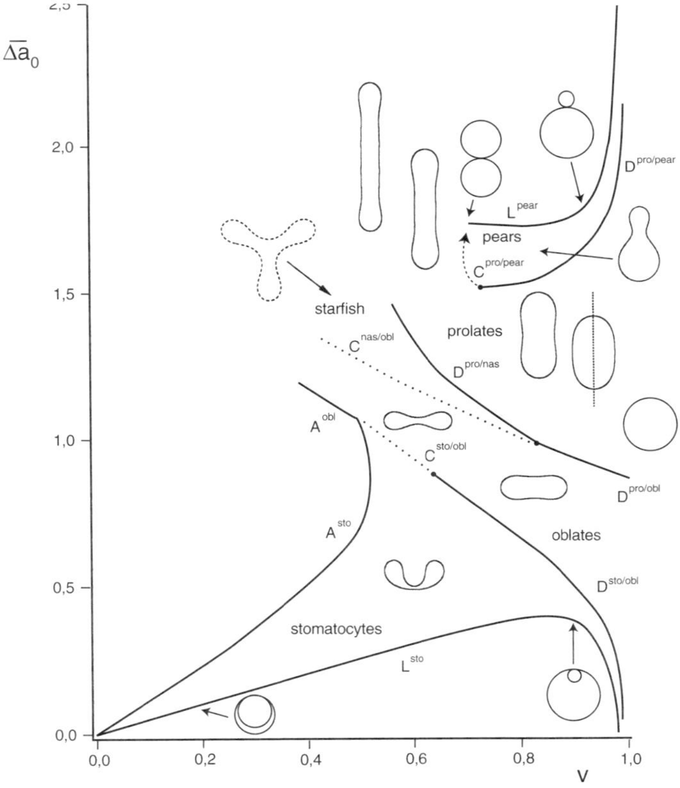


**Figure 10.4** Phase diagram of the ADE model at $\alpha = 1.4$. For each region, lowest-energy shapes are indicated with their centers of gravity located at the corresponding position in the phase diagram. The horizontal axis gives a measure of the volume-to-area ratio expressed by the reduced volume $v$, whereas the vertical axis shows the effective differential area $\overline{\Delta a_0}$, i.e. the preferred curvature of the vesicles.

**Figure 10.5** Example of a fluctuating tubular vesicle belonging to the prolate phase [73].

differential area, vesicles show generally an outward curved morphology, such as the budded shapes consisting of two spheres, whereas for small values of the effective differential area, shapes are predominantly curved inward, as for discocytic or stomatocytic vesicles. Lowest-energy shapes are illustrated for each region. As indicated for the prolate, all rotational symmetry axes are along the vertical $\overline{\Delta a_0}$ axis, except for the starfish-like vesicles which have a nonaxisymmetric shape. This region intervenes between the prolate and oblate phase. Note that for medium to small reduced volume, the oblate vesicles resemble the biconcave form of the red blood cell. First-order boundaries (D on Figure 10.4), i.e. those where shapes change discontinuously, are indicated by solid lines; second-order boundaries (C), where a particular shape symmetry is broken continuously, are dashed lines. Further, the pear and stomatocyte regions are bounded by limiting lines (L) that consist of two spheres connected by a neck with an infinitesimal radius. Beyond these lines and for small reduced volume, there is a region that has not been studied in detail. Vesicle shapes touch along special self-adhesion lines (A) where membrane–membrane interactions have to be taken into account in addition to bending energy [74].

All the shapes and shape classes shown in Figure 10.4 have by now been seen experimentally [22,37,38,71,75]. However, only the prolate phase and its boundaries have been studied systematically. Figure 10.6 is a more detailed phase diagram of the prolate region. In addition to the location of the phase transitions discussed above, the spinodals (M) of the prolate phase are given and denoted by dotted lines. The spinodal lines delimit the regions of metastable shapes. At its respective spinodal, a particular shape becomes unstable to small perturbations and decays into the globally stable configuration. Due to thermal fluctuations, a spinodal line becomes somewhat fuzzy at finite temperatures, and vesicles decay even within a finite distance of the spinodal whenever the activation energy to overcome the elastic barrier is on the order of *kT*.

Prolate shapes are locally stable between the upper spinodal line $M^{pro}_{odd}$ and the lower spinodal line $M^{pro}_{even}$. The lines $D^{pro/pear}$, $C^{pro/pear}$, $D^{pro/obl}$, and $D^{pro/nas}$ are boundaries to the region where prolates are the lowest-energy shapes. In the region immediately above $D^{pro/pear}$, lowest-energy shapes are pear-like. In the region immediately below the prolates, oblate and nonaxisymmetric (nas) shapes have

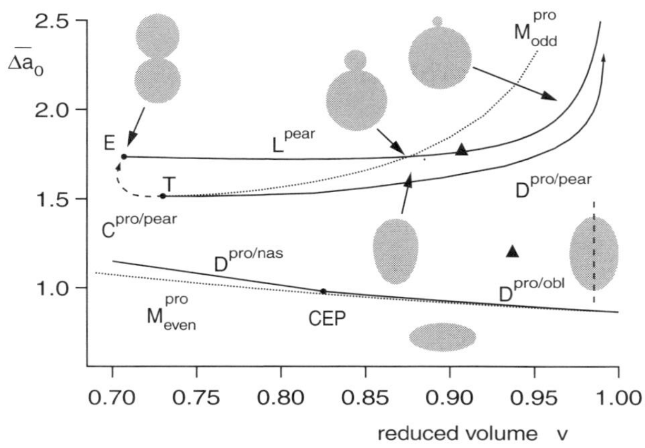


**Figure 10.6** Prolate region of the ADE phase diagram for $\alpha = 1.4$. Dotted lines correspond to the upper and lower stability boundary of prolate shapes. The two triangles mark the positions of the vesicle depicted in Figures 10.7 and 10.8, where the one with the smaller effective differential area $\overline{\Delta a_0}$ corresponds to Figure 10.7.

lowest energy. The endpoint (E) of the limiting line $L^{pear}$ represents two spheres of equal radii connected by a infinitesimal neck. The point (T) on the prolate–pear phase boundary is a tricritical point. In analogy to thermodynamics, this tricritical point separates a line of first-order ($D^{pro/pear}$) and second-order ($C^{pro/pear}$) phase transitions. At the first-order transition, there is a jump in the slope of the elastic bending energy as a function of the control parameter, for example $\overline{\Delta a_0}$. This corresponds to a discontinuous shape change. In contrast, at the second-order phase transition, there is a continuous change in slope, and the vesicles change their shape smoothly. CEP denotes a special kind of critical end point. At this point, the first-order transition $D^{pro/pear}$ disappears, revealing a pair of transitions, one of first order ($D^{pro/nas}$) and one of second order ($C^{nas/obl}$), previously hidden beneath it. The following section describes the theoretical features of the transitions out of the prolate phase and their experimental signatures.

## 5.2 Budding Transition

Budding, where a small satellite is expelled from a larger vesicle, is a paradigmatic example of a morphological transition. During the last decade, it has attracted a lot of interest [22,32,36,37,40,76]. The theoretical and experimental efforts to describe and characterize this transition have considerably improved our understanding of vesicle shapes and the theory of bending elasticity in general. Starting from a prolate vesicle, budding can be induced by raising temperature [22,37] or by increasing spontaneous curvature directly via appropriate exchange of the external solution

[41]. The transition is first order on the right of the tricritical point (T) and second-order on the left. Indeed, for budding at small reduced volume, one observes a continuous symmetry breaking [36,52]. Prolate vesicles loose their equatorial mirror symmetry during budding and become gradually more pear-shaped. In contrast, for large reduced volumes, prolate shapes become only instable at the spinodal line $M^{pro}_{odd}$ and spontaneously expel a small bud, see Figure 10.1 [22,76]. The approach to this instability is characterized by a strong increase in the pear-like fluctuations of the vesicles [40]. These spinodal fluctuations have long relaxation times and can easily be mistaken as stable pear shapes [36,37]. In Figures 10.7 and 10.8, a fluctuating prolate vesicle is shown at two different points in the phase diagram approaching the spinodal line of the budding transition, see Figure 10.6.

The first sequence of fluctuating shapes (Figure 10.7) is located within the stable prolate regime ($v = 0.937$ and $\overline{\Delta a_0} = 1.21$) [22]. At this point, the vesicle fluctuates around a prolate shape; subsequent frames are not visibly correlated. This situation changes near the budding instability (Figure 10.8). The shape parameters ($v = 0.907$ and $\overline{\Delta a_0} = 1.78$) [22] now place the vesicle within the metastable prolate region. The vesicle has crossed the first-order line, but keeps the prolate shape due to a finite activation energy of the budding transition in this region of the phase diagram. The approach to the spinodal line, where the prolate shapes become unstable, is characterized by the development of a soft mode. There are clearly strong pear-shape fluctuations visible with a long relaxation time. These spinodal fluctuations have been quantitatively characterized [40]. Both the mean-square amplitudes and the relaxation times of this soft pear-like mode scale with the inverse distance in reduced volume to the spinodal line. The final budding is caused by a particular strong thermal fluctuation which overcomes the residual energy barrier near the spinodal. An example of such an event can be seen in Figure 10.9. Note the quasistable appearance of the pear-like shape as it diffuses over the saddle point of the bending energy landscape.

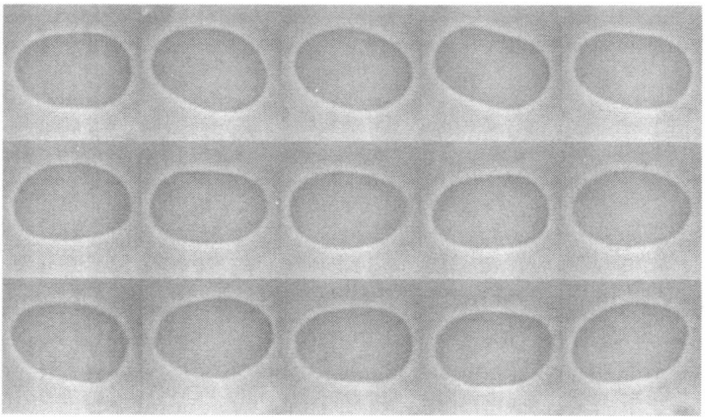

**Figure 10.7** Prolate vesicle fluctuations at $v = 0.937$ and $\overline{\Delta a_0} = 1.21$. The elapsed time between snapshots is 6.3 s [22].

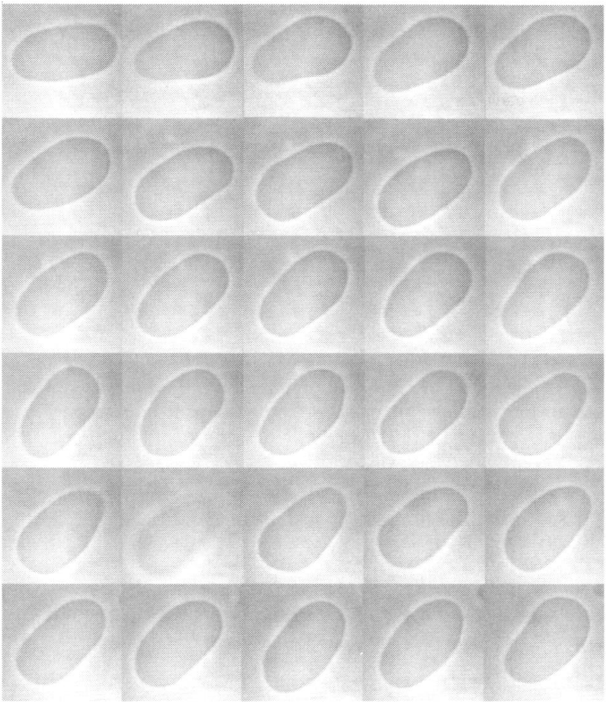

**Figure 10.8** Pear fluctuations of a prolate vesicle near the spinodal line of the budding transition. The same vesicle as in Figure 10.7 is shown, but now with parameter values $v = 0.907$ and $\overline{\Delta a_0} = 1.78$ which are much closer to the budding instability. The snapshots are taken at equal time intervals of 6.3 s.

Due to the crossing of the spinodal line $M^{pro}_{odd}$ with the limiting line $L^{pear}$, budding at large reduced volume misses the pear phase completely. Vesicles assume a double-spherical shape after the budding instability. With single-component membranes, the bud stays connected to the parental vesicle. However, in a multi-component sphingomyelin system, fission has been found to occur after the budding event [76]. Separation of the vesicles can also be observed in polymer-induced budding. Here, water-soluble polymers are grafted from the external side to the membrane via hydrophobic anchors, see Section 3. In this case, the two vesicles stay connected by an umbilical cord [41]. In both systems, there is probably lateral inhomogeneity in the molecular components within the membrane. In this respect, one has to distinguish between thermodynamically driven phase separation and molecular segregation due to the coupling of molecular density to membrane curvature. In the first case, the vesicles exhibit domain-induced budding [77], whereas in the second case it is the budding transition itself that allows segregation into a region of high curvature [78]. The latter scenario is only effective for

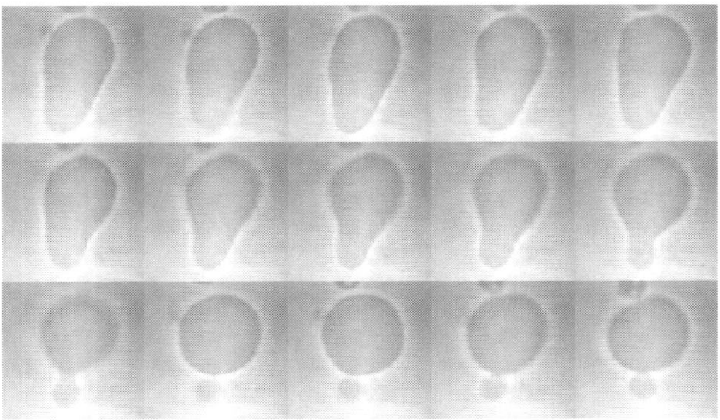

**Figure 10.9** Time sequence of the final budding instability with constant control parameters $(v, \overline{\Delta a_0})$ very close to the spinodal line [22]. The initial pear-like shape results from a giant thermal fluctuation of a metastable prolate vesicle onto the saddle point of the bending energy landscape. The elapsed time between snapshots is 1.2 s. First, a slow diffusion occurs over the saddle point region corresponding to the pear-like shapes until, finally, the vesicle relaxes towards the budded configuration by quickly closing down the neck between the two quasispherical parts of the shape.

sufficiently small vesicles where the bending energy becomes comparable to the entropy of mixing.

## 5.3 Prolate–oblate Transition

The previous section showed that thermal fluctuations play a major role in morphological transitions of vesicles. At the lower phase boundary of the prolate shapes, thermal fluctuations become even more dominant. The prolate–oblate transition, to the right of the critical end point (CEP), is weakly first order [70,72]. Thus, the energy barriers between the metastable shape and the shape of lowest energy can be crossed thermally in both directions. Indeed, prolate–oblate fluctuations have been observed experimentally [75]. Figure 10.10 shows a time sequence of a vesicle near the prolate–oblate transition. The vesicle is seen to alternate between prolate and oblate shapes. The latter shape appears as a quasi-spherical vesicle in the focal plane due to observation from below. A statistical analysis of the bistable shape distribution allowed extraction of an effective potential for the transition. Further, mean-escape times out of the prolate and oblate shapes, respectively, have been measured [75]. This dynamic quantity depends on the energy dissipation mechanism. In future experiments, it may be possible to distinguish between bulk and surface hydrodynamic drag and intermonolayer viscous coupling. The latter friction effect depends on the degree of molecular interdigitation and could be used for a characterization.

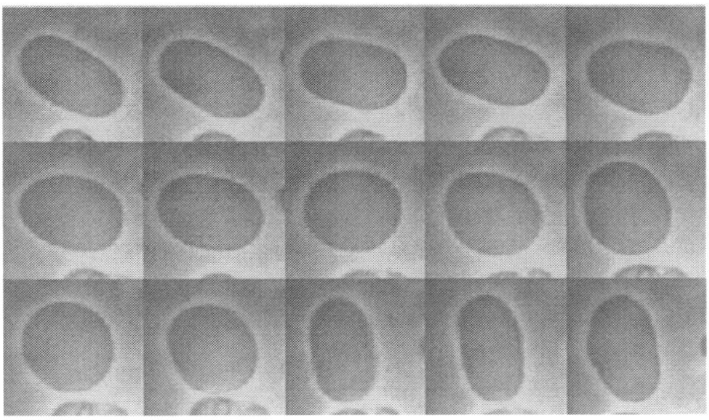

**Figure 10.10** Prolate–oblate fluctuations of a vesicle [75]. The long axis of the prolate vesicle is located within the focal plane. The oblate vesicle appears quasispherical because its rotational symmetry axis is normal to the focal plane. The elapsed time between snapshots is 6.3 s. A bistable behavior occurs where time sequences of fluctuating prolate and oblate shapes are separated by comparatively short-lived transitional shapes. Such a scenario confirms the theoretical expectations of a weakly first-order transition between prolate and oblate shapes. A statistical analysis of the shape distribution allows identification of the prolate configuration as the meta-stable shape in this example.

To the left of the point CEP (see Figure 10.4), the upper spinodal of the oblate phase turns into a real second-order transition from oblates to nonaxisymmetric shapes [72]. These shapes resemble starfish at small reduced volume and connect the prolate and oblate vesicle phases. They will be described in the next section.

## 5.4 Starfish Vesicles

A fascinating class of nonaxisymmetric shapes exist for medium values of the effective differential area. Figure 10.11 compares the experimental and theoretical shapes [71]. Starfish vesicles are found in several subclasses including $n$-armed stars, H-shapes, and shapes with even lower symmetry (Figure 10.12). These figures reveal the general building principle of this particular shape class. Whenever possible, the morphology is locally axisymmetric and consists of tubular sections connected via nonaxisymmetric joints. Increasing the spontaneous curvature drives budding of the tubular arms [41]. Note that starfish vesicles can be formed by homogeneous membranes. No segregation of lipids into regions of lower and higher curvature is necessary. Indeed, starfish have been found in single-component membranes [71]. The branched tubular structure can be understood from the requirement to accommodate a large area-to-volume ratio at medium values of differential area. Theoretical shapes have been obtained via direct minimization of the energy with a finite element method [71]. The vesicle shape shown in Figure 10.12 exhibits diffusion-

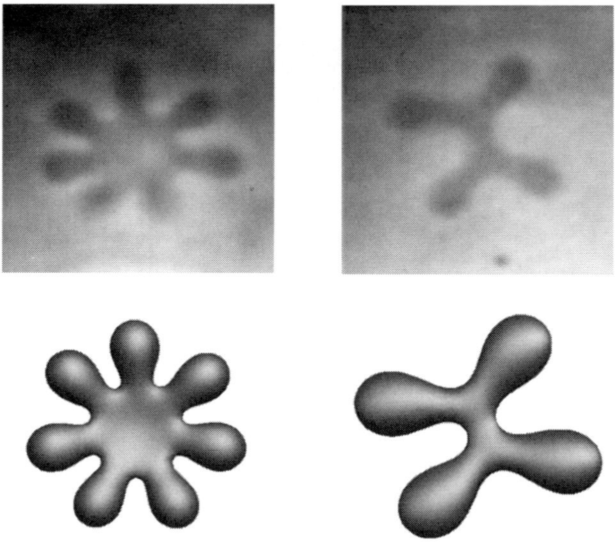

**Figure 10.11** Comparison of theoretical and experimental nonaxisymmetric starfish vesicles [71]. A seven-armed star and an H-shape are shown. These vesicles are swollen from the single phospholipid SOPC. Thus, highly branched shapes can be formed by homogeneous membranes.

like relative rotations of the arms with respect to each other. Thus, one concludes that the energy difference of the various configurations is well below 1 $kT$. As a consequence, the shape exhibiting two mirror planes depicted in Figure 10.12 is visible in the microscope only for a small fraction of time.

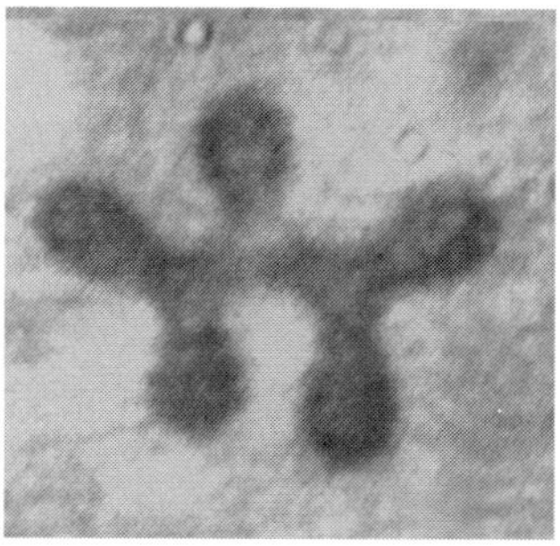

**Figure 10.12** Complex starfish vesicle with minimal symmetry [79]. The mean shape is mirror symmetric to the focal plane of the microscope and a plane orthogonal to it through the center of gravity of the vesicle.

## 6. CONCLUSION AND PERSPECTIVES

An overview has been given of vesicle shapes and shape transitions, and it is found that there is a fascinating interplay of the physical chemistry and the statistical physics of membranes. The detailed study of the bending elasticity of fluctuating membranes has brought order into the zoo of vesicle shapes and now enables these shapes to be used as morphological probes for general interactions of the membrane with its aqueous environment. In turn, this activity will spur further developments in membrane elasticity. Interfacial chemical reactions can couple to and/or change the internal structure of the bilayer. Further, it will be interesting to study the template role of soft interfaces in (bio)mineralization [80]. Continuing investigations into the various mechanisms for spontaneous curvature of amphiphilic interfaces will improve our ability to control the morphology of supramolecular surfactant aggregates.

## 7. REFERENCES

1. *Physics of Amphiphilitic Layers*. Proceedings of the workshop, Les Houches, France, February 10–19, 1987: (International Winter School on the Physics of Amphiphilic Layers). (Eds) J. Meunier, D. Langevin and N. Boccara, Springer-Verlag, Berlin (1987) Springer Proceedings in Physics.
2. R. Lipowsky, *Nature*, 349, 475 (1992).
3. *Statistical Mechanics in Physics and Biology*. Symposium held December 2–5, 1996, Boston, Massachusetts, U.S.A., (eds.) D. Wirtz and T. C. Halsey, Pittsburgh, Pennsylvania. Materials Research Society, **463**. Based on the Symposium *Statistical Mechanics in Physics and Biology*, **42**.
4. S. Svetina and B. Žekš, in *Handbook of Nonmedical Applications of Liposomes*, vol 1, (eds) D. D. Lasic and Y. Barenholz, CRC Press, Boca Raton, 1996, p. 1
5. U. Seifert, *Adv. Phys.*, **46**, 13 (1997).
6. G. Gompper and M. Schick, in *Phase transitions and critical phenomena*, vol 16, (eds) C. Domb and J. L. Lebowitz, Academic Press, 1994.
7. S. A. Safran, *Adv. Phys.*, **48**, 395 (1999).
8. M. B. Schneider, J. T. Jenkins and W. W. Webb, *J. Phys. France*, **45**, 1457 (1984).
9. H. Engelhardt, H. Duwe and E. Sackmann, *J. Phys. Lett.*, **46**, L395 (1985).
10. S. T. Milner and S. A. Safran, *Phys. Rev. A*, **36**, 4371 (1987).
11. J. Faucon, M. Mitov, P. Méléard, I. Bivas and P. Bothorel, *J. Phys. France*, **50**, 2389 (1989).
12. H. P. Duwe, J. Käs and E. Sackmann, *J. Phys.*, (France), **51**, 945 (1990).
13. E. Evans and W. Rawicz, *Phys. Rev. Lett.*, **64**, 2094 (1990).
14. U. Seifert, *Z. Phys. B*, **97**, 299 (1995).
15. V. Heinrich, F. Sevšek, S. Svetina and B. Žekš, *Phys. Rev. E*, **55**, 1809 (1997).
16. E. Sackmann, *FEBS Lett.*, **346**, 3 (1994).
17. R. Schekman and L. Orci, *Science*, **271**, 1526 (1996).
18. N. Mohandas and E. A. Evans, *Annu. Rev. Biophys. Biomol. Struct.*, **23**, 787 (1994).
19. U. Seifert, *Biophys. J.*, **75**, 1141 (1998).
20. S. K. Boey, D. H. Boal and D. E. Discher, *Biophys. J.*, **75**, 1573 (1998); D. E. Discher, D. H. Boal and S. K. Boey, *Biophys. J.*, **75**, 1584 (1998).

21. R. J. Mashl and R. F. Bruinsma, *Biophys. J.*, **74**, 2862 (1998).
22. H.-G. Döbereiner, E. Evans, M. Kraus, U. Seifert and M. Wortis, *Phys. Rev. E*, **55**, 4458 (1997).
23. W. Helfrich, *Z. Naturforsch.*, **28c**, 693 (1973).
24. W. Helfrich, *Z. Naturforsch.*, **29**, 510 (1974).
25. E. Evans, *Biophys. J.*, **14**, 923 (1974).
26. D. Marsh, *Handbook of Lipid Bilayers*, CRC Press, Boca Raton, 1990.
27. H. J. Deuling and W. Helfrich, *Biophys. J.*, **16**, 861 (1976).
28. S. Svetina, M. Brumen and B. Žekš, *Stud. Biophys.*, **110**, 177 (1985).
29. S. Svetina and B. Žekš, *Eur. Biophys. J.*, **17**, 101 (1989).
30. L. Miao, B. Fourcade, M. Rao, M. Wortis and R. K. P. Zia, *Phys. Rev. A*, **43**, 6843 (1991).
31. U. Seifert, K. Berndl and R. Lipowsky, *Phys. Rev. A*, **44**, 1182 (1991).
32. L. Miao, U. Seifert, M. Wortis and H.-G. Döbereiner, *Phys. Rev. E*, **49**, 5389 (1994).
33. M. P. Sheetz and S. Singer, *Proc. Natl Acad. Sci.*, USA, **71**, 4457 (1974).
34. E. A. Evans, *Biophys. J.*, **30**, 265 (1980).
35. R. E. Waugh, J. Song, S. Svetina and B. Žekš, *Biophys. J.*, **61**, 974 (1992).
36. K. Berndl, J. Käs, R. Lipowsky, E. Sackmann and U. Seifert, *Europhys. Lett.*, **13**, 659 (1990).
37. J. Käs and E. Sackmann, *Biophys. J.*, **60**, 825 (1991).
38. F. Farge, P. Devaux, *Biophys. J.*, **61**, 347 (1992).
39. B. L.-S. Mui, H.-G. Döbereiner, T. D. Madden and P. R. Cullis, *Biophys. J.*, **69**, 930 (1995).
40. H.-G. Döbereiner, E. Evans, U. Seifert and M. Wortis, *Phys. Rev. Lett.*, **75**, 3360 (1995).
41. H.-G. Döbereiner, O. Selchow and R. Lipowsky, *Eur. Biophys. J.*, **28**, 2 (1998).
42. J. Israelachvili, *Intermolecular and Surface Forces*, 2nd edn., Academic Press, London, 1991.
43. R. Homa and H. J. Pownall, *Biochim. Biophys. Acta*, **938**, 155 (1988).
44. R. D. Kornberg and H. M. McConnell, *Biochemistry*, **10**, 1111 (1971).
45. M. R. Raphael and R. Waugh, *Biophys. J.*, **71**, 1374 (1996).
46. T. E. Redelmeier, M. J. Hope and P. R. Cullis, *Biochemistry*, **29**, 3046 (1990).
47. M. Winterhalter and H. Helfrich, *J. Phys. Chem.*, **92**, 6865 (1988); *J. Phys. Chem.*, **96**, 327 (1992).
48. D. J. Mitchell and B. W. Ninham, *Langmuir*, **5**, 1121 (1989).
49. M. Winterhalter and H. Helfrich, *J. Phys. Chem.*, **96**, 327 (1992).
50. S. May, *J. Chem. Phys.*, **105**, 8314 (1996).
51. T. Chou, M. V. Jaric and E. D. Siggia, *Biophys. J.*, **72**, 2042, (1998).
52. J. B. Lee, P. G. Petrov and H.-G. Döbereiner, *Langmuir*, **15** (1999).
53. H.-G. Döbereiner, A. Lehmann, W. Goedel, O. Selchow and R. Lipowsky, in *Materials Science of the Cell*, (eds) B. Mulder, V. Vogel and C. Schmidt, *Mat. Res. Soc. Symp. Proc.*, **489**, 101, 1998.
54. R. Lipowsky, *Europhys. Lett.*, **30**, 197 (1995).
55. C. Hiergeist and R. Lipowsky, *J. Phys. II France*, **6**, 1465, (1996).
56. C. Hiergeist, V. Indrani and R. Lipowsky, *Europhys. Lett.* **36**, 491 (1996).
57. E. Eisenriegler, A. Hanke, S. Dietrich, *Phys. Rev. E*, **54**, 1134 (1996).
58. C. Dietrich, M. Angelova and B. Pouligny, *J. Phys II*, **7**, 1651 (1997).
59. K. Yaman, P. Pincus and C. M. Marques, *Phys. Rev. Lett.*, **78**, 4514 (1997).
60. A. D. Dinsmore, D. T. Wong, P. Nelson and A. G. Yodh, *Phys. Rev. Lett.*, **80**, 409 (1998).
61. R. Lipowsky and H.-G. Döbereiner, *Europhys. Lett.*, **42**, 219 (1998).
62. C. R. Safinya, I. Koltover and J. O. Rädler, *Current Opinion in Colloid Interface Science*, **3**, 69 (1998).
63. M. J. Hope, B. Mui, S. Ansell and Q. F. Ahkong, *Molecular Membrane Biology*, **15**, 1 (1998).

64. J. O. Rädler, I. Koltover, T. Salditt and C. R. Safinya, *Science*, **275**, 810 (1997).
65. R. Wick, M. I. Angelova, P. Walde and P. L. Luisi, *Chem. Bio.*, **3**, 105 (1996).
66. M. Sokabe, F. Sachs and Z. Q. Jing, *Biophys. J.*, **59**, 722 (1991).
67. S. L. Keller, S. M. Bezrukov, S. M. Gruner, M. W. Tate, I. Vodyanoy and V. A. Parsegian, *Biophys. J.*, **65**, 23 (1993).
68. W. Harbich, H. J. Deuling and W. Helfrich, *J. Phys.*, **38**, 727 (1977).
69. P. Méléard, J. Faucon, M. Mitov and P. Bothorel, *Europhys. Lett.*, **19**, 267 (1992).
70. V. Heinrich, S. Svetina and B. Žekš, *Phys. Rev. E*, **48**, 3112 (1993).
71. W. Wintz, H.-G. Döbereiner and U. Seifert, *Europhys. Lett.*, **33**, 403 (1996).
72. M. Jarić, U. Seifert, W. Wintz and M. Wortis, *Phys. Rev. E*, **52**, 6623 (1995).
73. L. Xu, PhD Thesis, Universität Potsdam (1999).
74. M. Kraus, U. Seifert and R. Lipowsky, *Europhys. Lett.*, **32**, 431 (1995).
75. H.-G. Döbereiner and U. Seifert, *Europhys. Lett.*, **36**, 325 (1996).
76. H.-G. Döbereiner, J. Käs, D. Noppel, I. Sprenger and E. Sackmann, *Biophys. J.*, **65**, 1396 (1993).
77. R. Lipowsky, *J. Phys. II France*, **2**, 1825 (1992).
78. U. Seifert, *Phys. Rev. Lett.*, **70** (1993).
79. O. Selchow, Diploma Thesis,. Technische Universität Müchen (1996).
80. S. Mann and G. A. Ozin, *Nature*, **382**, 313 (1996).

Chapter 11

# Oblate–Prolate Transition of Ellipsoidal Giant Magnetoliposomes: Experiments showing an Anisotropic Spontaneous Curvature

*Olivier Sandre, Christine Ménager, Jerôme Prost, Valérie Cabuil, Jean-Claude Bacri and Andrejs Cebers*

*Universités Paris 6 et Paris 7, France*

## 1. INTRODUCTION

The shape of fluid lipid vesicles is governed by the bending elasticity of their membrane, characterized by a bending modulus $K_c$ and a spontaneous curvature $c_0$ [1]. The latter exists only for bilayers separating nonsymmetric aqueous media. Döbereiner *et al.* described an experimental method based on the thermal fluctuations of giant vesicles to measure $c_0$ precisely and use it as a control parameter in phase diagrams [2]. The following discussion relates to particular nonsymmetric giant vesicles called magnetoliposomes, used as a tool for the observation of new shape transitions. Magnetoliposomes are giant liposomes filled up with a colloidal dispersion of magnetic particles. As with normal liposomes, magnetoliposomes exhibit thermal fluctuations. The effect of a magnetic field is to flatten the fluctuations and to induce a shape deformation even at low field intensity. Two types of deformations are reported here, depending on the characteristics of the

*Giant Vesicles* Edited by P. L. Luisi and P. Walde

encapsulated magnetic solution. They are explained by a modification of $c_0$, induced by the application of the magnetic field.

## 2. LIPOSOME PREPARATION

Magnetoliposomes are obtained by filling normal giant liposomes with an aqueous magnetic fluid (ferrofluid). They are suspended in a nonmagnetic aqueous phase. The ferrofluid is a colloidal suspension of magnetic nanoparticles (grains of $\gamma$-$Fe_2O_3$) [3]. The particle diameter is about 10 nm and the iron oxide surface is coated by trisodium citrate species in order to have a negative surface charge at pH 7. This ensures the colloidal stability even in physiological media [4]. There is an equilibrium between the citrate species adsorbed on the grains and the nonadsorbed trisodium citrate in the bulk. This electrolyte has a concentration $C_s$ which is monitored by dialysis and controlled by measuring the conductivity of the ferrofluid. The phospholipid constituting the membrane of the magnetoliposomes is 1,2-dioleoyl-*sn*-glycero-3-phosphocholine (DOPC from Avanti Polar Lipids, USA, used as purchased).

For the preparation of giant liposomes, the spontaneous swelling of a dense phospholipidic film was used because it enables encapsulation of the ferrofluid inside giant vesicles with a high efficiency. The ferrofluid is firstly prepared and its ionic strength $C_s$ is adjusted by dialysis. A small amount of lipid powder (1 mg) is mixed with 10 μL of this aqueous ferrofluid (pH = 7, volume fraction in magnetic particles is 0.05) and sheared on a glass support to obtain a dense lamellar phase. Then tridistilled water is added in excess (4 mL) to allow the spontaneous swelling of the liposomes during 1 h incubation at 40°C. Most of the liposomes prepared by this way are spherical with diameters ranging from 10 to 100 μm (Figure 11.1). They are red colored and they exhibit thermal fluctuations.

## 3. OBSERVATIONS

Cells were made of two glass slides separated by a 200 μm spacer and filled by aliquots of the magnetoliposomes suspension, then observed under an optical microscope (Leica ×40, NA 0.65). Pictures are taken with a CCD camera and digitized on a computer.

The magnetoliposomes are sensitive to an applied magnetic field of low intensity, typically 400 G. Two types of deformations have been observed. An initially spherical magnetoliposome is either elongated by the field (Figure 11.2) or, on the contrary, compressed at the poles (Figure 11.3). The shape is always described by an ellipsoid with rotational symmetry along the field direction: $a$ is the semi-axis parallel to the magnetic field $\vec{H}$ and $b$ is the value of the two other semi-axes perpendicular to $\vec{H}$. The type of deformation and its amplitude are described by the

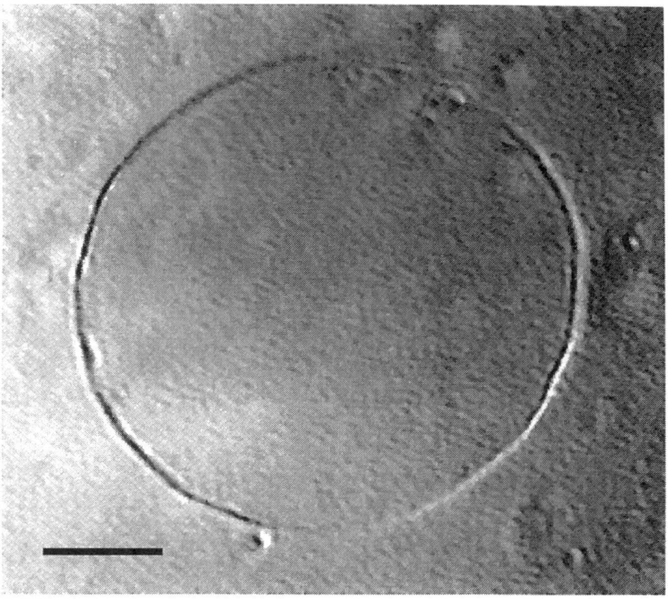

**Figure 11.1** A giant magnetoliposome without any magnetic field. The membrane is fluctuating as observed with differential interference contrast microscopy (DIC). Length of the bar is 20 μm.

ellipsoid eccentricity: $e^2 = 1 - b^2/a^2$. For an elongated liposome (prolate ellipsoid) $e^2$ is positive and for a compressed one (oblate ellipsoid) $e^2$ is negative. A combination of the two deformations is also observed and described as a spinning top shape (Figure 11.4).

## 4. ANALYSIS OF DEFORMATIONS UNDER A MAGNETIC FIELD

The experimental parameters relevant to the observed shape transitions are (a) the initial radius $R_0$ of the liposome measured on the video frame and (b) the two characteristics of the encapsulated solution, namely its magnetic susceptibility $\chi$ and the free salt concentration $C_s$.

An *in situ* measurement of $\chi$ is provided by magnetophoresis, i.e. the migration of a liposome submitted to a controlled gradient of magnetic field. A magnetic field gradient is performed perpendicularly to an applied magnetic field with the set-up described in Figure 11.5. The field lines of two coils in serial are deviated by polar pieces edged at 45°. The sample is in the middle region where the field lines are parallel to the O–$x$ axis and the field intensity gradient is in the O–$y$ direction. The

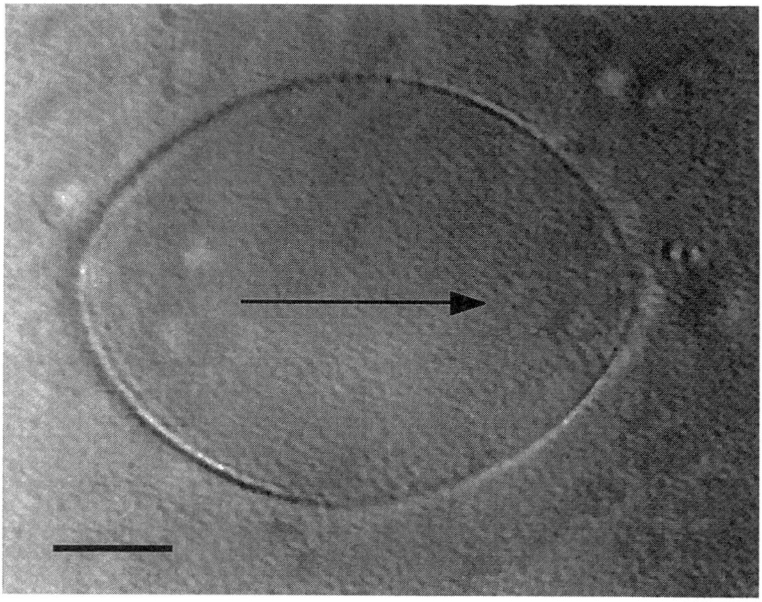

**Figure 11.2** The liposome shown in Figure 11.1 submitted to a magnetic field $\vec{H}$ (direction given by the arrow). The deformation is prolate and the membrane fluctuations vanish. Note the high ionic strength ($C_s = 85\,\text{mM}$). Length of the bar is 20 μm.

gradient is $110 \pm 2\,\text{G}\,\text{cm}^{-1}$ (measured with Hall effect probes) with a field equal to $410 \pm 2\,\text{G}$ in the sample place.

The magnetic force that causes the migration of a magnetoliposome submitted to a magnetic field gradient is

$$\vec{F}_\text{m} = \chi V H_x(\vec{\nabla}_y H_x) \tag{1}$$

where $V = \frac{4}{3}\pi ab^2$ is the volume of the liposome and $\vec{\nabla}$ is the gradient operator. It is balanced by a viscous force, which has been calculated [5] in the case of an ellipsoid elongated in the O–$x$ direction and moving along O–$y$ in a medium of viscosity $\eta$:

$$\vec{F}_\text{v} = 16\pi\eta \frac{a}{2\ln\left(\dfrac{2a}{b}\right) + 1}\vec{v} \tag{2}$$

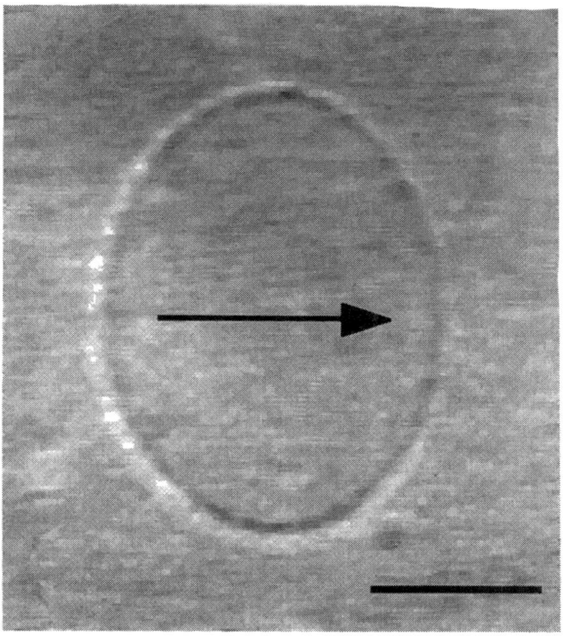

**Figure 11.3** A magnetoliposome filled with a low ionic strength magnetic fluid ($C_s = 7.75$ mM) and submitted to a magnetic field $\vec{H}$ (direction given by the arrow). The deformation is oblate. Length of the bar is 10 μm.

The viscosity of the suspending solution of the vesicles is $\eta = 10^{-2}$ P. The magnetophoretic velocity results from the balance of the two forces:

$$\vec{v} = \frac{b^2\left[2\ln\left(\frac{2a}{b}\right)+1\right]}{8\eta}\chi H_x(\vec{\nabla}_y H_x) \tag{3}$$

The measurement of the liposome velocity leads to the value of $\chi$ characterizing the ferrofluid encapsulated in a given liposome. The concentration $C_p$ of nanoparticles inside the liposome is then calculated using Langevin's law of superparamagnetism:

$$\chi = \frac{C_p V_p^2 m_s^2}{3kT} \tag{4}$$

where $m_s = 360$ G is the magnetization at saturation of $\gamma$-$Fe_2O_3$ and $V_p$ is the volume of a nanoparticle [6]. The following uses $V_p$ corresponding to a particle diameter of 12 nm, taking into account that the ferrofluid is polydisperse and that $\chi$ is ruled by the tail of the distribution.

Please note that $R_0$ and $C_p$ vary a lot with the population of liposomes due to the swelling process. Their values are measured thereafter. The concentration of trisodium citrate salt in the ferrofluid is adjusted to a known value by dialysis through a cellulose membrane (Spectra/Por MWCO 12-14000 from ROTH, France). Final concentrations $C_s$ ranging from 5–85 mM are measured from the electrical conductivity of the ferrofluid.

The experimental values of $R_0$ and $\chi$ are measured on large amounts of magnetoliposomes prepared with ferrofluids of different $C_s$. Some are given in Table 1. The experimental data in Figure 11.6 illustrate the influence of the three parameters $R_0$, $\chi$, and $C_s$ on the observed deformations. Vesicles that are larger, more magnetic, and filled with a high ionic strength magnetic fluid exhibit prolate deformation under a magnetic field, whereas smaller vesicles filled with a low ionic strength magnetic fluid and less magnetic become oblate. Two domains corresponding to both types of deformation are well defined on this topological phase diagram, the oblate one being in the region of low values of $R_0$, $\chi$ and $C_s$.

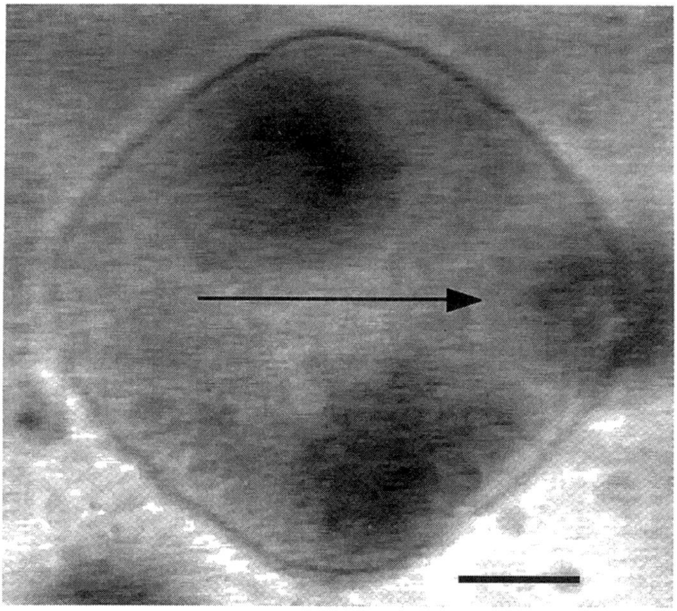

**Figure 11.4** Magnetoliposomes inside which the salt concentration is intermediate ($C_s = 25$ mM) and submitted to a magnetic field $\vec{H}$ (direction given by the arrow). The liposomes are always axisymmetric around the field direction but combine elongations both at the poles and at the equator (spinning-top shapes). The values of $e^2 = 1 - b^2/a^2$ can be (a) nearly zero, (b) negative or (c) positive. In that last case the sharp shape is only transient and relaxes to a more rounded one, but with (d) membrane buckling. Length of the bar is always 10 µm.

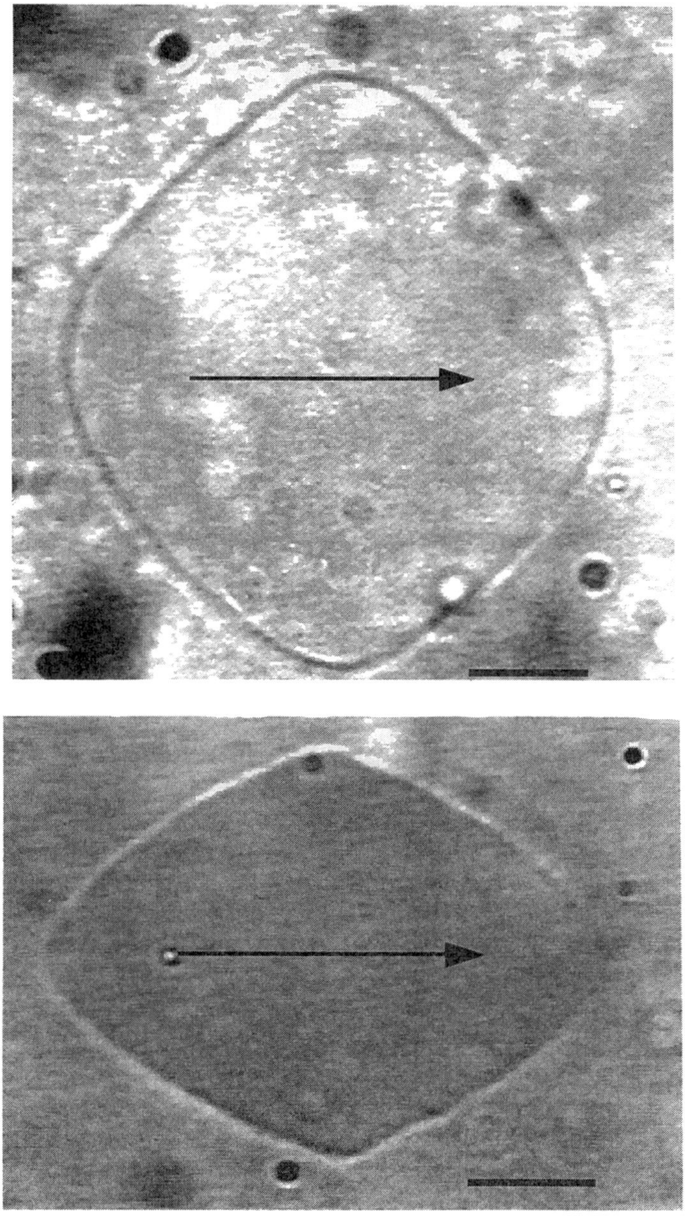

**Figure 11.4** (*continued*)

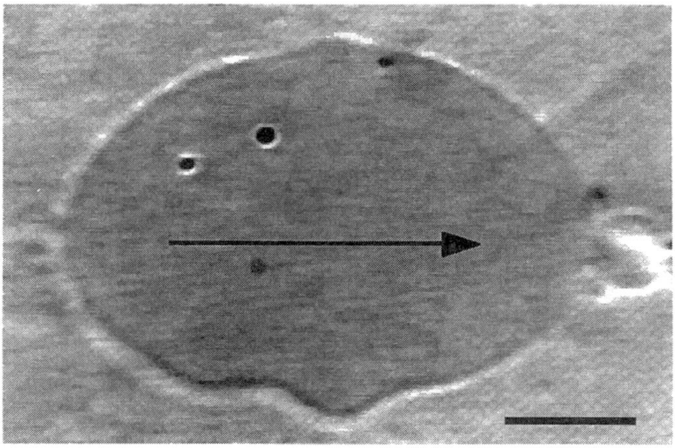

**Figure 11.4** (*continued*)

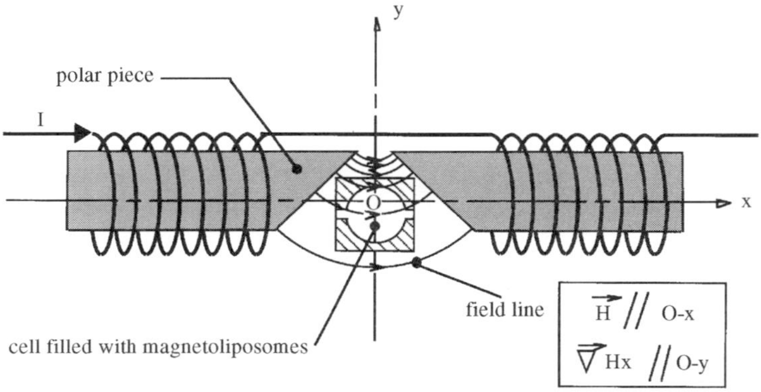


**Figure 11.5** Experimental set-up to produce a controlled magnetic field gradient (the field lines get closer towards increasing *y* coordinate).The sample containing the magnetoliposomes is placed between the two coils in the cell as described in the text.

## 5. INTERPRETATION

When submitted to an applied magnetic field, a liposome encapsulating a ferrofluid of susceptibility $\chi$ gets a bulk magnetization (proportional to $R_0^3$); as a consequence, the interface between inner magnetic and outer non magnetic media (i.e. the bilayer) is stressed. The deformation of an initially spherical object by a magnetic stress should always be a prolate ellipsoid, whatever the sign of $\chi$, so that magnetic droplets and magnetic holes (such as air bubbles) have the same behavior [7]. This was

**Table 11.1** Examples of experimental data about magnetoliposomes. $C_s$ is the concentration of trisodium citrate electrolyte in the ferrofluid, $R_0$ is the radius of the initial spherical shape, $e$ is the eccentricity of the ellipsoidal deformation and $\chi$ the magnetic susceptibility. The concentration $C_p$ of magnetic particles *in situ* is calculated from $\chi$ using eq. (4).

| $C_s$ (mol L$^{-1}$) | $\chi$ | $C_p$ (cm$^{-3}$) | $R_0$ (μm) | $e^2 = 1 - \frac{b^2}{a^2}$ |
|---|---|---|---|---|
| $5 \times 10^{-3}$ | $7.1 \times 10^{-5}$ | $8.2 \times 10^{13}$ | 25.2 | −0.397 |
| $7.75 \times 10^{-3}$ | $1.1 \times 10^{-4}$ | $1.3 \times 10^{14}$ | 13.3 | −0.596 |
| $7.75 \times 10^{-3}$ | $7.5 \times 10^{-5}$ | $8.7 \times 10^{13}$ | 21.4 | −1.17 |
| $2.5 \times 10^{-2}$ | $7.3 \times 10^{-5}$ | $8.4 \times 10^{13}$ | 24.6 | −0.845 |
| $2.5 \times 10^{-2}$ | $1.2 \times 10^{-4}$ | $1.4 \times 10^{14}$ | 22.5 | 0.496 |
| $2.5 \times 10^{-2}$ | $6.5 \times 10^{-5}$ | $7.6 \times 10^{13}$ | 50.4 | 0.343 |
| $3.5 \times 10^{-2}$ | $3.6 \times 10^{-5}$ | $4.2 \times 10^{13}$ | 21.1 | −1.31 |
| $4.5 \times 10^{-2}$ | $5.5 \times 10^{-5}$ | $6.4 \times 10^{13}$ | 31.6 | 0.466 |
| $6.5 \times 10^{-2}$ | $1.1 \times 10^{-3}$ | $1.2 \times 10^{15}$ | 14.1 | 0.536 |
| $8.5 \times 10^{-2}$ | $1.4 \times 10^{-4}$ | $1.7 \times 10^{13}$ | 13.0 | 0.586 |

observed for magnetoliposomes and used [8] in order to determine the bending modulus $K_c$ of the bilayer from the analysis of the liposomes deformation as a function of the magnetic field intensity.

Therefore the observation of an oblate deformation by a slight change of the experimental parameters is a tricky result. Note that the inner compartment of the magnetoliposomes is a complex colloidal solution. It includes both a 3 : 1 electrolyte (trisodium citrate) with concentration $C_s$ and a colloidal suspension of nanometric $\gamma$-$Fe_2O_3$ particles (giving $\chi$). The latter are both nanoscopic magnets and macro-ions, the net charge per nanoparticle being $Z \approx -20$ [4]. The importance of decreasing $C_s$ to obtain an oblate deformation suggests that there are electrostatic interactions between the particles and the bilayer. Indeed some electrophoresis experiments on large unilamellar liposomes (diameters about 200 nm) made of phospholipids with a phosphocholine head group have reported values of the $\zeta$-potential ranging from −4 mV for natural egg–lecithin [9] to about −10 mV for synthetic lipids such as 1,2-dimysristoyl-*sn*-glycero-3-phosphocholine (DMPC) [10]. The $\zeta$-potential varies with temperature, ionic strength and specific adsorption of anions [11]. It is assumed in the present study that the bilayer and the magnetic nanoparticles are both negatively charged and so repel each other.

The shape of magnetoliposomes submitted to a magnetic field results from a competition between the magnetic energy of the bulk and the bending energy of the bilayer. The magnetic energy always favors the prolate shape but, as shown below, the bending energy can stabilize the oblate deformation. The contribution to the bending energy of the bilayer due to a surface charge density has been extensively studied theoretically [12–14], but to our knowledge only experiments on small unilamellar vesicles (diameters about 10 nm) have been reported [15]. The shape of

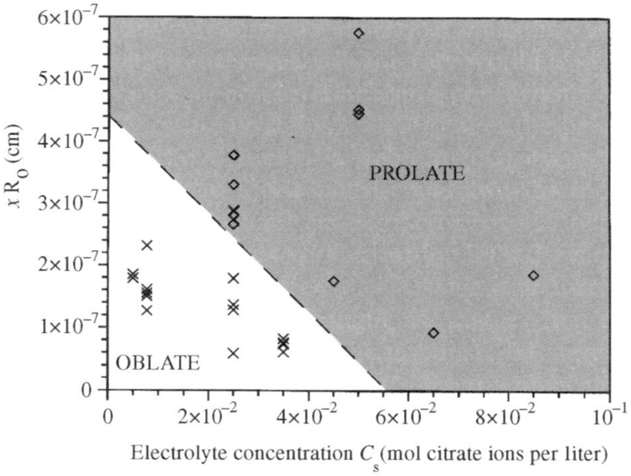


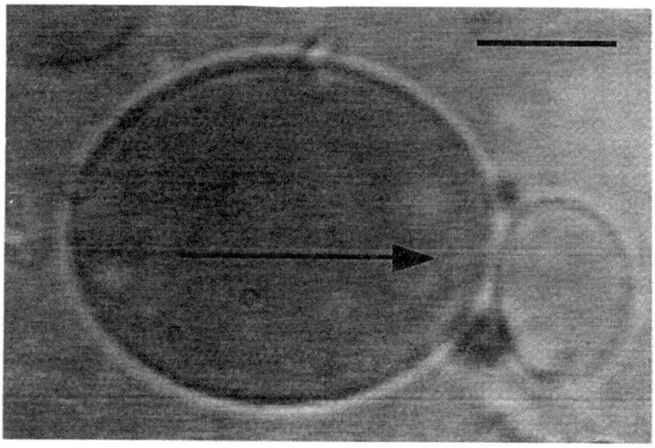

**Figure 11.6** (a) Topological phase diagram of the magnetoliposomes under a magnetic field where the product $\chi R_0$ is plotted versus trisodium citrate salt concentration $C_s$ in the ferrofluid. The dotted line between the two domains of prolate and oblate deformations is only a guide for the eye. (b) Effect of product $\chi R_0$ at given salt concentration ($C_s = 25\,\text{mM}$) on two neighboring liposomes with opposite deformations under a magnetic field. The prolate one is large and highly colored (high $\chi R_0$) whereas the oblate one is smaller and appears more dilute in magnetic particles (low $\chi R_0$). Length of the bar is 10 μm. The arrow indicates the direction of the field.

liposomes is known to be strongly dependent on the spontaneous curvature of the bilayer ($c_0$). In the case of magnetoliposomes, $c_0$ originates from the asymmetry between the aqueous compartments (ferrofluid inside, nonmagnetic solution outside). Winterhalter and Helfrich [12] have introduced a method to derive the elasticity constants of a bilayer bearing two different charge densities on both its sides. By using the same approach, considering that the screening length of the electrostatic potential is $\kappa_i^{-1}$ inside the liposome and $\kappa_o^{-1}$ outside, the spontaneous curvature can be written as

$$c_0 = \frac{\kappa_o^3}{3}\left(\frac{1}{\kappa_o^2} - \frac{1}{\kappa_i^2}\right) \tag{5}$$

The effect of the magnetic field (intensity $H_0$) on the spontaneous curvature $c_0$ is due to the modification it implies on the particles concentration profile in the vicinity of the bilayer. Qualitatively the depletion in nanoparticles is reduced near the magnetic poles because of magnetic attraction. The effect of this phenomenon on $c_0$ can be quantitatively described by introducing an angular dependence in the expression of the decay factor of the electrostatic potential away from the bilayer:

$$\kappa_i^2(\theta) = \kappa_o^2\left(1 + \frac{d}{1 + \frac{4\pi\chi^2 H_0^2}{C_p kT}\cos^2(\theta)}\right) \tag{6}$$

where $\theta$ is the angle between the field direction and the normal to the bilayer and $d$ is the ratio $Z^2 C_p/12C_s$ [16]. Thus, when a magnetic field is applied, $c_0$ becomes lower near the poles ($\theta = 0$ and $\theta = 180^\circ$) than at the equator. But, according to eq. (6), the angular contribution of $\kappa_i^{-1}$ depends on the amount of magnetic materials through $C_p$ and $\chi$, and on the ionic strength through $d$. This explains the driving role of these parameters in the experiments (Figure 11.6). For example a decrease of $C_s$ (by dialysis of the ferrofluid) raises up the anisotropy of the spontaneous curvature. Hence the oblate shape minimizes the bending energy because it develops a less curved bilayer at the magnetic poles where $c_0$ is decreased. If the concentration of magnetic oxide is not too high, so that the bending energy prevails over the magnetic energy, then the oblate shape is favored. This explains the tendency of oblate deformation under a magnetic field for magnetoliposomes encapsulating a ferrofluid diluted both in electrolyte and in nanoparticles.

This article describes a new shape transition for magnetoliposomes. This experimental result is interpreted as a coupling between magnetostatics and electrostatics. A complete theoretical model is being developed. The main finding of our results so far is that even a small charge density on the phospholipidic membrane can induce a coupling with charged encapsulated species.

## 6. REFERENCES


1. W. Helfrich, *Z. Naturforsch.*, **28c**, 693–703 (1973).
2. H. G. Döbereiner, E. Evans, M. Kraus, U. Seifert and M. Wortis, *Phys. Rev. E*, **55**, 4458–4474 (1997).
3. R. Massart *IEEE Trans. Magn.* **MAG-17**, 1247–1248 (1981).
4. J. C. Bacri, R. Perzinski, D. Salin, V. Cabuil and R. Massart, *J. Magn. Magn. Mat.*, **85**, 27–32 (1990).

5. J. Perrin, *J. Phys. Radium.*, **7**, 33 (1936).
6. J. C. Bacri, D. Salin, R. Perzinski, V. Cabuil and R. Massart, *J. Magn. Magn. Mat.*, **62**, 36–46 (1986).
7. A. T. Skjeltorp, *J. Magn. Magn. Mat.*, **65**, 195 (1987); A. T. Skjeltorp, *Phys. Rev. Lett.*, **51**, 2306 (1984); A. T. Skjeltorp, *J. Appl. Phys.*, **55**, 2587 (1984).
8. J. C. Bacri, V. Cabuil, A. Cebers, C. Ménager and R. Perzinski, *Europhys Lett.*, **33**, 235–240 (1996).
9. F. J. Carrion, A. De La Maza and J. L. Parra, *J. Colloid Interf. Sci.*, **164**, 78–87 (1994).
10. K. Makino, T. Yamada, M. Kimura and T. Oka *Biophys. Chem.*, **41**, 175–183 (1991).
11. S. A. Tatulian, *Biochim. Biophys. Acta*, **736**, 189–195 (1983).
12. M. Winterhalter and W. Helfrich, *J. Phys. Chem.*, **92**, 6865–6867 (1988); M. Winterhalter W. Helfrich, *J. Phys. Chem.*, 396, 327–330 (1992).
13. P. Pincus, J. F. Joanny and D. Andelman, *Europhys. Lett.*, 763–768 (1990).
14. D. Bensimon, S. David, S. Leibler and A. Pumir, *J. Phys. France*, **51**, 689–695 (1990).
15. J. Oberdisse and G. Porte, *Phys. Rev. E*, **56**, 1965–1975 (1997).
16. O. Sandre, C. Ménager, J. Prost, V. Cabuil, J. C. Bacri and A. Cebers (submitted for publication).

# Chapter 12

# Micromanipulation of Tubular Vesicles

*LIYU XU AND HANS-GÜNTHER DÖBEREINER*

*Max-Planck-Institut für Kolloid-und Grenzflächenforschung, Golm, Germany*

## 1. MICROMANIPULATION OF VESICLES

Micromechanical measurements on giant (fluid) vesicles using micrometer-size glass pipettes have been used extensively to extract membrane material parameters. Mutual adhesion energies, area thermal expansivities and elastic compressibilities are routinely measured [1,2]. At high levels of membrane tension, rupture strengths of membranes [1] and the edge energy of stabilized pores [3] can be obtained. Measurements in the intermediate and entropic tension regimes give the membrane bending moduli [4–7]. In addition to parameters characterizing static behavior, dynamic phenomena such as forced flip-flop between the membrane monolayers [8,9] and intermonolayer drag [8] have been monitored. However, the dynamics of membranes is generally much less studied than their static properties [10]. Recently, a series of particular interesting papers on the manipulation of tubular vesicles with optical tweezers found a Rayleigh-like pearling instability of the cylindrical tube after application of laser light [11–13].

This amazing versatility and usefulness of the controlled application of forces to vesicles motivated us to build a micropipette station as a standard tool to measure membrane material parameters (Figure 12.1). The technique has been improved by implementing real-time analysis of the morphological vesicle response to applied forces within an arrangement of computer-controlled components.

Also of interest are the dynamics of tubular membranes and the microrheology of the lipid bilayer under defined boundary conditions. The next section presents the

*Giant Vesicles* Edited by P. L. Luisi and P. Walde

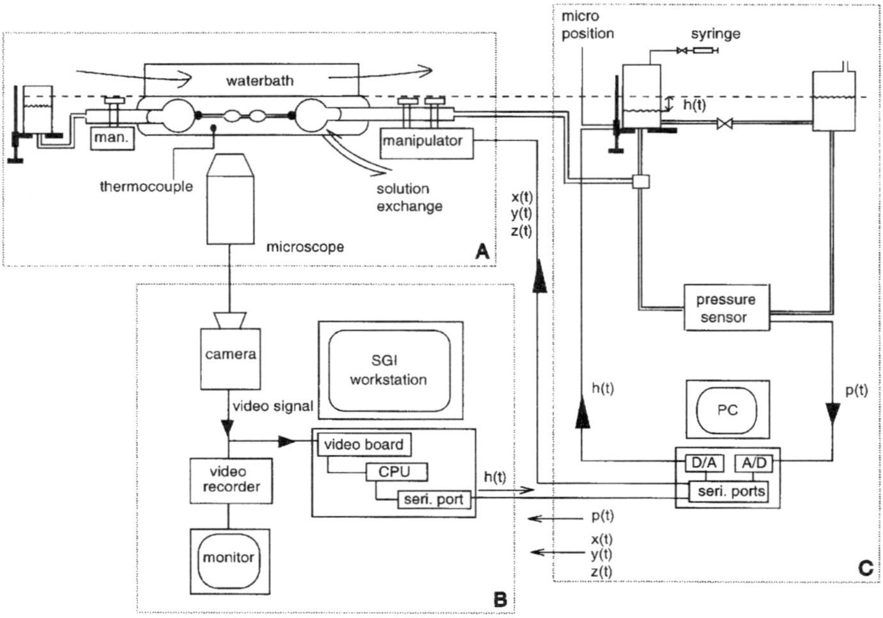


**Figure 12.1** Micromanipulation system consisting of a micropipette chamber mounted on an inverted phase-contrast microscope equipped with a sensitive CCD camera (A), real-time image analysis running on a UNIX workstation (B), and a set-up controlling pressure and manipulator position by a personal computer (C). The pipette suction pressure is created hydrostatically. We can generate controlled membrane tensions down to $10^{-4}$ dyn cm$^{-1}$ corresponding to 1 μm height difference between two linked water reservoirs. The micropipettes are mounted on two three-axis manipulators which can be moved manually or by computer. Image analysis is integrated with the automated experimental device control.

response of tubular vesicles with given membrane area and enclosed volume to pulling and pushing forces on their poles. A prototype experiment for probing various mechanisms of energy dissipation during forced membrane movement is also described.

## 2. PULLING AND PUSHING OF TUBES

Vesicles were swollen from 1,2-dioleoyl-*sn*-glycero-3-phosphocholine (Avanti Polar Lipids, USA) and *N*-((6-(biotinoyl)amino)hexanoyl)-1,2-dihexadecanoyl-*sn*-glycero-3-phosphoethanolamine (biotin-X DHPE, Molecular Probes, The Netherlands) at a ratio of 99 : 1. Tubes were grown out of lipid globules in shear flow and cut to the desired length with a pipette. Streptavidin-coated beads (Dynal, Germany) were used

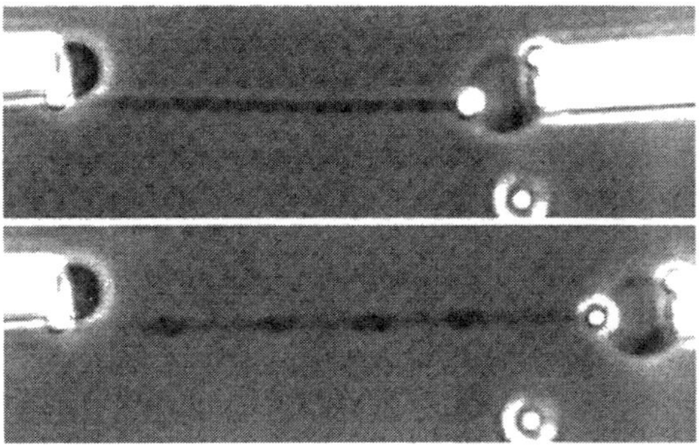

**Figure 12.2** Pulling of the tubes results in a Rayleigh-like pearling instability of the cylindrical membrane. The pole-to-pole distance is increased quickly ($20\,\mu m\,s^{-1}$) from $57.4\,\mu m$ to $75.4\,\mu m$.

to hold the tubules. Probe vesicles and beads (Figures 12.2 and 12.3) were picked up with two opposing pipettes. Beads were then allowed to adhere to the tubular vesicle under investigation and placed at a distance approximately equal to the free length of the cylindrical vesicle. Thermal fluctuations were clearly visible. A sudden increase of bead distance leads to the well-known pearling instability (Figure 12.2) [11–13]. The control parameter is the axial length of the tubule which can be set and changed with a given speed. Note that the vesicle volume and membrane area are kept constant, whereas the boundary conditions given in references [11,13] are less clear. With time, three of the four spindles decay to produce a stable spindle–tether–bead configuration [14]. In contrast, decreasing the length of the tubule leads to a buckling instability of the cylindrical membrane (Figure 12.3). One observes a distinct

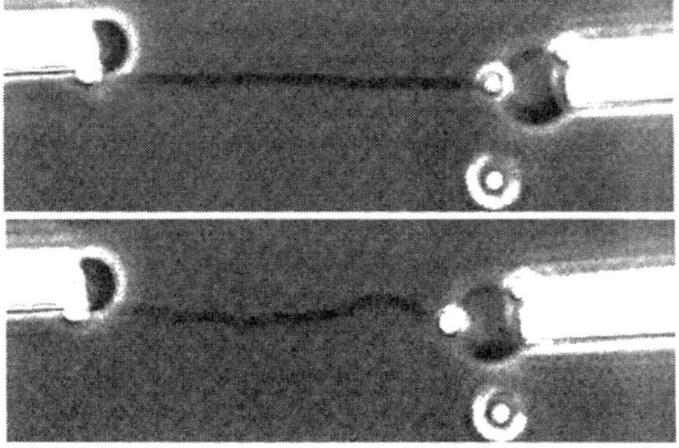

**Figure 12.3** Pushing of the tubes leads to a buckling instability. The pole-to-pole distance is decreased quickly ($20\,\mu m\,s^{-1}$) from $58.7\,\mu m$ to $54.5\,\mu m$.

selection and relaxation of the bending modes which depend on the amplitude and velocity of imposed length changes. Pulling and pushing can be repeated automatically several times with the same vesicle to establish reproducibility and good data quality.

## 3. MICRORHEOLOGY OF MEMBRANES

By fine-tuning of boundary conditions, i.e., allowing the tubule to obtain different adhesion areas with the respective beads at the two ends, one can achieve a stretched quasi-equilibrium configuration where a single spindle moves along a conelike tether with constant speed [15,16]. Measurement of this velocity together with an analysis of the elastic force on the spindle allows deduction of the kinetic coefficient(s) of energy dissipation. Depending on the tubular geometry and spontaneous curvature of the vesicle [17,18] different microrheological regimes, including bulk water and bilayer lipid viscous flow, as well as intermonolayer drag [8] can be studied.

## 4. ACKNOWLEDGMENTS

We thank E. Evans, W. Fenzl, R. Lipowsky, and M. Neese for useful discussions, and V. Heinrich and U. Seifert for their helpful collaborations.

## 5. REFERENCES

1. E. Evans and D. Needham, *J. Chem. Phys.*, **91**, 4219 (1987).
2. E. Evans, in *Structure and Dynamics of Membranes*, (eds), R. Lipowsky and E. Sackmann, Elsevier, Amsterdam 1995.
3. D. V. Zhelev and D. Needham, *Biochim. Biophys. Acta.*, **1147**, 89 (1993).
4. E. Evans and W. Rawicz, *Phys. Rev. Lett.*, **64**, 2094 (1990); **79**, 2379 (1997).
5. R. E. Waugh, J. Song, S. Svetina and B. Zeks, *Biophys. J.*, **61**, 974 (1992).
6. D. V. Zhelev, D. Needham and R. M. Hochmuth, *Biophys. J.*, **67**, 720 (1994).
7. V. Heinrich and R. E. Waugh, *Ann. Biomed. Eng.*, **24**, 595 (1996).
8. E. Evans and A. Yeung, *Chem. Phys. Lipids*, **73**, 39 (1994).
9. S. Svetina, B. Zeks, R. E. Waugh and R. M. Raphael, *Eur. Biophys. J.*, **27**, 197 (1998).
10. U. Seifert, *Adv. Phys.*, **46**, 13 (1997).
11. R. Bar-Ziv and E. Moses, *Phys. Rev. Lett.*, **73**, 1392 (1994).
12. P. Nelson, T. Powers and U. Seifert, *Phys. Rev. Lett.*, **74**, 3384 (1995).
13. R. Bar-Ziv, E. Moses and P. Nelson, *Biophys. J.*, **75**, (1998).
14. V. Heinrich, B. Bozic S. Svetina and B. Zeks, *Biophys. J.*, **76**, (1999).
15. J. L. Goveas, S. T. Milner and W. B. Russel, *J. Phys. II France*, **7**, 1185 (1997).
16. This observation might explain, why pearls continue to move along tethers after shutting-off laser light [13,15].
17. H.-G. Döbereiner, E. Evans, M. Kraus, U. Seifert and M. Wortis, *Phys. Rev. E*, **55**, 4458 (1997).
18. H.-G. Döbereiner, O. Selchow and R. Lipowsky, *Eur. Biophys. J.*, **28**, 174 (1999).

# Chapter 13

# Electromechanical Properties of Model Membranes and Giant Vesicle Deformations

*PHILIPPE MÉLÉARD, CLAIRE GERBEAUD, AND TANJA POTT*

*Centre de Recherche Paul Pascal, Pessac, France*

*MARIN D. MITOV*

*Bulgarian Academy of Sciences, Sofia, Bulgaria*

## 1. INTRODUCTION

Many different phenomena involving membrane morphological changes occur naturally during the growth of a cell. Such shape deformations involve molecular size to cell size scales. For example, the normal shape of the erythrocyte is discocyte with a radius of 7–8 μm and a thickness close to 2 μm. When migrating into small vessels, it has to deform into a stomatocyte shape to adapt its larger size to the capillary. Such deformations are controlled by the cytoskeleton network lying under the membrane but they could not occur without involving also some fundamental mechanical properties of the membrane.

Since the mid-1970s, much work has been done to determine the role of these mechanical properties on the overall morphological behavior of cell membranes. Canham [1], Helfrich [2] and Evans [3] have identified the different characteristics responsible for the spontaneous shape of a membrane or its resistance to deformation. Using phenomenological parameters, they gave descriptions of the membrane deformation valid on a scale where membranes can be considered as a continuum material (i.e. on a scale larger than the membrane thickness). Following this idea

*Giant Vesicles* Edited by P. L. Luisi and P. Walde

Brochard and Lennon demonstrated that the bending rigidity of the red blood cell membrane can be obtained from the analysis of its flickering [4]. Despite this and other work [5–8], most experimental studies on the mechanical properties of natural component systems have been done on bilayers composed of phospholipids only. Among the different structures that can be formed when mixing lipids and water, giant vesicles ($R \geq 10\,\mu m$) are studied because their size is large enough to be clearly visible using videomicroscopy. Such simple models can mimic some of the morphological changes previously observed for cell membranes [9,10].

The following discussion briefly introduces the physical parameters that describe the membrane behavior when a deformation occurs. The focus is on the bending elasticity measurement from the analysis of thermal fluctuations of giant vesicles. The experimental part deals first with the dependence between the bending elasticity and the bilayer content. On one hand, the only membrane molecules that spontaneously associate and form large surface are amphiphilic lipids such as phosphatidylcholines. Their phase behavior is essentially thermotropic at a water weight ratio larger than $\sim 40\%$, whereas some lyotropism can be noted at lower water content [11]. Giant unilamellar vesicles behave as lamellar phases due to their dilution. They can thus be studied to measure the relationship between the bending elasticity and the lipid phase behavior. Single-component bilayers from three different synthetic saturated phosphatidylcholines were chosen, to compare their mechanical behaviors as a function of temperature and chain length. On the other hand, it is well-known that natural membranes are composed of several molecules with different chemical structures. Using mixtures, the membrane bending elasticity is demonstrated to be strongly dependent on very small membrane incorporates. All these studies indicate that some correlations can be made between the mechanical behavior and the chemical structure of the molecules within the bilayer.

The last experimental study presents some recent experiments on the deformation of giant vesicles induced by electric fields. These observations represent the first part of a work to study the electric interaction with lipid bilayers, as a function of the frequency of the applied electric field (from 100 Hz to 10 kHz). This paper, focuses on the low frequency regime ($<1$ kHz) when pure water can be considered as a conductive medium while the membrane is an insulator. The observed large shape changes of giant vesicles are related to their bilayer bending properties.

## 2. DESCRIPTION OF MEMBRANE DEFORMATIONS

Membrane thickness can generally be neglected when it is compared to any cell size. To study the role of the membrane mechanical properties in cell deformability, bilayers or membranes are described as mathematical surfaces, that are homogeneous and of vanishing thickness, whose small shape changes are simply understood using three different elementary deformations characterized by their elastic moduli (Figure 13.1).

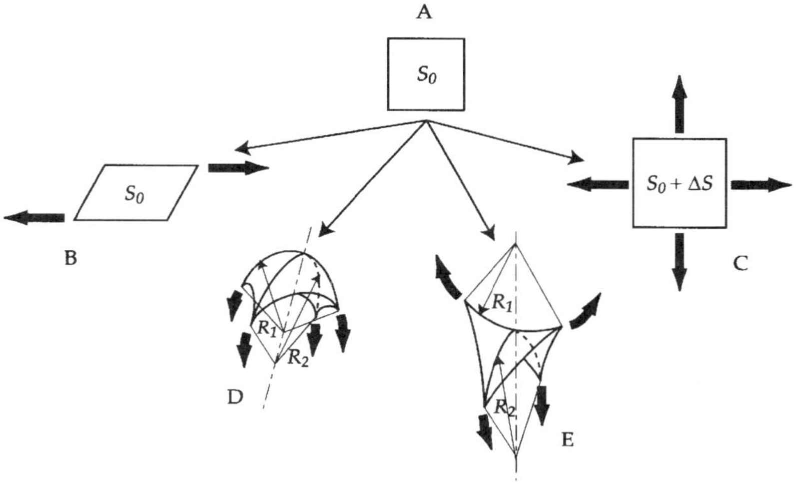


**Figure 13.1** Three basic deformations for a membrane element considered as a mathematical surface. (A) Initially, the undeformed state of the membrane element is assumed to be flat. (B) A shear deformation occurs when tangential forces parallel to the side of the membrane are applied at constant area. (C) Tangential forces perpendicular to the membrane sides increase or decrease the membrane area at constant curvature. The two last sketches represent bending deformations occurring when the forces applied to the membrane are out of the plane, in the cases of (D) positive or (E) negative Gaussian curvature.

## 2.1 Elementary Deformations

Shear deformation does not contribute to energy changes when considering fluid bilayers, i.e. membranes at a temperature higher than the main phase transition $T_m$ [12]. However, when the bilayer is not fluid, shear deformation has to play a role (Figure 13.1 A→B) [13].

When considering small enough stretching deformations (Figure 13.1 A→C), the energy cost per unit area is written as a quadratic expression [2,14]:

$$f_\kappa = \frac{\kappa}{2}\left(\frac{\Delta A}{A_0}\right)^2 = \frac{\sigma}{2} \times \left(\frac{\Delta A}{A_0}\right) \tag{2}$$

where $\kappa$ (in $\mathrm{N\,m^{-1}}$ or $\mathrm{J\,m^{-2}}$) is the stretching modulus, $A_0$ and $A$ are the areas before and after the deformation ($\Delta A = A - A_0$), and $\sigma$ is the membrane tension defined as usual by $\sigma = A_0 \times \partial f_\kappa / \partial A$. It can be noted that the reference area $A_0$ corresponds to the zero tension state. The membrane tension $\sigma$ is very similar to the interfacial tension commonly used in colloidal chemistry. However for phospholipid bilayers, the number of molecules in a membrane cannot change rapidly. As a consequence, any dilatation of bilayers containing phospholipids leads generally to an increase of the area per molecule, unless the bilayer contains some soluble amphiphilic chemicals [15,16].

Deformations inducing curvature changes without shear or stretching are finally considered (Figure 13.1 A$\rightarrow$D and A$\rightarrow$E). The energy cost per unit area for small curvature deformations is expressed as [2]

$$f_c = \frac{k_c}{2}(c_1 + c_2 - c_0)^2 + \bar{k}_c c_1 c_2 \tag{3}$$

In this expression, $k_c$ and $\bar{k}_c$ are the bending and the saddle-splay moduli respectively (in Joules), $c_0$ is the spontaneous curvature, $c_1$ and $c_2$ are the principal curvatures ($c_i = 1/R_i$ where $R_i$ is the radius of curvature, Figure 13.1). As for the expansion energy, eq. (2), the splay term $k_c(c_1 + c_2 - c_0)^2/2$ is very similar to the energy cost of a spring deformation (Hooke's law). Note that a deformation defined by $c_1 = -c_2 \neq 0$, $c_0 = 0$ (Figure 13.1 E) does not cost any energy if one only considers the first squared term in eq. (3), explaining the requirement of the saddle-splay term $\bar{k}_c c_1 c_2$.

Any membrane deformation has to increase the energy compared to that at equilibrium. Accordingly the elastic moduli $\mu$, $\kappa$ and $k_c$ should be positive. It can also be demonstrated that $2k_c + \bar{k}_c > 0$ (assuming that flat bilayers correspond to the stable state). Yet the sign for $\bar{k}_c$ is not determined *a priori* and depends on the membrane composition. When $\bar{k}_c > 0$, the membrane prefers the negative Gaussian curvature shape (i.e. $c_1 c_2 < 0$, Figure 13.1 E). On the contrary, the case when $\bar{k}_c < 0$ characterizes a membrane that prefers spherical deformations (i.e. $c_1 c_2 > 0$, Figure 13.1 D).

## 2.2 From Open to Closed Objects

Equations (1), (2), and (3) give the contributions of shear, stretching, and bending elasticities when a local deformation is applied on a membrane element. For mesoscopic objects defined by their membrane, these equations have to be added and integrated over the whole surface of the bilayer to evaluate the total free energy cost for any deformation. However, this interpretation is not complete if these three mechanical contributions only are taken into account. Two other theoretical occurences have to be considered: (i) the case of open objects where the bilayer is not continuous, i.e. there exists one or several mesoscopic holes through the membrane, and (ii) the case of close objects where the membrane is now continuous.

Consider now a large-scale object ($R > 0.1\ \mu$m) defined by its bilayer. When the bilayer contains pores with dimension larger than the molecular size (Figure 13.2), the free energy cost of these edges must be considered:

$$E_\gamma = \gamma \times \delta \tag{4}$$

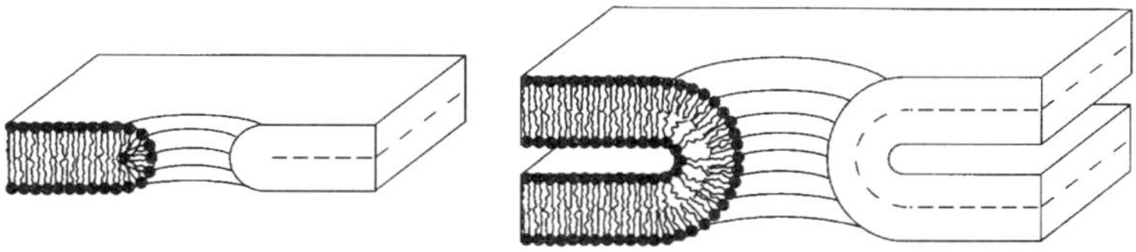

(a) (b)

**Figure 13.2** (a) An open vesicle possesses border molecules that arrange as a line defect where molecules bend on a molecular length. (b) The fusion of two bilayers yields a passage connecting two adjacent bilayers (adapted from Ref 66).

where $\gamma > 0$ (in $\mathrm{J\,m^{-1}}$) is the line energy and $\delta$ is the total border length. Physically, the pore results from the fusion of the two monolayers of a given bilayer (Figure 13.2(a)). Then, the molecules within the border are strongly curved, which is an unfavorable situation for phosphatidylcholines due to their hydrophobic packing. This is no more the case after a bilayer fusion (Figure 13.2(b)). The typical membrane curvature along this passage is now roughly dependent on the water thickness separating the two bilayers.

Finally, consider the case of a similar large-scale object ($R > 0.1\ \mu$m) where the membrane is now continuous. For such a close bilayer, the Gauss–Bonnet theorem states that the integration on the liposome total surface of the Gaussian curvature, $c_1c_2$, depends only on the topological characteristics of the vesicle [17], i.e. $\oint c_1c_2\,\mathrm{d}S = 4\pi(1-g)$, where $g$ is the genus of the surface. For example, any deformation from a spherical to an ellipsoidal shape implies $\mu$, $\kappa$, and $k_c$ only, but not the saddle-splay modulus $\bar{k}_c$, because the topology is conserved during the transformation. Considering a fluid phospholipid membrane ($\mu = 0$), it can be demonstrated that the stretching modulus $\kappa$ is too high to give measurable area fluctuations at thermal equilibrium [18]. As a consequence and because lipid molecules are usually poorly soluble in water solutions, the total area of a bilayer is fixed for short time experiments. Considering also the volume enclosed by the bilayer as a constant, due to the low water permeability (more precisely, giant vesicles are mainly studied in isoomostic solutions), this gives two geometric constraints that depend on the vesicle formation process. It should be noted that the total vesicle area and the vesicle volume may vary considerably from one liposome to another.

## 3. MEASURING THE MECHANICAL PARAMETERS

Giant unilamellar vesicles (GUV) are one of the models that can be used to determine the different mechanical parameters of a bilayer because their large size

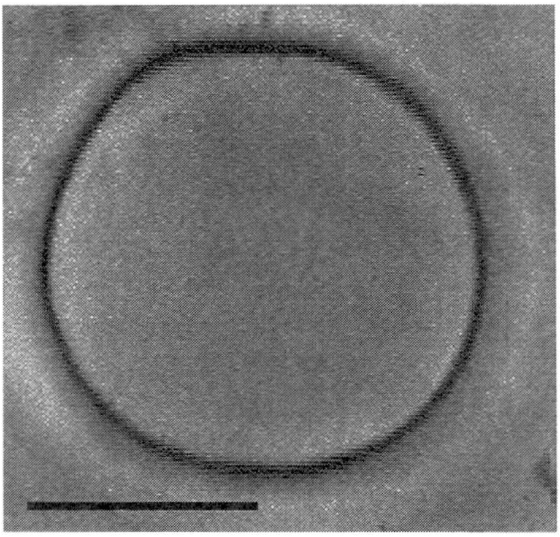

(a)

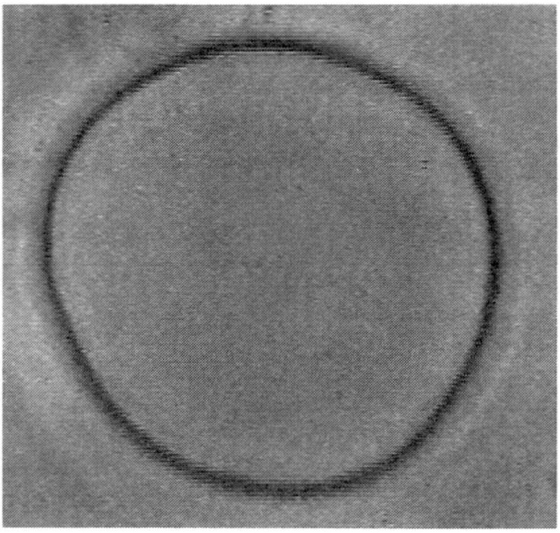

(b)

**Figure 13.3** Parts (a) and (b) are two characteristic pictures of the same fluctuating giant vesicle at different times (the bar represents 10 μm). They are obtained using a phase contrast microscope and after the digitization of the video images obtained from a contrast enhancement camera. The membrane position is clearly seen as a black curve.

is easily seen under an optical microscope. As lipid bilayers do not absorb light, different contrast enhancement techniques are used: Zernicke phase contrast, Nomarski differential interference contrast (DIC), Hoffman contrast, or epifluorescence. Figure 13.3 shows characteristic images of a GUV that can be obtained using phase contrast microscopy. The equatorial cross-section is clearly visible from the

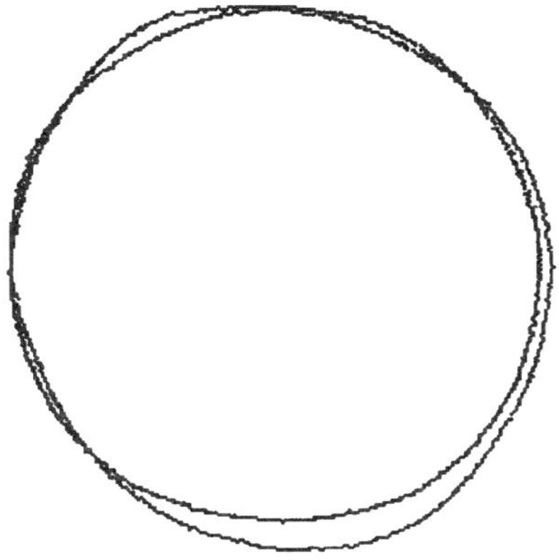

(c)

**Figure 13.3** Part (c) is the superposition of the two contours obtained from the image analysis of the previous pictures.

surrounding environment and can easily be used to extract the contour line after the digitization of the video signal (Figure 13.3(c)), [19]. The GUV is quasispherical, i.e. it is locally deformed around a mean spherical shape. The observed deformations are time and space dependent and originate from the Brownian motion of the membrane lipids and the surrounding water molecules. Their amplitudes are closely related to the bending elasticity $k_c$, introduced in the previous section.

## 3.1 Bending Elasticity $k_c$

Historically, the first measurement of a mechanical property of natural membranes was made by Brochard and Lennon [4]. The flickering phenomenon of red blood cells was analyzed as being due to light interferences resulting from thickness oscillations of the cell. This flickering is physically analogous to the thermal fluctuations of giant vesicles (Figures 13.3(a) and (b)). Nevertheless, the first bending elasticity measurement on model systems was made on EPC tubular liposomes [20]. Schneider, Jenkins and Webb [21] and Engelhardt, Duwe and Sackmann [22] proposed two different methods to measure $k_c$ from the analysis of thermal fluctuations of giant quasispherical vesicles. Later, Milner and Safran obtained the proper theoretical description of the thermal fluctuation phenomenon of quasispherical objects, where the constraint of the constant liposome area was first taken into account [23]. This latter description was used to analyze the data and the experimental limitations of such a procedure [24]. The main improvements of this approach were the use of a large number of images to limit the noise contribution, the decomposition of the autocorrelation function of the relative thermal fluctuations into Legendre polynomials, and finally a correct understanding of the role of the time integration effect from the video camera [25]. This, together with the development of the GUV electroformation method [26] gave an opportunity to follow the behavior of the bending elasticity $k_c$ as a function of different parameters (see below).

Other methods are given in the literature that lead to bending elasticity measurements of phospholipid bilayers and which do not directly use the analysis of thermal fluctuations of giant vesicles. The first was introduced by Bo and Waugh [27], using micropipette manipulation. It can be applied to estimate the nonlocal bending elasticity [28] and was refined recently by Heinrich and Waugh [29]. A second pipette method was published a short time later by Evans and Rawicz [30]. Finally, Helfrich and co-workers followed the interaction of fluctuating liposomes with an alternative electric field [31,32].

### 3.2 Other Mechanical Properties

The previous section showed that the bending elasticity measurement of lipid bilayers has led to the development and improvement of different sophisticated methods based on the observation or the manipulation of giant vesicles. Stretching elasticity can be obtained by micromanipulation using suction pressures that are definitely much larger than for $k_c$ determination, leading in principle to independent $k_c$ and $\kappa$ measurements depending on the pressure range [30,33]. A similar experiment has been successfully applied to the estimation of the failure limit, i.e. the mechanically induced opening of GUVs, for different natural membranes and model systems or to study the area–pressure relationship close to the main phase transition of DMPC bilayers [34]. More recently, Zhelev and Needham have combined high amplitude electric pulsed experiments with the micromanipulation of giant vesicles to obtain the line energy $\gamma$ [35]. For SOPC bilayers and SOPC containing 50 mol % cholesterol, $\gamma$ is found to be $(0.92 \pm 0.07) \times 10^{-11}$ N and $(3.05 \pm 0.12) \times 10^{-11}$ N, respectively (the latter value has been corrected recently by Moroz and Nelson [36]). Of note is that the saddle-splay elasticity $\bar{k}_c$ has never been obtained from direct measurements of giant vesicle deformations or other systems.

## 4. MATERIALS AND METHODS

1,2-Dilauroyl-*sn*-glycero-3-phosphocholine (DLPC), 1,2-dimyristoyl-*sn*-glycero-3-phosphocholine (DMPC), 1,2-dipalmitoyl-*sn*-glycero-3-phosphocholine (DPPC), 1,2-dilauroyl-*sn*-glycero-3-phosphatidic acid (DLPA) and cholesterol (Chol) were obtained from Fluka; 1-stearoyl-2-oleoyl-*sn*-glycero-3-phosphocholine (SOPC) was obtained from Avanti Polar Lipids, 1,2-diacyl-3-digalactosyl-*sn*-glycerol (DGDG, from whole wheat flour) and gramicidin D were obtained from Sigma. Egg-yolk 1,2-diacyl-*sn*-glycero-3-phosphocholine (EPC) was obtained following a published procedure [37]. Giant vesicles were produced using the electroformation method [38] in MilliQ water (Millipore mQ) or in a Tris buffer at 1 $mmol\,l^{-1}$ (pH = 7.4, Trizma Base, Sigma). The pH is controlled by adding concentrated NaOH solution. The electroformation cell and the procedure used to obtain giant vesicles have been described elsewhere [39]. The chamber is made of a 1 mm thick glass cell (similar to a fluorescence cell) where two parallel and horizontal platinum wires of 0.8 mm diameter have been introduced and connected to the generator (Hewlett Packard HP 3324 A). Fluctuating vesicles are observed using a phase contrast microscope (Axiovert 135 with a water immersion $\times 40/0.75$ objective, Zeiss) and a CCD video camera (C2400-77, Hamamatsu). Video images are recorded on a U-matic recorder (VO-7630, Sony), digitized by a AV300 AA module (Digital) and analyzed on an AlphaStation (Dec 3000, Digital). The details of the procedure used for the experimental determination of bending elasticity from thermal fluctuation analysis of giant vesicles can be found in previously published papers [19,24,39]. To study the interaction of giant vesicles with an alternating electric field, the electroformation cell was used. Accordingly, the applied electric field is perpendicular to the optical axis and is thus parallel to the observation plane.

## 5. RESULTS AND DISCUSSION

The following section presents some experimental results obtained in the study of giant vesicles using optical microscopy, beginning with bending elasticity measurements of different model membrane systems. Observations made on giant vesicle deformations induced by low-frequency electric fields are also discussed.

### 5.1 Bending Elasticity as a Function of Bilayer Content

#### *(a) One-Component Bilayers*

To study the relationship between phase behavior and mechanical properties at full hydration, the bending modulus was measured as a function of temperature, in the region of the gel–fluid phase transition temperature $T_m$, for three different saturated phosphatidylcholines (PC) of increasing chain length: dilauroyl-PC (DLPC), dimyristoyl-PC (DMPC), and dipalmitoyl-PC (DPPC) [40]. The results are summarized in Figure 13.4(a). In the low-temperature region, i.e. when $T$ is decreased close to $T_m$ for DMPC and DPPC bilayers, a sharp reduction of $k_c$ was observed, as for multibilayer systems [41]. This behavior can be understood by considering local deformations [40]. Another important result of this work was to demonstrate that the bending modulus of symmetric and saturated PCs is roughly independent of the

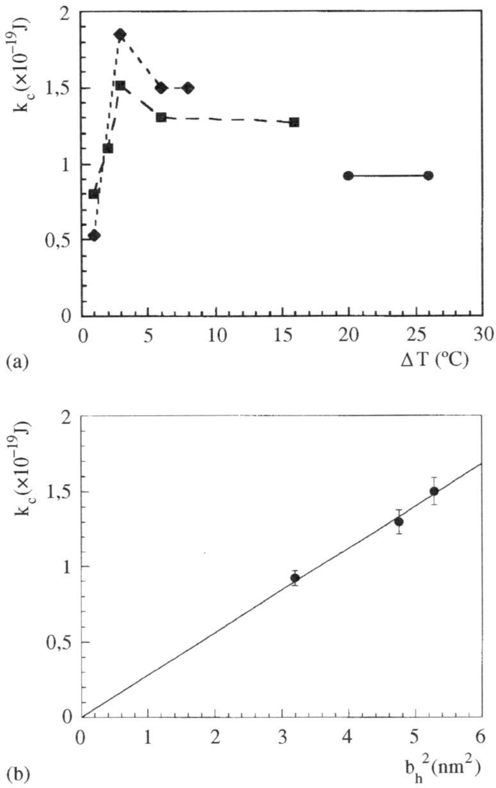


**Figure 13.4** (a) Bending elasticity measurements $k_c$ for DLPC (●), DMPC (■) and DPPC (◆) as a function of the temperature difference to the main phase transition, $\Delta T = T - T_m$. (b) Comparison of the bending elastic moduli $k_c$, for DLPC, DMPC and DPPC in the high-temperature region as a function of the square of the hydrophobic length (adapted from Reference 40).

temperature when $T$ is larger than $\sim T_m + 6°C$, contrary to the lower temperature region [40]. When comparing the bending moduli of DLPC, DMPC, and DPPC bilayers in this high-temperature region as a function of the square of their hydrophobic chain length, a linear dependence is obtained, Figure 13.4(b) [40]. This behavior can be interpreted by referring to molecular models [42]. For one-component bilayers, the bending elasticity $k_c$ can be simply expressed as a function of the lipid area $a$ and some bilayer thickness $d$ by:

$$k_c \approx \frac{d^2}{a}$$

Intrinsically, $d$ is not equal to the hydrophobic thickness of the bilayer $b_h$. It is related to the distance spanning between the neutral surfaces close to the interfacial regions. It is very close to $b_h$ for long enough lipid molecules. As for saturated PCs the molecular area seems to be almost independent of the chain length in the fluid state [11], a linear increase of $k_c$ with $b_h^2$ was obtained (Figure 4b). It can be noted that such a simple scale dependence of $k_c$ as a function of the membrane thickness is the first clear evidence that mechanical properties are closely related to the molecular dimensions.

Most studies concerning natural systems have to be done in solutions at controlled pH and ionic strength. This is necessary to limit protein denaturation or osmotically induced stress of the cell membrane. For model systems like closed liposomes, one has also to maintain some osmotic equilibrium to ensure mechanical stability (see Section 3.2). The pH solution also has to be controlled when proteins are incorporated. This is why giant vesicles are produced in a buffer solution (Tris 1 mmol $l^{-1}$, pH = 7.4). As it can be seen in Table 1, such an environmental change strongly increases the bending elasticity of SOPC bilayers by about 40% compared to deionized water (pH ~ 5.5). This is surprising because the pH change in the bulk (from 5.5 to 7.4) remains very small by comparison to published p$K$ data for phosphatidylcholines in a bilayer [43] and cannot change the surface charge density of the lipid bilayer. Therefore, one cannot expect any contribution from such charge

**Table 1** Bending elasticities of different bilayers at room temperature. Bilayer mixtures are usually studied in deionized water at equilibrium with the normal $CO_2$ pressure concentration (pH ~ 5.5).

| Bilayer composition (mol % ratio)[a] | $k_c (\times 10^{-19}$ J) |
|---|---|
| DLPC | $0.91 \pm 0.05$ |
| DLPC/DLPA (98/2) | $0.92 \pm 0.05$ |
| DLPC/DLPA (90/10) | $1.05 \pm 0.06$ |
| DLPC/DLPA (80/20) | $1.06 \pm 0.06$ |
| SOPC | $1.27 \pm 0.07$ |
| SOPC[b] | $1.81 \pm 0.08$ |
| SOPC/gramicidin D (99.5/0.5)[b] | $1.32 \pm 0.06$ |
| SOPC/gramicidin D (97/3)[b] | $1.38 \pm 0.02$ |
| SOPC/gramicidin D (95/5)[b] | $1.51 \pm 0.03$ |
| EPC (1 day) | $0.66 \pm 0.06$ |
| DGDG | 0.08–0.1 |

[a]DLPC = dilauroylphosphatidylcholine, DLPA = dilauroylphosphatidic acid, SOPC = stearoylolephophosphatidylcholine, EPC = egg yolk phosphatidylcholine, DGDG = digalactosyldiacylglycerol from whole wheat flour.

[b]bilayers have been studied in a buffer solution (pH ~ 7.4).

effects to the bending stiffness (see the following discussion about DLPC–DLPA mixtures). However, this pH change can induce some subtle change within the hydrophilic region such as some reorientation of the phosphocholine head group. Another possible explanation might be that Tris buffer shows some interaction with the bilayer. If such an assumption is true, the measured effect should be dependent on the Tris concentration or on the pH in the buffer solution (these measurements are in progress).

### *(b) Two-component Bilayers*

Evidently, natural membranes are complex systems where the perturbation due to a given lipid or protein on the overall behavior is hardly distinguishable. To understand how the chemical nature of different membrane molecules governs the mechanical stability of 2-dimensional structures, systematic measurements were made of the bending elasticity of bilayers as a function of their composition. Four perspectives were chosen. First, the role of sterol was investigated by studying DMPC–cholesterol mixtures as a function of the temperature and the sterol content. The resulting behavior was compared with that of pure DMPC and that of red blood cell lipid extracts. Second, the influence of the surface charge density was studied on mixtures containing a synthetic phosphatidylcholine and the corresponding phosphatidic acid (DLPC–DLPA mixtures). Third, the role of hydrophobic chain polydispersity for natural lipid extracts from different origins was investigated. Finally, some preliminary results were also given on the influence of gramicidin D, a natural hydrophobic peptide from *B. brevis* [44], on the bending elasticity of stearoyloleoylphosphatidylcholine (SOPC) bilayers characterized by a monolayer hydrophobic thickness slightly larger than that of gramicidin. All these studies indicated that some correlations can be made between the mechanical behavior and the chemical structure of the molecules within the bilayer.

Mechanical properties of cholesterol containing bilayers have already been studied extensively by several authors. It is known that cholesterol increases strongly the stretching elasticity [45–47]; some information exists about its influence on bending elasticity [48]. A recent report discusses the role of cholesterol content and temperature on the bending elasticity of DMPC bilayers, which were extensively examined and compared to lipid extracts from red blood cell membranes [39]. The main results are presented in Figure 13.5. At low sterol content (molar ratio $\leq 10\%$) and when $T \geq T_m$, DMPC–Chol mixtures behave similarly to pure DMPC bilayers, particularly close to the main phase transition temperature $T_m$. At higher cholesterol content (molar ratios of 30% and 50%), $k_c$ increases continuously when $T$ decreases and the values of $k_c$ can be obtained for $T \leq T_m$. This can be understood by referring to the phase behavior of DMPC–Chol systems [49]. At 30 mol % in cholesterol and in the temperature range studied herein, DMPC–Chol forms a $\beta$-phase that is also known as a liquid ordered phase (explaining the existence of thermal fluctuations

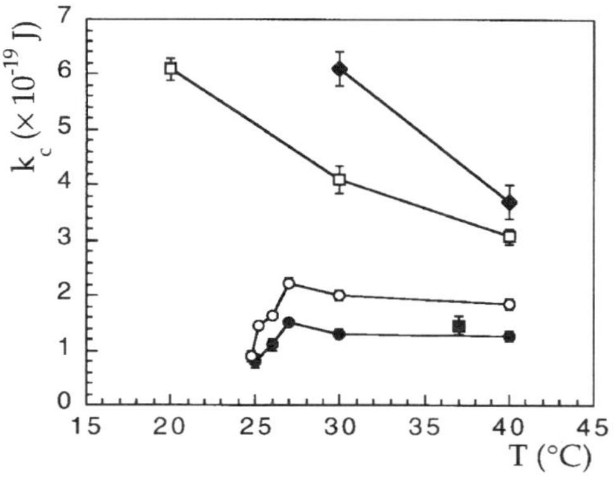


**Figure 13.5** Bending elasticity $k_c$ of cholesterol containing membranes as a function of the sterol content and the temperature $T$: (●) DMPC bilayers; (○) 10 mol % cholesterol in DMPC; (□) 30 mol % cholesterol in DMPC; (◆) 50 mol % cholesterol in DMPC, and (■) total lipid extracts of red blood cell membrane at 37°C. Lines are drawn to help reading the figure (adapted from Reference 39).

and thus the measurement of $k_c$). Interestingly, the total extracts from red blood cell membranes are characterized by a very low bending constant (Figure 5), comparable to that of pure DMPC bilayers in the same temperature region. One can argue that a high cholesterol ratio ($\sim$40%) is maintained to regulate the red blood cell membrane thermotropism (due to the high sphingomyelin content [50]) but not for mechanical reasons [39].

The role of surface charge on the mechanical behavior of model membranes has been studied on DLPC–DLPA mixtures. In that case, the perturbation is introduced by a variable surface charge that depends on the PA molecular ratio and on the environmental characteristics of the surrounding medium while the lipid region of the bilayer is not changed. As seen from Table 1, the bending modulus of the DLPC–DLPA mixtures increase with the percentage of PA within the bilayer. This behavior agrees with the theoretical works reviewed by Andelman [51] that predict an increase for $k_c$ whose rate has to depend on the Debye–Hückel screening length $\lambda_D$ and on the surface charge density $\chi$. Nevertheless, these results show only a 15% augmentation in the saturation region. This is definitely too small by comparison to what is expected for the calculated $\lambda_D$ (if the only ions present in the medium are $HCO_3^-$ and $H^+$) and $\chi$ (deduced from fitting procedures and taking the published acidic constant for PA lipids [43]).

The bending elasticities of membranes containing egg-yolk phosphatidylcholine (EPC) and digalactosyl diglyceride (DGDG) from wheat flour were also studied. In both cases, the bilayer is an extract of naturally occurring chain mixture with a chemically defined head group. The results are summarized in Table 1. It is interesting to compare the bending elasticity of freshly prepared EPC vesicles

($k_c \approx 0.7 \times 10^{-19}$ J [38]) to synthetic lipid with a similar hydrophobic length (for DPPC bilayers at high temperature $k_c \approx 1.5 \times 10^{-19}$ J [40], and for SOPC bilayers in deionized water $k_c \approx 1.3 \times 10^{-19}$ J). EPC bilayers are much less rigid than the synthetic DPPC and SOPC membranes. For DGDG bilayers (Table 1), the bending elasticity of synthetic lipids with the same head region have not yet been measured. Nevertheless, $k_c$ is in that case very small, its modulus being in the same range as that of mixed surfactant systems [52]. Natural membrane lipids such as EPC are characterized by aliphatic chains with different lengths and degrees of unsaturations from one molecule to the other. As a consequence, the lipid mixture has to be characterized by a higher disorder, resulting in a smaller aliphatic packing density. As demonstrated previously when comparing DLPC, DMPC, and DPPC at temperature higher than the corresponding $T_m$, $k_c$ is strongly dependent of the characteristics of this aliphatic region. Consequently, this change in the packing density leads to a decrease of $k_c$. This effect is strongly enhanced for DGDG extracts due to its unusual high content in polyunsaturated fatty acids where $\alpha$-linolenic acid represents the main part [53].

Finally, this presentation of the relationship between bending elasticites and membrane composition concludes with a study of SOPC bilayers containing different amounts of gramicidin D. Gramicidin D is a mixture containing mainly three different peptides with a very similar structure [44]. Gramicidin is a 15-hydrophobic amino acid. It can arrange as a double-stranded helix or $\beta$-helix depending on the experimental conditions [44]. Herein gramicidin adopts the $\beta$-helix conformation and inserts directly into the aliphatic region of the membrane, its length being comparable to half the hydrophobic length of natural membranes. Its activity as a cation-selective transmembrane channel is known to be due to short-time dimerization of gramicidin molecules located in two monolayers of the same bilayer [44]. Due to its hydrophobic nature, gramicidin is first cosolubilized in chloroform–methanol following a procedure described elsewhere [39]. We succeeded in making giant vesicles containing different amounts of gramicidin in SOPC bilayers. It was observed that the bending rigidity of SOPC bilayers is first modified by a very small amount of gramicidin (1 mol %, Table 1). However, at higher molar ratios (3 and 5 mol %), $k_c$ is practically constant (Table 1). Unfortunately, vesicles do not form for peptide ratios higher than 5 mol %. This might be due to the appearance of the inverted hexagonal phase occurring at larger peptide ratio [44], that may be induced by a hydrophobic length for SOPC monolayers [11] larger than the hydrophobic thickness of gramidicin molecules [44]. Nevertheless it is clear from this study that the bending elasticity of SOPC bilayers is not strongly modified by the addition of gramicidin. On one hand, such behavior is unexpected precisely because gramicidin has to modify the packing arrangement of the lipids to induce $H_{II}$ formation, although these experiments were made at concentrations below that necessary for $H_{II}$ formation. On the other hand, the observation that a transmembrane protein has little influence on the membrane mechanical properties can be reasonably anticipated.

## 5.2 Vesicle Shape Transformations Induced by Alternating Electric Fields

The previous section shows that giant vesicles are easy to deform due to the very low bending elasticity of their bilayers ($\sim 2k_BT \leq k_c \leq \sim 150k_BT$, where $k_B$ is the Boltzmann constant and $T$ is the temperature). Consequently, they are highly susceptible to any perturbation in the environment, such as Brownian motion, solvent flow, or mechanical shock.

Vesicles interact also with electrical fields [54]. Experimentally, it is found that giant liposomes behave differently depending on the frequency of the applied electric field [55] as shown in Figure 13.6. This is due to a change of the water behavior with respect to the electric field frequency $\nu$. At frequency smaller than a critical threshold $\nu_c$ water is a conductive material, whereas at higher frequency it becomes a dielectric. At $\nu < \nu_c$, the electric problem is then equivalent to a dielectric thin shell in a conductive media, whereas the electric field propagates through a dielectric bilayer in another dielectric media (water) when $\nu > \nu_c$. If $\kappa_w$ is the conductivity of deionized water ($\sim 5 \times 10^{-6}\,\mathrm{S\,m^{-1}}$) and $\varepsilon_w = \varepsilon_r\varepsilon_0$ is the water dielectric constant in the low frequency range ($\sim 6.9 \times 10^{-10}\,\mathrm{C^2\,J^{-1}m^{-1}}$), the critical frequency, $\nu_c = \kappa_w/2\pi\varepsilon_w$, was experimentally and theoretically found to be close to 1 kHz.

Thus, a general solution explaining how the electric field interacts with the vesicle wall is not easy to find due to the strong difference in the water nature around $\nu_c$. It has lead to independent theoretical approaches for $\nu < \nu_c$, [31,32] and for $\nu > \nu_c$, [55], where the bending elasticity is the only mechanical property that controls the vesicle deformation in an electric field. This is true because (i) the electric energy is much smaller than the energy required to increase the vesicle area, and (ii) the vesicle does not generally change its genus during the deformation (Figure 13.6). Consequently, the electromechanical model describing the interaction takes into account the electrical properties of the surrounding water and the membrane, the bending elasticity $k_c$ of the bilayer, and the vesicle geometry, i.e. the area to volume ratio [31,32,54,55].

In the low frequency regime (at $\nu = 100\,\mathrm{Hz}$), vesicles generally behave as in Figure 13.6(b). However, when applying high-amplitude electric fields (square waves at $20\,\mathrm{kV\,m^{-1}}$), some vesicles (not all of them) orient in the electric field direction then open at the poles (Figure 13.7), similarly to previously published results [56]. These passages appear at a speed that depends on the amplitude of the applied electric field. Their final size is vesicle dependent, i.e. it is a function of the apparent bilayer area. It is also found that the passage size becomes smaller if the amplitude of the electric field is reduced (not shown). They finally close (at microscope resolution) under a critical amplitude (or if the electric field is switched off). This phenomenon is not universal as neighboring vesicles with similar size, lamellarity and shape do not open.

One can interpret the images shown in Figure 13.7 as a topology change of the vesicle when submitted to high intensity electric fields. Such interpretation was

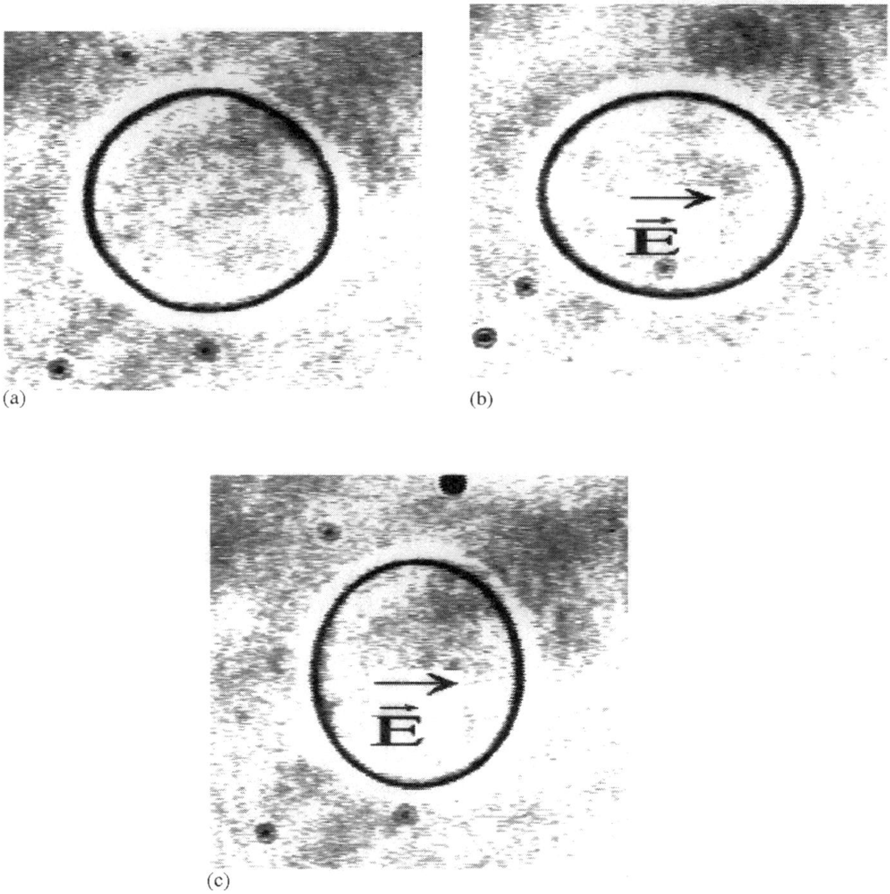


**Figure 13.6** (a) A fluctuating liposome (radius ~ 18.1 μm) deforms itself into (b) a prolate ellipsoid when the water surrounding the bilayer can be considered as a conducting medium (low frequency regime), whereas it transforms into (c) an oblate ellipsoid when the water is a dielectric (high-frequency regime). The frequency threshold characterizing this prolate to oblate transformation when a liposome interacts with an electric field is experimentally found to be close to 1 kHz for dionized water (pH ~ 5.5).

proposed by Harbich and Helfrich to analyze the vesicle transformation and its equilibrium shape [56]. Following this approach, we have tried to find how the different mechanical properties of such a liposome can control the shape observed at given electric field amplitude. Developing a theoretical model including the bending moduli $k_c$ and $\bar{k}_c$ and the line energy $\gamma$ and assuming the local electric energy per unit area to be proportional to $n_z^2$ ($\vec{n}$ being the local bilayer normal), the differential

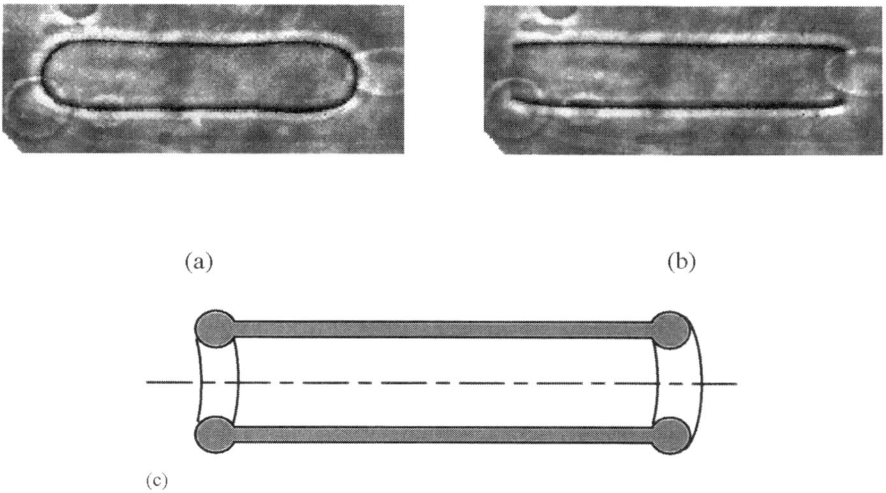


**Figure 13.7** Two shapes of the same SOPC giant vesicle (a) in the absence and (b) in the presence of a horizontal electric field (100 Hz, 20 kV m$^{-1}$). The long axis is close to 50 μm. (c) Diagram of the model we use to analyze the shape of open vesicles.

equations that describe the observed situation do not lead to reasonable values for the two unknown mechanical parameters $\bar{k}_c$ and $\gamma$ (results not shown). This is particularly true for the $\gamma$ value of $\sim 2 \times 10^{-14}$ J m$^{-1}$, compared to the experimental measurements from Zhelev and Needham ($\gamma = (0.9\text{–}3) \times 10^{-11}$ J m$^{-1}$ [35,36]). Consequently, the experimental shapes needed re-examination, such as those shown in Figure 13.7. It is well-known that pore formation has to occur at the pole positions where the voltage difference through the bilayer is the highest [57,58]. However, it was found that passage formation does not appear systematically at a pole position. It can appear anywhere on the bilayer if the amplitude of the electric field is reduced to the value inducing the opening of a passage (not shown). This indicates that the vesicle opening might be due to a preexisting defect, such as a connection between two neighboring bilayers (Figure 13.2(b)). In that case, the electric field orients first the vesicle bilayer. Then, the defects have to move to the poles where the obstruction of the current lines is reduced (Figure 13.7(c)). Following this ideal sketch, it was concluded that the central part of the vesicle is essentially two tubular bilayers with the same axis, the openings resembling to tori connecting the tubular bilayers (Figure 13.7(c)). The water thickness $e$ that separates the two bilayers and the smaller radius of the torus $r$ are related due to (i) the interactions in the central part of the vesicle and (ii) the curvature energy stored in the torus regions. The interactions originate from thermal fluctuations. The concentric bilayers in the middle of the vesicle still fluctuate even if these undulations are not visible with a microscope. A reduction of $e$ increases the pressure exerted by one bilayer on the other as previously proposed by Helfrich [55]. Consequently, the bilayer separation is given by the equilibrium of attractive van der Waals forces and repulsive undulation forces. The attractive part is a material feature whose amplitude is given by the Hamacker constant, whereas the repulsions depend on the bending elasticity $k_c$ (explaining the limited swelling of PC multilayer systems when the bending elasticity is larger than $k_B T$ [59]). Strictly speaking, $r$ and $e$ are given by the minimization of the vesicle free energy at fixed total area and entrapped volume, i.e. the interaction energy in the tubular part and the curvature energy of the vesicle (at both ends and in the tubular region). However, $r$ is sufficiently small in these experiments to remain unresolved with a phase-contrast microscope. Using a simple model, where $e$ is assumed to be close to the water thickness ($\sim 25$ Å), $r$ is roughly given by the water volume entrapped by the vesicle wall. The vesicle area is also roughly twice the area of one of the tubular bilayer. This leads to an entrapped volume in the order of 1 $\mu m^3$ and $r \leq 0.1$ μm.

## 6. CONCLUSIONS

This paper shows how the mechanical properties of model membranes can be measured by the study of giant vesicle deformations. This presentation is by no means exhaustive. Other studies demonstrate how closed objects, such as giant vesicles, can change their shape as a function of a precise tuning of their geometric characteristics and their molecular distribution [60–63].

The shape transformations of giant vesicles can result from spontaneous deformation (thermal fluctuations), from interactions with a glass pipette (micromanipulation experiments) or alternating electric fields. The mechanical parameters that characterize such morphological changes are strongly dependent on the bilayer content. This is especially the case for $k_c$ which was found to vary from $2\,k_B T$ to $150\,k_B T$. However, the results obtained on model systems have to be carefully interpreted when compared to natural membranes. This is true for DMPC bilayers containing cholesterol at a ratio similar to that in mammal membranes when $k_c$ is related to that of red blood cell lipid extracts. Most studies do not consider the contribution of the membrane proteins on the mechanical stability of the bilayer. Finally, all the experiments presented have used vesicles with a symmetric bilayer of controlled composition. Thus, their monolayers are chemically identical and the inside and outside aqueous solutions are the same, these conditions being achieved during the vesicle formation. Most of the natural membranes, however, work in an asymmetric environment where both monolayers are composed of different constituents. (The consequences of induced asymmetry on the giant vesicle morphology have been investigated by Devaux and co-workers [64,65]).

The interaction of electric fields with membranes is interesting mainly for potential applications of electroporation and electropermeation for cell transfection [57,58]. The observations on large-scale openings through giant vesicles is similar to that of Harbich and Helfrich, but the interpretation is in contradiction with theirs [56]. These authors proposed that such shapes result from pore formation. In that case, the deformation induced by the electric field is dependent of the bending elasticity $k_c$, the Gaussian curvature elasticity $\bar{k}_c$ and the line energy $\gamma$. In our interpretation, the final shape has the symmetry of a torus (genus = 1) with a very small volume compared to the total area. It remains a closed bilayer object during the application of the electric field. The induced deformation is then controlled by the single bending elasticity $k_c$ (through the curvature energy of the vesicle and the interaction energy) and a material feature (the lipid Hamacker constant, [59]). Although very crude, this model suggests the interpretation to be possible. The transmembrane voltages used are in the same range than those currently applied in electrofusion studies [58]. Thus, one can not reject a preceding fusion process in some bilamellar vesicles, thereby creating the necessary passages that are seen under a microscope when applying the electric field.

## 7. ACKNOWLEDGMENTS

This work was the result of a collaboration within the French Bulgarian Laboratory, supported by the Centre National de la Recherche Scientifique and the Bulgarian Academy of Science. It was also sponsored by grant 7BUPJ048478 from the Swiss National Science Foundation.

## 8. REFERENCES

1. P. B. Canham, *J. Theoret. Biol.*, **26**, 61 (1970).
2. W. Helfrich, *Z. Naturforsch.*, **28c**, 693 (1973).
3. E. A. Evans, *Biophys. J.*, **13**, 926 (1973).
4. F. Brochard and J.-F. Lennon, *J. Physique*, **36**, 1035 (1975).
5. E. Evans, *Biophys. J.*, **43**, 27 (1983).
6. E. Evans and A. Yeung, *Biophys. J.*, **56**, 151 (1989).
7. A. Yeung and E. Evans, *Biophys. J.*, **56**, 139 (1989).
8. H. Strey, M. Peterson and E. Sackmann, *Biophys. J.*, **69**, 478 (1995).
9. M. Bloom, E. Evans and O. G. Mouritsen, *Quart. Rev. Biophys.*, **24**, 293 (1991).
10. R. Lipowsky and E. Sackmann (eds), *Structure and Dynamics of Membranes*, Elsevier, Amsterdam, 1995.
11. D. Marsh, *Handbook of Lipid Bilayers*, CRC Press, Boca Raton, 1990.
12. E. Evans and D. Needham, *J. Phys. Chem.*, **91**, 4219 (1987).
13. E. Evans and R. Skalak, *Mechanics and Thermodynamics of Biomembranes*, CRC Press, Boca Raton, 1980.
14. E. Evans and S. Simon, *J. Colloid Interface Sci.*, **51**, 266 (1975).
15. D. Needham and D. V. Zhelev, *Annals Biomed. Engineering*, **25**, 287 (1995).
16. D. V. Zhelev, *Biophys. J.*, **71**, 257 (1996).
17. G. Darboux, *Leçons sur la théorie générale des surfaces et les applications géométriques*

*du calcul infinitésimal. Third edition*, Chelsea Publishing Company, Chelsea, 1972.
18. R. Kwok and E. Evans, *Biophys. J.*, **35**, 637 (1981).
19. M. D. Mitov, J.-F. Faucon, P. Méléard and P. Bothorel, in *Advances in Supramolecular Chemistry*, vol. II, (ed.) G. W. Gokel, Jai Press, Greenwich, 1992, p. 93.
20. R. M. Servuss, W. Harbich and W. Helfrich, *Biochim. Biophys. Acta*, **436**, 900 (1976).
21. M. B. Schneider, J. T. Jenkins and W. W. Webb, *J. Physique*, **45**, 1457 (1984).
22. H. Engelhardt, H. P. Duwe and E. Sackmann, *J. Phys. Lett.*, **46**, L395 (1985).
23. S. T. Milner and S. A. Safran, *Phys. Rev. A*, **36**, 4371 (1987).
24. J. F. Faucon, M. D. Mitov, P. Méléard, I. Bivas and P. Bothorel, *J. Physique*, **50**, 2389 (1989).
25. P. Méléard, J.-F. Faucon, M. D. Mitov and P. Bothorel, *Europhys. Lett.*, **19**, 267 (1992).
26. M. I. Angelova, S. Soléau, P. Méléard, J.-F. Faucon and P. Bothorel, *Springer Proc. Phys.*, **66**, 178 (1992).
27. L. Bo and R. E. Waugh, *Biophys. J.*, **55**, 509 (1989).
28. R. E. Waugh, J. Song, S. Svetina and B. Zeks, *Biophys. J.*, **61**, 974 (1992).
29. V. Heinrich and R. E. Waugh, *Annals of Biomedical Engineering*, **24**, 595 (1996).
30. E. Evans and W. Rawicz, *Phys. Rev. Lett.*, **64**, 2094 (1990).
31. M. Kummrow and W. Helfrich, *Phys. Rev. A*, **44**, 8356 (1991).
32. G. Niggemann, M. Kummrow and W. Helfrich, *J. Phys. II France*, **5**, 413 (1995).
33. D. Needham and D. Z. Zhelev, in *Vesicles*, (ed.) M. Rosoff, Marcel Dekker, New York, 1996, p. 373.
34. E. Evans and R. Kwok, *Biochemistry*, **21**, 4874 (1982).
35. D. V. Zhelev and D. Needham, *Biochim. Biophys. Acta*, **1147**, 89 (1993).
36. J. D. Moroz and P. Nelson, *Biophys. J.*, **72**, 2211 (1997).
37. W. S. Singleton, M. S. Gray, M. L. Blown and J. L. White, *J. Am. Oil Chem. Soc.*, **42**, 53 (1965).
38. M. I. Angelova, S. Soléau, P. Méléard, J.-F. Faucon and P. Bothorel, *Prog. Colloid Polym. Sci.*, **89**, 127 (1992).
39. P. Méléard, C. Gerbeaud, T. Pott, L. Fernandez-Puente, I. Bivas, M. D. Mitov, J. Dufourcq and P. Bothorel, *Biophys. J.*, **72**, 2616 (1997).
40. L. Fernandez-Puente, I. Bivas, M. D. Mitov and P. Méléard, *Europhys. Lett.*, **28**, 181 (1994).
41. T. Hönger, K. Mortensen, J. H. Ipsen, J. Lemmich, R. Bauer and O. G. Mouritsen, *Phys. Rev. Lett.*, **72**, 3911 (1994).
42. A. G. Petrov and I. Bivas, *Prog. Surf. Sci.*, **16**, 389 (1984).
43. J. F. Tocanne and J. Teissié, *Biochim. Biophys. Acta*, **1031**, 111 (1990).
44. J. A. Killian, *Biochim. Biophys. Acta*, **1113**, 391 (1992).
45. E. Evans and D. Needham, *Faraday Discuss. Chem. Soc. USA*, **81**, 267 (1986).
46. D. Needham, T. J. McIntosh and E. Evans, *Biochemistry*, **27**, 4668 (1988).
47. D. Needham and R. S. Nunn, *Biophys. J.*, **58**, 997 (1990).
48. H. P. Duwe, J. Käs and E. Sackmann, *J. Physique*, **51**, 945 (1990).
49. P. F. F. Almeida, W. L. C. Vaz and T. E. Thompson, *Biochemistry*, **31**, 6739 (1992).
50. Y. Barenholz and T. E. Thompson, *Biochim. Biophys. Acta*, **604**, 129 (1980).
51. D. Andelman, in *Structure and Dynamics of Membranes*, (eds) R. Lipowsky, and E. Sackmann, Elsevier, Amsterdam, 1995, p. 603.
52. F. Nallet, D. Roux and J. Prost, *J. Phys. France*, **50**, 3147 (1989).
53. F. D. Gunstone, J. L. Harwood and F. B. Padley, *The Lipid Handbook, Chapman and Hall*, London, 1986.
54. M. Winterhalter and W. Helfrich, *J. Colloid Interface Sci.*, **122**, 583 (1988).
55. M. D. Mitov, P. Méléard, M. Winterhalter, M. I. Angelova and P. Bothorel, *Phys. Rev. E*, **48**, 628 (1993).

56. W. Harbich and W. Helfrich, *Z. Naturforsch.*, **34a**, 1063 (1979).
57. E. Neumann, A. E. Sowers and C. A. Jordan, *Electroporation and Electrofusion in Cell Biology*, Plenum Press, 1989.
58. M. P. Rols and J. Teissié, *Bioelectrochem. Bioenerg.*, **24**, 101 (1990).
59. W. Helfrich and S. R.M., *Nuovo Cimento*, **3D**, 137 (1984).
60. J. Käs and E. Sackmann, *Biophys. J.*, **60**, 1 (1991).
61. X. Michalet and D. Bensimon, *J. Phys. II France*, **5**, 263 (1995).
62. U. Seifert, *Advances in Physics*, **46**, 13 (1997).
63. H. G. Döbereiner, E. Evans, M. Kraus, U. Seifert and M. Wortis, *Phys. Rev. E*, **55**, 4458 (1997).
64. E. Farge and P. F. Devaux, *Biophys. J.*, **61**, 347 (1992).
65. L. Mathivet, S. Cribier and P. F. Devaux, *Biophys. J.*, **70**, 1112 (1996).
66. J. M. Seddon, *Biochim. Biophys. Acta*, **1031**, 1 (1990).

# Chapter 14

# Mechanical Properties of Lipid Bilayers Containing Grafted Lipids

*ISAK BIVAS, VICTORIA VITKOVA AND MARIN D. MITOV*
*Bulgarian Academy of Sciences, Sofia, Bulgaria*

*MATHIAS WINTERHALTER*
*Biozentrum, Basel, Switzerland*

*ROSSITSA G. ALARGOVA*
*University of Sofia, Bulgaria*

*PHILIPPE MÉLÉARD AND PIERRE BOTHOREL*
*Centre de Recherche Paul Pascal, Pessac, France*

## 1. INTRODUCTION

One of the most promising applications of the liposomes is their use as drug carriers [2–4]. Usually, when drug carrying liposomes circulate in a blood stream, they are taken up by the mononuclear phagocitic system in several minutes. Their circulation time could be increased by adding modified (PEGylated, grafted, stealth – named after the military plane because the supplied vesicles, circulating in the blood stream, are invisible to the immune system) lipids to their lipid membrane. These are ordinary lipids with a polymer chain (usually poly(ethylene glycol), PEG), cova-

*Giant Vesicles* Edited by P. L. Luisi and P. Walde

lently attached to their hydrophilic heads. The polymer layer screens the vesicles and increases their average lifetime up to several days. The modified lipids change the mechanical properties of the membranes. The aim of the present work is to investigate both theoretically and experimentally the mechanical properties of a lipid bilayer with modified lipids. To achieve this aim, phenomena related to the out-of-plane fluctuations of the membranes of giant vesicles were exploited.

In what follows, a lipid bilayer is discussed, with or without modified lipids in its matrix and only in its liquid crystal state, so that it can be considered as a two-dimensional liquid. The mechanical properties can thus be characterized by stretching and bending elastic moduli. Let $S_0$ be the area of a flat tension-free membrane. If $\Delta S$ is the change of $S_0$, the density of the stretching elastic energy $g_s$ and its tension $\sigma(\Delta S)$ is then defined as

$$\begin{aligned} \sigma(\Delta S) &= k_s \frac{\Delta S}{S_0} \\ g_s &= \frac{1}{2} k_s \times \left( \frac{\Delta S}{S_0} \right)^2 \end{aligned} \tag{1}$$

where $k_s$ is the stretching elasticity modulus of the membrane.

If the tension-free membrane is bent, its shape can be characterized locally by its principal curvatures, $c_1$ and $c_2$. Then the density of the curvature elastic energy $g_c$ is given by [5]

$$g_c = \frac{k_c}{2}(c_1 + c_2 - c_0)^2 + \bar{k}_c c_1 c_2 \tag{2}$$

where $k_c$ and $\bar{k}_c$ are the bending and saddle bending elastic moduli and $c_0$ is the spontaneous curvature of the bilayer. Symmetrical membranes with $c_0 = 0$ are considered later.

Two different bending elastic moduli exist: $k_c^{fr}$, when the exchange of lipid molecules between the monolayers of the bilayer is free, and $k_c^{bl}$, when it is blocked. When the exchange is forbidden, the number of the molecules in each monolayer of the bilayer is constant. At free flip-flop, the bending elasticity energy is lower because it has been minimized with respect to the difference between the number of molecules in each monolayer and, consequently, $k_c^{fr} < k_c^{bl}$. For all phenomena related to the out-of-plane fluctuations of membranes, the relevant quantity is $k_c^{fr}$ [6–8]. These phenomena include the thermal fluctuations of quasispherical vesicles [9,10], as well as vesicle suction in micropipettes at very low suction pressures [1].

## 2. THEORETICAL DESCRIPTION OF THE MECHANICAL PROPERTIES OF LIPID BILAYERS CONTAINING GRAFTED LIPIDS

For the theoretical description of the free energy of the PEG-chains of any modified lipid, the results of Milner *et al.* [11,12] were used. Assume that identical polymer chains are grafted on a flat surface, and let $f_p(s_p, N)$ be the free energy per chain in the brush regime at mean surface area $s_p$ and the number of segments per chain be $N$. If $k_B$ denotes the Boltzmann constant, $T$ is the absolute temperature, $a$ is the length of a single segment of the polymer chain, and $\xi$ is the persistence length of the polymer chain, the free energy $f_p(s_p, N)$ can be expressed in the following way [11,12]

$$f_p(s_p, N) = \frac{9}{10}\left(\frac{\pi^2}{12}\right)^{1/3} N k_B T \frac{a^2 \xi^{-2/3}}{(s_p)^{2/3}} \tag{3}$$

If a layer containing grafted polymers is curved as a part of a cylindrical surface with curvature $c$, the free energy per chain $F_p(s_p, c, N)$ is

$$F_p(s_p, c, N) = f_p(s_p, N)\left\{1 - \frac{5}{24}h(s_p, N)c + \frac{5}{64}[h(s_p, N)]^2 c^2 + \cdots\right\} \tag{4}$$

where $h(s_p, N)$ is the thickness of the polymer layer, given by

$$h(s_p, N) = \left(\frac{12}{\pi}\right)^{1/3} N \frac{a\xi^{2/3}}{(s_p)^{1/3}} \tag{5}$$

As previously shown [13,14], the results of Milner *et al.* [11,12] can be used for calculating the stretching $K_s(C, N)$, bending $K_c^{fr}(C, N)$, and saddle bending elasticities $\bar{K}_c(C, N)$ of a lipid bilayer, containing a modified lipid with molar concentration $C$ and number of monomers per polymer chain $N$. In these theoretical investigations it was assumed that the lipid part of the modified lipid molecules is identical to that of the basic lipid. It was also supposed that the effects of protrusion of the molecules of the modified lipid are negligible [15]. Retain the symbols $k_s$, $k_c^{bl}$, and $k_c^{fr}$ for the elastic moduli of the lipid bilayer without modified lipid. Let $s_0$ be the mean area per molecule in the monolayer of a flat tension-free bilayer without PEGylated lipid and $s(C, N)$ – the mean area per lipid molecule (including modified lipid) in the decorated bilayer. Then $K_s(C, N)$ and $s(C, N)$ satisfy the equations

$$\left[\frac{K_s(C, N)}{k_s} - 1\right]\left[\frac{3}{8}\frac{K_s(C, N)}{k_s} + \frac{5}{8}\right]^{5/3} = \frac{16}{5}\left(\frac{\pi^2}{12}\right)^{1/3} \frac{k_B T N a^2 \xi^{-2/3}}{k_s (s_0)^{5/3}} C^{5/3} \tag{6}$$

$$\frac{s(C, N) - s_0}{s_0} = \frac{3}{8}\left[\frac{K_s(C, N)}{k_s} - 1\right] \tag{7}$$

The bending elastic modulus at free flip-flop, $K_c^{fr}(C, N)$, depends on the elastic moduli $k_c^{\rm fr}$, $k_c^{\rm bl}$ and $k_s$ of the lipid bilayer without modified lipid and on the thickness $d_{\rm ch}$ of the hydrophobic core of the monolayer of the flat tension-free bilayer without modified lipid:

$$K_c^{\rm fr}(C, N) = \frac{s_0}{s(C, N)} k_c^{\rm bl} + \left[\frac{s_0}{s(C, N)}\right]^3 (k_c^{\rm fr} - k_c^{\rm bl}) + k_s \frac{[s(C, N) - s_0](s_0)^2}{[s(C, N)]^3} d_{\rm ch}\left[d_{\rm ch} - 2\varepsilon\sqrt{\left(\frac{k_c^{\rm fr} - k_c^{\rm bl}}{k_s}\right)}\right] \tag{8}$$

where $\varepsilon$ is equal to $+1$ or $-1$ depending on the transport direction of the molecules between the monolayers of a cylindrically curved bilayer when flip-flop is permitted. If $\varepsilon = +1$, the molecules move from the monolayer whose hydrophobic heads are nearer to the center of the curvature to the other monolayer (equivalently, they move within one monolayer from the higher-density to the lower-density polymer chain regions). On the contrary, when $\varepsilon = -1$, the molecular flow between the monolayers is in the opposite direction.

The saddle-bending elasticity is [14,16]

$$\bar{K}_c(C, N) = \frac{s_0}{s(C, N)} \bar{k}_c - \frac{6}{35}\left(\frac{12}{\pi^2}\right)^{1/3} k_B T \frac{a^4 \xi^{2/3} N^3}{[s(C, N)]^{7/3}} C^{7/3} \tag{9}$$

where $K_s(C, N)$ and $K_c^{\rm fr}$ depend only on the product $CN^{3/5}$. At a given value of $C$, $\bar{K}_c(C, N)$ is proportional to $N^3$, which could lead to negative values of great magnitude. One of the criteria for stability of the lamellar phase is

$$2K_c^{\rm fr}(C, N) + \bar{K}_c(C, N) \geq 0 \tag{10}$$

which means that the flat membrane is energetically more favorable in comparison to a vesicles' suspension. For high enough values of $C$ and/or $N$ this inequality is not valid and the lamellar phase, built of such a lipid system, will be destabilized.

## 3. SOME LIMITATIONS OF THE MICROPIPETTE METHOD FOR DETERMINATION OF THE ELASTIC MODULI OF A LIPID BILAYER

There are several methods for the experimental measurement of $k_c^{\rm fr}$. One of these, [1] consists of determination of the effective stretching modulus of a membrane fluctuating at very low tensions. The principles of the method are shown on Figure 14.1.

A vesicle of radius $R_v$ is sucked in a micropipette with inner radius $R_p (R_v > R_p)$. The hydrostatic pressures inside and outside the pipette and inside the vesicle are $p^{\rm in}$,

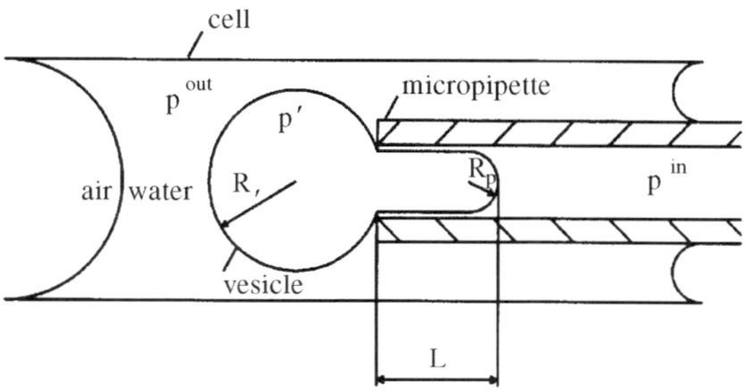


**Figure 14.1** Experimental set-up for bending elasticity $k_c^{fr}$ measurements. A vesicle with a radius $R_v$ is sucked in a micropipette with a radius $R_p < R_v$. The hydrostatic pressures inside and outside the micropipette and inside the vesicle are $p^{in}$, $p^{out}$, and $p'$. The length of the membrane sucked in the micropipette is $L$. The bending elasticity is determined from the dependence of $L$ on $\Delta p = p^{out} - p^{in}$.

$p^{out}$, and $p'$. Denote $\Delta p$ as $p^{out} - p^{in}$. Let $L$ be the length of the membrane sucked in the micropipette. For low enough $\Delta p$, the following relationship between $L$ and $\Delta p$ exists [1]:

$$\frac{1}{2}\left[\left(\frac{R_p}{R_v}\right)^2 - \left(\frac{R_p}{R_v}\right)^3\right]\frac{L}{R_p} = \frac{k_B T}{8\pi k_c^{fr}}\ln(\sigma) + \frac{\sigma}{k_s} + \text{constant} \tag{11}$$

where $\sigma$ is related to $\Delta p$ via the Laplace law:

$$\sigma = \frac{\Delta p R_v \cdot R_p}{2(R_v - R_p)} \tag{11a}$$

These relationships permit the measurement of $k_c^{fr}$ when $\sigma$ is low enough, and $k_s$ when $\sigma$ is high enough.

The following investigates some limitations of the micropipette method for determination of the elastic moduli. Equation (11) is obtained assuming that the volume of the vesicle and the number of molecules in its membrane do not change. The volume of the vesicle remains constant only if the membrane is completely impermeable, which is not exactly the case. Taking into account the permeability of the membrane and assuming that it is permeable only for the water and not for the solute, there will be a flow of water from the cell to the inner part of the vesicle, as well as from the inner part of the vesicle to the canal of the micropipette. The vesicle will be in equilibrium if the quantity of water, penetrating into the vesicle, is equal to the quantity leaving it. The equilibrium state of the vesicle does not depend on the permeability of the membrane. This quantity determines only the period of time for which the equilibrium state at given value of $\Delta p$ is reached. Let $C^a$ be the concentration of admixtures in the water. Assuming that the solution is ideal (which is true when the activity of dissolved admixtures is equal to 1), the following dependency is obtained as a generalization of eq. (11):

$$\frac{1}{2}\left[\left(\frac{R_{\rm p}}{R_{\rm v}}\right)^2 - \left(\frac{R_{\rm p}}{R_{\rm v}}\right)^3\right]\frac{L}{R_{\rm p}} = \frac{k_{\rm B}T}{8\pi k_{\rm c}^{\rm fr}}\ln\sigma + \frac{\sigma}{k_{\rm s}} + \frac{2}{3}\frac{\Delta p}{C^{\rm a}R^{\rm gas}T}\left[\frac{R_{\rm p}}{R_{\rm p}-R_{\rm v}} + \frac{1}{2}\frac{R_{\rm p}^2}{R_{\rm v}^2}\right] + \text{constant} \tag{12}$$

where $R^{\rm gas}$ is the gas constant and the relationship between $\sigma$ and $\Delta p$ is given by eq. (11a).

In such experiments the accuracy of measurement is usually of the order of 10%. Using the estimations given below, the conditions are now specified for when the error in the calculation of $k_{\rm c}^{\rm fr}$, due to the permeability of the membrane and its stretching elasticity, is negligible. The stretching elasticity effects are negligible if the following inequality is satisfied ($k_{\rm s} \sim 200$ dyn cm$^{-1}$, $k_{\rm c}^{\rm fr} \sim 20k_{\rm B}T$):

$$\sigma^{\max} < 0.1\frac{k_{\rm B}T}{8\pi k_{\rm c}}k_{\rm s} \sim 0.05 \text{ dyn cm}^{-1} \tag{13}$$

The effects with respect to volume changes are negligible if:

$$C^{\rm a} > \frac{320\pi}{3}\frac{k_{\rm c}}{(k_{\rm B}T)^2}\frac{1}{R_{\rm v}A}\sigma^{\max} \sim 10^{-2} \text{ mol l}^{-1} \tag{14}$$

where $A$ is Avogadro's number. To estimate eq. (14) take $R_{\rm v}$ as $10^{-3}$ cm and $\sigma^{\max}$ as 0.05 dyn cm$^{-1}$.

When $k_{\rm s}$ is measured the effects due to the permeability of the membrane can be disregarded when

$$C^{\rm a} > \frac{40}{3}\frac{k_{\rm s}}{Ak_{\rm B}TR_{\rm v}} \tag{15}$$

If $k_{\rm s} \sim 200$ dyn cm$^{-1}$ $C^{\rm a}$ has to be greater than 0.1 mol l$^{-1}$. If $k_{\rm s} \sim 1500$ dyn cm$^{-1}$ (e.g. for membranes containing cholesterol) $C^a$ has to be of the order of 0.7 mol l$^{-1}$. These estimations must be taken into account in the measurements.

## 4. MATERIALS AND METHODS

Liposomes were made from 1-stearoyl-2-oleoyl-*sn*-glycero-3-phosphocholine (SOPC, Avanti Polar Lipids Inc., USA). The modified lipid was 1-palmitoyl-2-oleoyl-*sn*-glycero-3-phosphoethanolamine-*N*-[poly(ethylene glycol)-2000] (PEG2000 lipid, Avanti Polar Lipids Inc., USA). The average molecular weight of the polymer chain was 2000 Da. This corresponds to a mean number of $N$ of 45 monomers per chain. Fluorescent dye, 1.5 mol % (lyssamine™ rhodamine B 1,2-dihexadecanoyl-

*sn*-glycero-3-phospho-ethanolamine, triethylammonium salt (rhodamine DHPE, L-1392), (Molecular Probes Inc., USA)) was added together with SOPC and PEG2000 lipid. The quantity of the PEG2000 lipid in the aqueous phase was controlled to be much less than this in the lamellar phase, as described in the next section. Liposomes with 5, 10, and 15 mol % PEG2000 lipid in their membrane were prepared. To obtain giant vesicles, about 3–7 mg of SOPC/PEG2000 lipid mixture (depending on the molar concentration of the PEG2000 lipid) were dissolved into 1 ml of 2 : 1 (v : v) chloroform : methanol solvent. The organic solution was placed in a 10 ml flask and the solvent was removed by vacuum for about 5 h. The flask was then filled with solution of 0.18 mol $l^{-1}$ sucrose and 2.5 mmol $l^{-1}$ imidazol (to prevent proliferation of bacteria) into deionized water. The sample was kept at room temperature for 72 h. Probes were taken from the flask, taking care not to disturb the solution. Vesicles were chosen as a unilamellar objects without any visible defects on the bilayer, their diameter being of the order of 15–20 μm.

The experimental set-up was similar to that of Evans and Rawicz [1]. An inverted fluorescent microscope *Axiovert* 100 (Zeiss, Germany) was used. The pressure transducers (Sedeme Kistler, France) that measure the difference between hydrostatic pressure inside and outside the micropipette have a precision better than 0.1 N $m^{-2}$, corresponding to a water column height of 10 μm. The CCD camera was a C2400-77 (Hamamatsu Photonics K.K., Japan). The resolution of the whole optical system was 0.17 μm per pixel. The micropipettes were prepared from a borosilicate glass using a puller (Narishige, Japan). They had an inside diameter of the order of 6 μm.

The vesicle membrane containing the fluorescent marker, was observed under the microscope, by working in a fluorescent regime.

The experimental results show that under the conditions described above the vesicle membrane sucked into the micropipette attains its equilibrium position very quickly (5 min after the pressure was applied there was no change in the position of the membrane inside the micropipette). All the reported measurements were performed using the same micropipette.

## 5. EXPERIMENTAL RESULTS AND DISCUSSION

The modified PEG2000 lipid used is significantly more soluble in water than the ordinary lipids and, consequently, its quantity in the aqueous phase is not negligible. An estimation was therefore needed of the maximal concentration of the PEG2000 lipid in micellar solution in equilibrium with liquid crystal phase. Note that the swelling of the vesicles is maintained under these conditions. In fact, this is a ternary system, comprising water–lipid (SOPC)–PEG2000 lipid. The boundaries of the micellar region in the phase diagram of this system depend on the ratio PEG2000 lipid : lipid. The reasonable assumption was made that the quantity of PEG2000 lipid in the micellar phase will attain its maximal value in the system pure PEG2000 lipid–water. The light absorption of water solutions with various concentrations of PEG2000 lipid was investigated. The absorption maximum is at $\lambda = 200$ nm (Figure 14.2).

The concentration of $\sim 4 \times 10^{-6}$ mol l$^{-1}$ is the cross-point of the linear fit of the absorptions at low and high concentrations of the PEG2000 lipid. This concentration is assumed to be the boundary of the micellar phase. Consequently, the concentration of the PEG2000 lipid in the water subphase will not exceed $4 \times 10^{-6}$ mol l$^{-1}$.

It was noted that in each sample (see Materials and Methods) the quantity of the PEG2000 lipid in the water subphase was 10 times lower than the maximum quantity of the PEG2000 lipid in the entire volume of the sample. This guaranteed a better than 10% precision of the molar ratio between the lipid and the PEG2000 lipid in the membranes.

Most of the vesicles in the sample had defects inside and/or outside. These defects were either thin tubes (some having a length of tens of microns) or bright packages, connected to the membrane by such tubes. The defects were often absorbed by the membrane when a vesicle was sucked up with a high enough pressure into the micropipette. When decreasing the pressure, package-type defects almost always appeared (usually inside the vesicle). When a vesicle without observable defects was sucked into the micropipette, in the process of the decreasing of the pressure defects (packages of lipid) appeared. This meant we could not do any measurements by decreasing the pressure.

Only unilamellar vesicles without observable defects and diameters between 15 and 20 μm were studied. The criterion for the estimation of the number of bilayers of the vesicle was the brightness of its image observed under a microscope working in a

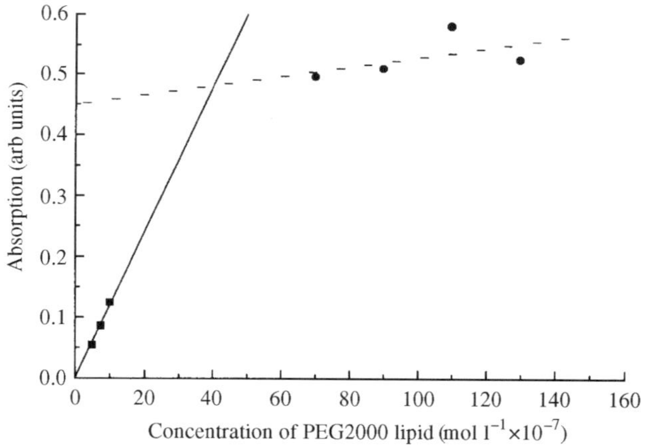


**Figure 14.2** Absorption of light with $\lambda = 200$ nm as a function of the PEG2000 lipid concentration in water: (■) experimental points at low concentrations; (——) linear fit of low concentrations; (●) experimental points at high concentrations; (– – –) linear fit of high concentrations.

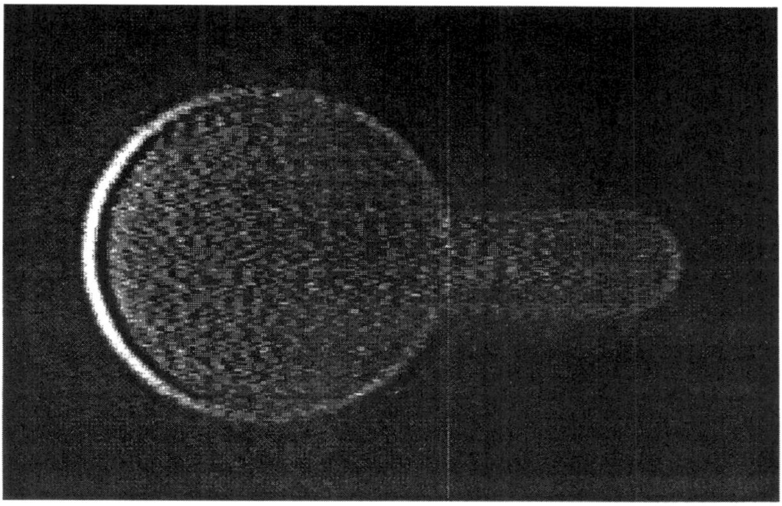

**Figure 14.3** A vesicle, sucked up in a pipette, as observed under a microscope working in a fluorescence regime. The diameter of the vesicle is ~20 μm, while the inner diameter of the pipette is ~6 μm. The vesicle matrix is SOPC, with 5 mol % PEG2000 lipid and 1.5 mol % rhodamine DHPE.

fluorescent regime [17]. Figure 14.3 shows a vesicle, sucked into the micropipette, as observed under the microscope.

Three molar concentrations $C$ of the PEG2000 lipid in the membrane were studied, namely 0.05, 0.10, and 0.15. Measurements of the length of the membrane sucked into the micropipette were made only for the cases when the applied pressure was consequently increased. The pressures were in the range from 20 dyn cm$^{-2}$ to 300 dyn cm$^{-2}$, corresponding to tensions from $\sim 5 \times 10^{-3}$ dyn cm$^{-1}$ to $\sim 7.5 \times 10^{-2}$ dyn cm$^{-1}$. The results are presented in Figure 14.4. The bending elasticity increases with $\sim k_B T$ each time the concentration of the modified lipid is increased with 5 mol %. The same dependence was measured by Evans and Rawicz [18] for a membrane consisting of the same PEG2000 lipid in a matrix of DGDG. They obtained the same change of $K_c$ with the concentration, but the value of $K_c$ for the pure lipid, measured by them [1], is about four times greater than the extrapolated value obtained from the results. The bending elasticity of a pure SOPC membrane, measured by analyzing the thermal form fluctuations of the vesicle, is up to six times greater than ours [19]. The following discussion presents a possible explanation for this discrepancy.

As previously mentioned, the experimental data were analyzed under the assumption that the stretching elasticity of the membrane is linear and of the order of 200 dyn cm$^{-1}$. Another assumption used implicitly is that for the studied ratios of lipid : PEG2000 lipid the membrane is a homogeneous planar structure. There is

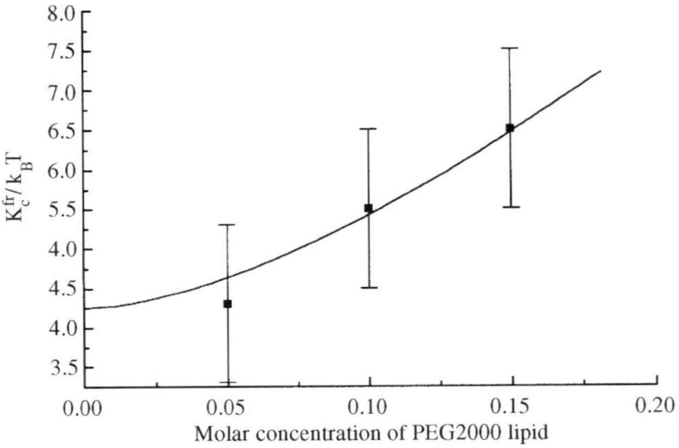


**Figure 14.4** Dependence of the bending elasticity of a bilayer of SOPC containing molar concentrations 0.05, 0.10, and 0.15 of the PEG2000 lipid. The measured bending elasticities are $(4.3 \pm 1)k_B T$, $(5.5 \pm 1)k_B T$, and $(6.5 \pm 1)k_B T$, respectively. The continuous line is the theoretical curve, calculated with eq. (9). The parameter values are $k_c^{fr} = 4.25 k_B T$, $k_c^{bl}/k_c^{fr} = 15$, $d_{ch} = 15$ Å, $s_0 = 70$ Å $^2$, $a = 3.5$ Å, $\xi = a$.

evidence that these two assumptions might not be fulfilled. In such a case the theory has to be modified.

Consider a flat tension-free membrane of SOPC and a cylindrical micelle consisting of SOPC, with PEG2000 lipid molar concentrations of $C1$ and $C2$ respectively. Let $\mu_l^1(C1)$, $\mu_p^1(C1)$, $\mu_l^2(C2)$ and $\mu_p^2(C2)$ be the chemical potentials of one lipid and PEG2000 lipid molecule in the membrane and in the micelle, respectively. The expressions for these quantities are:

$$\mu_l^1(C1) = 0.56 \frac{a^2 \xi^{-2/3} N k_B T}{[s_1(C1)]^{2/3}} (C1)^{5/3} \tag{16a}$$

$$\mu_p^1(C1) = 0.84 \frac{a^2 \xi^{-2/3} N k_B T}{[s_1(C1)]^{2/3}} \left(\frac{5}{3} - \frac{2}{3} C1\right)(C1)^{2/3} \tag{16b}$$

$$\mu_l^2(C2) = f^{mic} - 0.56 \frac{a^2 \xi^{-2/3} N k_B T}{[s_2(C2)]^{2/3}} \left\{ 1 - \frac{5}{24} \frac{h\left(\frac{s_2}{C2}, N\right)}{R^{out}} + \frac{5}{64} \left[ \frac{h\left(\frac{s_2}{C2}, N\right)}{R^{out}} \right]^2 \right\} (C2)^{5/3} \tag{16c}$$

$$\mu_p^2(C2) = f^{mic} + 0.84 \frac{a^2 \xi^{-2/3} N k_B T}{[s_2(C2)]^{2/3}} \times \left\{ 1 - \frac{5}{24} \frac{h\left(\frac{s_2}{C2}, N\right)}{R^{out}} + \frac{5}{64} \left[ \frac{h\left(\frac{s_2}{C2}, N\right)}{R^{out}} \right]^2 \right\}$$
$$\times \left(\frac{5}{3} - \frac{2}{3} C2\right)(C2)^{2/3} \tag{16d}$$

In eqs (16), $f^{\text{mic}}$ is the free energy of a lipid molecule in the cylindrical micelle, $s_1(C1)$ and $s_2(C2)$ are the mean areas per lipid molecule in the flat membrane and in the cylindrical micelle at its outer surface (where the polymer chains are grafted), and $R^{\text{out}}$ is the radius of curvature of this surface. All other quantities are those defined in the section on theory. To determine the value of $f^{\text{mic}}$, use the fact that the cylindrical micelle can be considered as two hydrophilic edges of a flat membrane glued together. Consequently, the free energy per unit length of the micelle is twice the edge energy $\gamma$. The edge energy of a membrane, made of egg-yolk lecithin was measured by Harbich and Helfrich [20]. They obtained $\gamma = 2.1 \times 10^{-6}$ dyn, assuming that $k_c$ for the membrane is $2.1 \times 10^{-12}$ erg. Taking for this membrane the more realistic value of $k_c = 0.55 \times 10^{-12}$ erg [9], and using their approach, gave $\gamma = 0.5 \times 10^{-6}$ dyn. Assuming the edge energy of the SOPC membrane to be the same. Then $f^{\text{mic}} = 2\gamma/M$, where $M$ is the number of the molecules per unit length of the micelle. To determine $M$, it was supposed that the volume of the hydrophobic part of one molecule remains unchanged, the radius of the core of the cylindrical micelle is equal to the length $l^{\text{at}}$ of the all-*trans* conformation of the hydrophobic chain, and that it is 50% greater than the thickness $d_{\text{ch}}$ of the hydrophobic part of a monolayer of the tension-free bilayer. The last assumption stems from the fact that at the main gel–liquid crystal transition of a lipid bilayer, its area increases by about 50% while the volume of the hydrophobic part remains practically constant [21]. A value of 10Å was used for the height of the hydrophilic head of the SOPC. Taking $d_{\text{ch}} = 15$Å, as the radius of the hydrophobic core of the vesicle gave $l^{\text{at}} = 22.5$Å, and the radius of the outer surface of the micelle as $R^{\text{out}} = 32.5$Å. The number of molecules $M$ per unit length of the micelle is then $\pi(l^{\text{at}})^2/(s_0 d_{\text{ch}})$. The condition for equilibrium between the flat membrane and the rod-like micelle is

$$\mu_1^1(C1) = \mu_1^2(C2)$$
$$\mu_p^1(C1) = \mu_p^2(C2)$$

This is a system of equations for the quantities $C1$ and $C2$. With the parameters values given above, the solution is $C1 \approx 0.1$, and $C2 \approx 0.4$. The exact values of $C1$ and $C2$ are not so important, because the numbers used are estimations, and not exact values. For example, if the value of the correlation length $\xi$ is assumed to be equal to $4a$ ($a$ being the length of the monomer unit) instead of $a$, $C1$ becomes of the order of 0.05. More important is the fact that $C1$ and $C2$ exist. If the total concentration of the PEG2000 lipid is less than $C1$, the planar membrane is stable. Only in this case are the theoretical results described in section Theory relevant. When the total concentration of the PEG2000 lipid is between $C1$ and $C2$, the membrane consists of planar parts with attached rod-like micelles. In the planar parts the concentration of the modified lipid is $C1$, whereas in the rod-like micelles it is $C2$. When planar parts and rod-like micelles coexist, the stretching elasticity of the membrane becomes very low. It is possible that the concentration range of PEG2000 lipids studied are within this domain. If this were the case, the studied vesicles, even without any observable defects, should contain some microscopic rod-like micelles, in equilibrium with the planar membrane. When the suction pressure increases, the membrane enters in the micropipette not only because the amplitudes of the fluctuation modes decrease, but because part of the rod-like micelle is transformed into planar membrane. This could be the reason why the analysis gave very low value of bending elasticity, and would also explain the observations previously described in this section. The theory presented in the beginning of the paper is developed under the assumption that the bilayer, containing modified lipids, can be considered as a relatively regular surface, whereas the last theoretical results show that it could be surface bearing whiskers of cylindrical micelles. In the latter case, the quantity of the material stored in the whiskers will depend on the concentration of the modified lipid and, therefore, on the tension of the bilayer. The theoretical description of the mechanical properties of such a complicated system have yet to be elaborated.

Some of these difficulties should be partially avoidable if the bending elasticity determined from the analysis is of the form fluctuations of a quasispherical vesicle.

Our theoretical predictions show, that $C1^{\max}$ (the maximal concentration of the PEG2000 lipid in the planar part of the membrane) decreases when the number of segments $N$ in the polymer chain increases and becomes zero at sufficiently high $N$. The theory of elasticity developed by the authors is valid only for the brush regime, where the surface concentration of the grafted chains is high and the statistical coils interact. For low enough C1 the brush regime is no longer valid and the mushroom regime with practically independent statistical coils becomes effective. Evidently, for high enough $N$, the maximal concentration of the PEGylated lipid that the membrane can absorb is of the order of $s_0/(a^2N^{6/5})$.

## 6. ACKNOWLEDGMENTS

This work was funded in part by project 7BUPJ048478 ‘Mechanical and electromagnetic properties of lipid monolayers and bilayers’, supported by the Swiss National Science Foundation and the Institute of Solid State Physics to the Bulgarian Academy of Sciences, Sofia, Bulgaria.

## 7. REFERENCES

1. E. A. Evans and W. Rawicz, *Phys. Rev. Lett.*, **64**, 2094 (1990).
2. G. Blume and G. Cevc, *Biochim. Biophys. Acta*, **1029**, 91 (1990).
3. D. Lasic, F. J. Marin, A. Gabizon, S. K. Huang and D. Papahadjopoulos, *Biochim. Biophys. Acta*, **1070**, 187 (1991).
4. D. Needham, K. Hristova, T. McIntosh, M. Dewhirst, N. Wu and D. Lasic, *J. Liposome Res.*, **2**, 411 (1992).
5. W. Helfrich, *Z. Naturforsch. C*, **29**, 693 (1973).

6. I. Bivas, P. Hanusse, P. Bothorel, J. Lalanne and O. Aguerre-Chariol, *J. Phys. France*, **48**, 855 (1987).
7. A. Yeung and E. Evans, *J. Phys. II*, **5**, 1501 (1995).
8. I. Bivas, P. Méléard, I. Mirtcheva and P. Bothorel, *Colloids and Surfaces A* (in press).
9. J.-F. Faucon, M. D. Mitov, P. Méléard, I. Bivas and P. Bothorel, *J. Phys. France*, **50**, 2389 (1989).
10. M. D. Mitov, J.-F. Faucon, P. Méléard and P. Bothorel, in *Advances in Supramolecular Chemistry*, (ed.) G. W. Gokel, Jai Press Inc., Greenwich, 1992, vol. II, p. 93.
11. S. T. Milner, T. A. Witten and M. E. Cates, *Macromolecules*, **21**, 2610 (1988).
12. S. T. Milner and T. A. Witten, *J. Phys. France*, **49**, 1951 (1988).
13. I. Bivas, D. Georgescauld, N. Jeandaine, M. Winterhalter, P. Méléard, G. Marinov and P. Bothorel, *Progr. Colloid Polym. Sci.*, **105**, 197 (1997).
14. I. Bivas, M. Winterhalter, P. Méléard and P. Bothorel, *Europhys. Lett.*, **41**, 261 (1998).
15. J. Majewski, T. L. Kuhl, M. C. Gerstenberg, J. N. Israelachvili and G. S. Smith, *J. Phys. Chem. B*, **101**, 3122 (1997).
16. C. Hiergeist, V. A. Indrani and R. Lipowsky, *J. Phys. II*, **6**, 1465 (1996).
17. K. Akashi, H. Miyata, H. Itoh and K. Kinosita Jr., *Biophys. J.*, **71**, 3242 (1996).
18. E. Evans and W. Rawicz, *Phys. Rev. Lett.*, **79**, 2379 (1997).
19. P. Méléard, C. Gerbeaud, P. Bardusco, N. Jeandaine, M. D. Mitov and L. Fernandez-Puente, *Biochimie (France)*, **80**, 401 (1998).
20. W. Harbich and W. Helfrich, *Z. Naturforsch.*, **34a**, 1063 (1979).
21. V. Luzzati and A. Tardieu, *Ann. Rev. Phys. Chem.*, **25**, 79 (1974).

# Chapter 15

# Motion of Particles Attached to Giant Vesicles: Falling Ball Viscosimetry and Elasticity Measurements on Lipid Membranes

*RUMIANA DIMOVA, CHRISTIAN DIETRICH AND BERNARD POULIGNY*

*Centre de Recherche Paul Pascal, Pessac, France*

## 1. INTRODUCTION

The material DMPC (1,2-dimyristoyl-*sn*-glycero-3-phosphocholine) is a very convenient lipid for studying the thermomechanics of lipid transitions in membranes due to its easily accessible phase transition temperature of about 24°C. Many studies have been carried out to obtain a better understanding of bilayer properties when the composing lipid freezes from the fluid $L_\alpha$ phase to the rippled gel $P_{\beta'}$ phase. Phospholipid phase transitions could be important, for example, in regulating the activities of membrane-associated proteins [1]. A set of experiments was performed based on the motion of micronsized latex particles in contact with giant DMPC vesicles. This work presents data for the temperature dependence of the shear surface viscosity in the phase transition region of the lipid. The experimental findings correspond well to results provided by the micropipette aspiration technique [2]. Preliminary results on the elasticity of the membrane in the gel phase are obtained by manipulating two particles on a single vesicle by means of optical trapping.

*Giant Vesicles* Edited by P. L. Luisi and P. Walde

## 2. EXPERIMENTAL PROCEDURES AND DATA ANALYSIS

Experiments were performed on giant liposomes prepared from DMPC (Avanti Polar Lipids) by electroformation [3] (see also Chapters 3 and 4). Vesicles (generally larger than 30 μm in diameter) are formed in pure water, in the $L_\alpha$ state (usually at 30°C), under electric field (about 1 V, 10 Hz). The investigation chamber (Figure 15.1) is equipped with a circulating water jacket connected to a cryothermostat (Lauda RM6) which allows working at temperatures down to about 14°C. Temperature is controlled by a thermocouple (±0.1°C).

After vesicles are formed, a dilute suspension of latex spheres is gently injected in the chamber away from the cluster of vesicles on the electrode. We use polystyrene spheres (Polyscience) with radii $a$ in the 1–10 μm range. The beads are manipulated by a long working-distance optical trap [4], basically a couple of counter-propagating laser beams (argon ion laser, 514 nm). In water, a trapped sphere equilibrates itself with its center on the laser beam axis. Off-centering the particle by a distance $d_{\mathrm{off}}$ produces a linear restoring force (due to radiation pressure) $F = k_{\mathrm{RP}} \times d_{\mathrm{off}}$, provided that $d_{\mathrm{off}}$ is small enough, say $d_{\mathrm{off}} \leq 0.6a$ [4]. The term $k_{\mathrm{RP}}$ can be regarded as the optical trap spring constant. Typical forces are in the 1–100 pN range [4,5]. No heating of the particles was detected [4]. When a particle is trapped, its sedimentation velocity in water is measured. Then the particle is brought into contact with a previously chosen vesicle which could be either free or connected to the cluster of surrounding vesicles at the electrode. The nature of the particle adhesion and the phenomena observed throughout particle encapsulation into the vesicle membrane is described elsewhere [6]. Generally, spheres stick at the vesicle surface attaining a finite contact angle which depends on the initial tension of the vesicle.

### 2.1 Viscosimetry ($L_\alpha$ Phase)

Once stuck to the membrane, the particle is brought near to the vesicle top and then released to glide down along the vesicle contour till it reaches the bottom. The

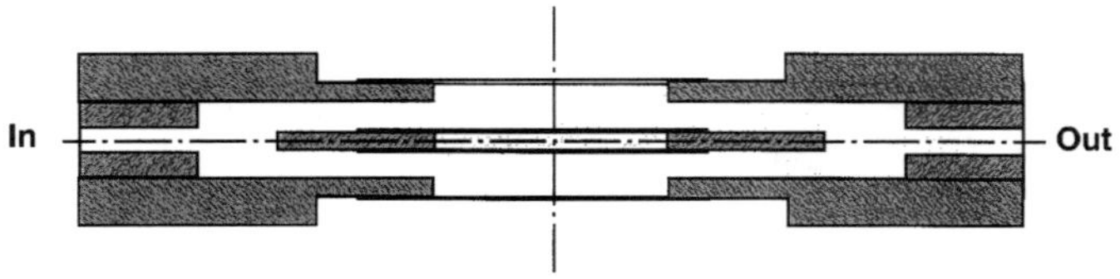


**Figure 15.1** Diagram of the temperature controlled chamber. The sample is contained inside the central area (40 × 9 × 1 mm$^3$). This zone contains two platinium wires (0.8 mm in diameter) for vesicle electroformation, a syringe needle for water and solid particles injection, and a thermocouple (not represented). The shaded zone is the volume flooded by the circulating water. The sample and flooded zones are bounded by microscope cover slips (170 μm in thickness). The thickness of the chamber at the level of the symmetry axis is about 3.4 mm. With the microscope objectives used in the set-up (Zeiss Epiplan, ×50, N.A. = 0.5), the entire sample volume is accessible to observation and optical manipulation.

particle trajectory is viewed from above. The position of the sphere is recorded every 0.2 s with a resolution of ±0.5 µm. The friction experienced by the particle sedimenting along the vesicle is influenced by the viscous properties of the membrane. Thus the vesicle–particle system is a viscosimeter working at the microscopic scale. The temperature dependence of the membrane viscosity near the phase transition region was studied.

The analysis of the particle track from the sedimentation experiment reduces to assessing the friction experienced by the sphere. Particle motion along the vesicle membrane is described by

$$\tilde{m}g \sin\theta = \zeta\tilde{R}\dot{\theta} \qquad (1)$$

where $\tilde{m}$ is the particle mass corrected by buoyancy, $g$ is gravity acceleration, and $\theta$ is the polar angle as defined in Figure 15.2. Equation (1) holds in the zero-temperature limit, i.e. when Brownian motion is neglected [7]. The factor $\zeta$ is the wanted friction coefficient and $\tilde{R}$ is the distance between the particle and vesicle centers. A more detailed analysis on the application of this equation is available elsewhere [7]. To avoid the entropic contribution due to Brownian motion to the measured friction ratio, particles were chosen to be large enough [7]. The solution of eq. (1) is

$$f[\theta(t)] = f[\theta_0] - \frac{\tilde{m}g}{\tilde{R}\zeta}t \qquad (2)$$

where $f[\theta] = \text{atanh}[\cos(\theta)]$ and $\theta_0$ defines the particle position at time $t = 0$. Thus the experimentally obtained time dependence of the function $f$, provides the value of the friction coefficient $\zeta$. Basically, $\zeta$ contains the information about the viscosity of the membrane. The theory of Danov and co-workers [8] was used to invert the friction coefficient data into shear surface viscosity $\eta_S$; for 1-stearoyl-2-oleoyl-*sn*-glycero-3-phosphocholine bilayers at room temperature, $\eta_S$ was found to be about $3 \times 10^{-6}$ sp surface poise, dyn.s/cm). The theory of Danov and co-workers deals

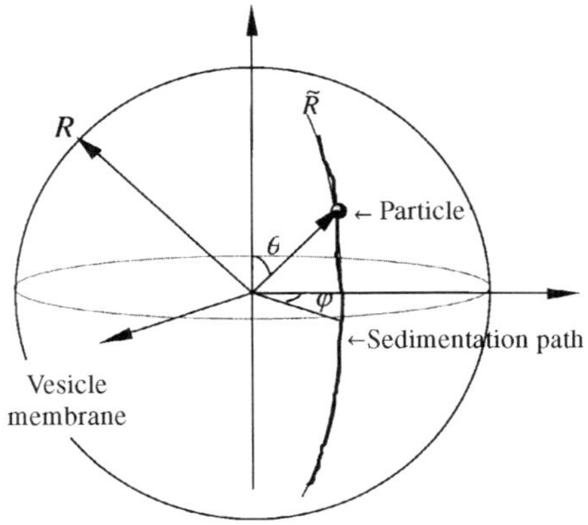


**Figure 15.2** Sedimentation experiment. Illustrative sketch of a particle attached to a spherical vesicle and bead trajectory when falling along the vesicle membrane.

with the motion of a spherical particle along a flat infinite film at the air–water interface. This theory can be adapted to a particle across a membrane (i.e. with water on both sides), provided that $a$ is much smaller than $R$, the vesicle radius, and that the contact angle is not too far from 90°. When $a$ is not small compared to $R$, the particle motion induces a circulation of the water internal to the vesicle. This increases $\zeta$ well above the value corresponding to $R/a \to \infty$. An improved theory that takes this effect into consideration was recently worked out [9].

### 2.2 Visco-elasticity (gel phase)

Two latex spheres of radii close to 5 μm (beads of similar sizes were chosen) are stuck either individually or simultaneously to a larger vesicle ($R > 30$ μm), in the $L_\alpha$ state. For the optical manipulation a double-trap configuration of the optical system was used, i.e. two pairs of laser beams [4,5]. One trap is fixed while the second one can be moved. The distance between the two traps is adjustable between 0 and about 35 μm.

#### *(a) Gel-phase static shear elasticity*

After particle adhesion, the membrane is cooled down to about 15°C, well in the gel state domain, while holding the two particles at fixed distance $d_0$. Then the trap separation was slowly increased to a distance $d_0 + e$. In response to this perturbation, the interparticle distance, $d$ (initially $d = d_0$) increases slightly. The new distance, $d = d_1$, is found $< d_0 + e$ because of the membrane elasticity. Here the membrane can be modeled as a spring of stiffness $k_M$ binding the two particles. The value of $d_1$ is simply found by balancing radiation pressure ($k_{RP}$) and membrane ($k_M$) forces:

$$d_1 - d_0 = \frac{k_{RP}}{2k_M + k_{RP}} e \tag{3}$$

The experimental procedure is to first measure $d_1$ at $T = 15$°C, then to increase $T$, which has the consequence of increasing $d_1$ slightly. The system is left for about 15 min to reach equilibrium after each temperature step. Thus a ($k_M$, $T$), graph is obtained, up to the main transition temperature $T_m$ (23°C $\leq T_m \leq$ 24°C). Afterwards, the system is cooled again, to repeat the experiment. Thus measuring the interparticle distance for each temperature provides a way to estimate $k_M$ as a function of $T$.

#### *(b) Gel-phase viscosity*

A simple extension of the above procedure allows the dynamics of membrane reaction to be studied. When the interparticle distance is increased to $d_1$, the laser beams are switched off. The system then relaxes, i.e. $d$ decreases down to a final value, in principle down to about $d_0$. In practice the membrane is stretched or shrunk by displacing one of the particles (again the double trap configuration is employed). After switching off the mobile trap only, the relaxation tracks are recorded over time.

If it is assumed that the particle friction in the gelified system can still be characterized by a constant coefficient $\zeta$, the above two-spring model predicts the simple exponential relaxation

$$d_1 = A + B\exp\left(-\frac{t}{\tau}\right) \tag{4}$$

where $A$ and $B$ are constants and $\tau = \zeta/(2\,k_M)$ is the characteristic time of the relaxation process. With the value of the spring constant $k_M$, available from the static elasticity experiments for the corresponding temperature, one can estimate approximately the friction coefficient $\zeta$. Thus the membrane shear viscosity can be assessed even for low temperatures where the membrane freezes and sedimentation experiments are not possible.

## 3. RESULTS

### 3.1 DMPC $L_\alpha$ phase viscosity

To reduce the vesicle finite-size effect as much as possible, systems for which the $R:a$ ratio is large enough are used, i.e. so that the bead experiences the membrane as being flat. A set of data for a single vesicle–particle system is shown in Figure 15.3. The results are shown in terms of the dimensionless friction ratio $\bar{\zeta} = \zeta/(6\pi\eta a)$, where $\eta$ is the viscosity of the surrounding media (bulk water). In Figure 15.3, $\bar{\zeta}$ is

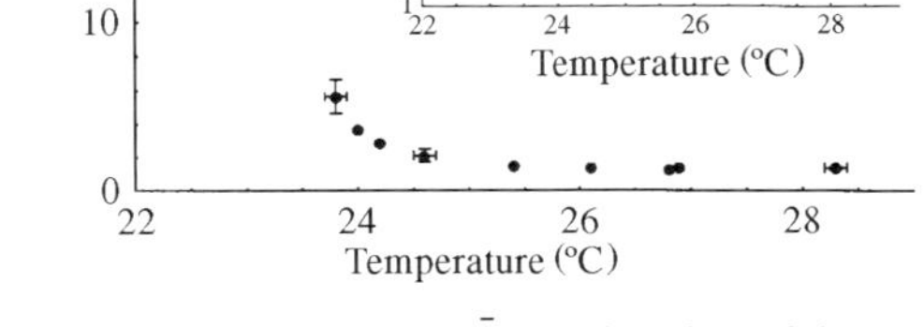


**Figure 15.3** The dimensionless friction ratio $\bar{\zeta}$ as a function of the temperature for a single vesicle–particle system. The insert presents the same data converted to shear surface viscosity ($\eta_S$).

plotted as a function of the temperature for a particle of radius $a = 2.7\ \mu m$ on a vesicle of radius $R = 46.7\ \mu m$. The values of $\eta_S$, found from those of $\bar{\zeta}$ using Danov and co-workers theory [7,8], are displayed in the inset to Figure 15.3, using a logarithmic scale. Sedimentation experiments were carried out down to 23°C. Below this temperature, the particles do not sediment further (within experimental time, about 1 h). This was the criterion used to decide that the membrane was gelified.

The main feature in Figure 15.3 is the considerable continuous increase (three orders of magnitude) of the viscosity when the phase transition temperature is approached.

### 3.2. DMPC gel phase elasticity

*(a) Statics*

The effective stiffness $k_M$ is an equilibrium (i.e. static) property of the membrane. The variation of $k_M$ as the temperature is raised is shown in Figure 15.4. The absolute value of $k_M$ was estimated from that of $k_{RP}$, using eq. (3). The value of $k_{RP}$ was computed using the generalized Lorenz–Mie theory [5] from the characteristics of the particles and knowing the powers of the laser beams in the sample chamber. This gave $k_{RP} \approx 10^{-3}$ dyn cm$^{-1}$ for the conditions used in the experiments. The different symbols in Figure 15.4 correspond to three individual experimental temperature cycles for one and the same vesicle–particles system. Reasonable reproducibility is demonstrated also when the system is altered. Although preliminary, these results show an approximately linear variation of $k_M$ over the studied temperature interval. When the lipid is in the $L_\alpha$ phase the bilayer looses its elastic properties, the membrane becomes fluid and $k_M = 0$. The dispersion of the results and the experimental error for lower temperatures increase because the membrane becomes stiffer and particle displacement is smaller. In principle, it should be

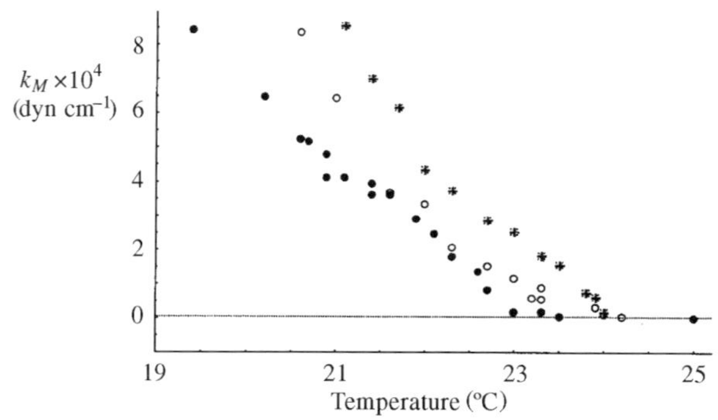


**Figure 15.4** Static elastic experiments data. Membrane effective stiffness constant $k_M$ plotted as a function of temperature. Different symbols correspond to three individual temperature runs.

possible to extract the value of the membrane shear modulus $\mu$ from that of $k_M$. A simple limit is that of a plane infinite membrane, with two spheres whose centers are located in the membrane plane (this corresponds to 90° contact angles for both particles). In this case, using Muskhelichvili's analysis for the elasticity of two-dimensional systems [10],

$$k_M \approx C\mu \tag{5}$$

where $C$ is a constant $= 8\pi/\ln(d_0/a)$. This result holds if $d_0 \gg a$ and if the membrane compression modulus $K$, is much larger than $\mu$ (for DMPC bilayers, $K$ is at least of the order of 100 dyn cm$^{-1}$). In our conditions $d_0/a \approx 5$.

Direct application of eq. (5) to the data, yields $\mu \approx 10^{-4}$ dyn cm$^{-1}$ at the lower end of the investigated temperature range. Notice that such a small value of $\mu$ is consistent with the above-mentioned $\mu \ll K$ condition. The estimated value of $\mu$ is surprisingly small for a supposedly ordered phase [11]. In fact, $k_M$ could be considerably smaller than $C\mu$, because the particles in the experiments were almost completely inside the vesicle; in other words, the contact angles were very small. Pulling the particles apart by means of the laser beams most probably makes them roll and induces an out-of-plane (more precisely out-of-sphere) deformation of the membrane. This effect needs to be taken into account for the value of $\mu$ to be correctly estimated. In any case, the above-mentioned value should be taken as a gross lower boundary.

### *(b) Dynamics*

The membrane relaxation dynamics were studied following the protocol explained in 2.2.b. An example of interparticle distance relaxation is shown in Figure 15.5. The

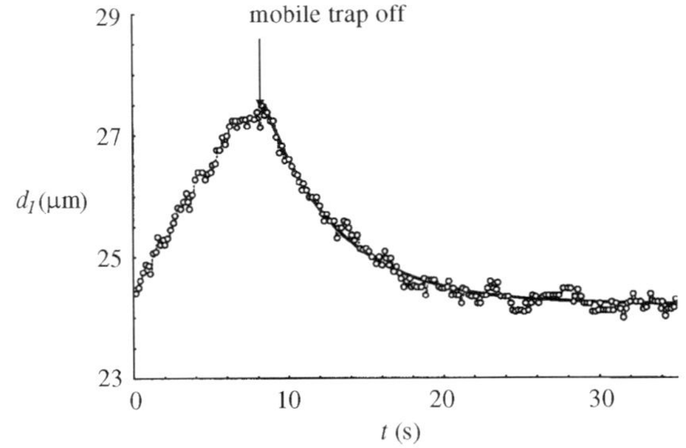


**Figure 15.5** Dynamic elasticity experiment at 19°C. The points represent the measured interparticle distance $d_1$. The solid line is an exponential fit for the relaxation part of the experiment according eq. (4), $\tau = 5$ s.

data is nicely fitted by a simple exponential function (solid line), with $\tau \approx 5$ s. For $k_M = 1.3 \times 10^{-3}$ dyn cm$^{-1}$, which is the value at $T = 19\,^\circ$C, we find $\zeta \approx 140$. Such a friction corresponds to about 0.01 sp in a fluid membrane, i.e. about 3000 times the viscosity of DMPC well above the main transition.

## 4. DISCUSSION AND CONCLUSIONS

The approach presented in this work gives an overview of both the viscous and elastic properties of vesicle membranes when the phase transition region is crossed. Three types of experiments were performed. Following the particle track in the sedimentation experiments provides a direct measurement of the membrane shear viscosity $\eta_S$. Unlike techniques as fluorescence recovery after photobleaching, electron spin resonance, or nuclear magnetic resonance, which are sensitive to molecular parameters, this method works on macroscopic scale; therefore, it does not require a theory relating molecular dynamics to macroscopic properties. The sedimentation experiments demonstrate a gradual change in the properties when the phase transition temperature is approached, which is the indication of strong pretransitional effects.

To our knowledge, these are the first measurements of a lipid bilayer linear elastic response, i.e. the basic macroscopic characteristic of a solid structure. It was found that the membrane effective stiffness decreases down to zero (within experimental error) at $T_m$, which is the indication of a continuous gel–fluid phase transition.

These results are in line with others obtained by different – but less direct – techniques applied to unilamellar systems. For instance, ultrasound propagation and attenuation were found continuous at $T_m$ [12] and a critical slowing down of membrane relaxation was observed [13]. Pretransitional phenomena in $L_\alpha$ phases near $T_m$ are correlated to structure fluctuations in the form of gel-like domains, of finite size and lifetime, and whose existence was experimentally detected by fluorescence spectroscopy [14].

## 5. ACKNOWLEDGMENTS

This work was supported by ‘Laboratoire Franco-Bulgare’ (CNRS/Bulgarian Academy of Sciences/University of Sofia) and by CNRS programme Ultimatech. We thank P. Martinoty for his interest and fruitful discussions.

## 6. REFERENCES

1. T. Hønger, K. Jørgensen, R. L. Biltonen and O. G. Mouritsen, *Biochemistry*, **35**, 9003 (1996); P. K. J. Kinnunen, *Chem. Phys. Lipids*, **57**, 357 (1991).

2. E. Evans and D. Needham, *J. Phys. Chem.*, **91**, 4219 (1987).
3. M. I. Angelova and D. S. Dimitrov, *Prog. Colloid Polymer Sci.*, **76**, 59 (1988); M. I. Angelova, S. Soléau, P. Méléard, J. F. Faucon and P. Bothorel, *Springer Proc. Phys.*, **66**, 178 (1992).
4. M. I. Angelova and B. Pouligny, *Pure Appl. Opt. A*, **2**, 261 (1993).
5. G. Martinot-Lagarde, B. Pouligny, M. I. Angelova, G. Gréhan and G. Gouesbet, *Pure Appl. Opt. A*, **4**, 571 (1995).
6. C. Dietrich, M. I. Angelova and B. Pouligny, *J. Phys. II France*, **7**, 1651 (1997).
7. K. Velikov, C. Dietrich, A. Hadjiisky, K. Danov and B. Pouligny, *Europhys. Lett.*, **40**, 405 (1997); K. Velikov, K. Danov, M. I. Angelova, C. Dietrich and B. Pouligny, *Colloids and Surfaces A*, **149**, 245 (1999).
8. K. Danov, R. Aust, F. Durst and U. Lange, *J. Colloid Interface Sci.*, **175**, 36 (1995).
9. K. Danov, unpublished results.
10. N. Muskhelishvili, in *Some basic problems of the mathematical theory of elasticity*, P. Noordhoff Ltd, Groningen, The Netherlands (1963).
11. C. Zakri, A. Renault, J.-P. Rieu, M. Vallade, B. Berge, J.-F. Legrand, G. Vignault and G. Grübel, *Phys. Rev. B*, **55**, 14163 (1997).
12. S. Mitaku, A. Ikegami and A. Sakanishi, *Biophys. Chem.*, **8**, 295 (1978).
13. S. Mitaku and T. Date, *Biochim. Biophys. Acta*, **688**, 411 (1982).
14. A. Jutila and P. K. J. Kinnunen, *J. Phys. Chem. B*, **101**, 7635 (1997); S. Pedersen, K. Jørgensen, T. R. Bækmark and O. G. Mouritsen, *Biophys. J.*, **71**, 554 (1996).

# Chapter 16

# Control of Fusion Between Giant Vesicles

*SEK WEN HUI, MEGHAN PERKINS AND PARTHAPRATIM CHANDAROY*
*Roswell Park Cancer Institute, Buffalo, USA*

## 1. INTRODUCTION

Membrane fusion, a fundamental cellular process, is an integral step in endocytosis, exocytosis, fertilization, infection, and cell division. The occurrence of membrane fusion in these cases is precisely controlled. In most cases, however; the molecular controlling mechanism is not well understood. Nevertheless, by knowing the basic physicochemical principles involved in lipid bilayer fusion, and more often, by trial and error, one may induce membranes to fuse solely by physicochemical manipulation. The most common artificial fusion methods are polyethylene glycol treatment and electrofusion. These artificial fusion methods are somewhat crude compared to biologically controlled fusion, in terms of selecting fusion partners and fusion sites. Yet artificial fusion techniques serve many useful purposes. Many modern biotechnology areas, such as monoclonal antibody production, drug delivery, gene transfer, and somatic cell hybrid formation, cannot develop without methods for artificially induced membrane fusion. Because of the need to improve efficiency and selectivity in artificially induced fusion, much effort is directed at understanding these fusion processes.

This article is concerned with the control of membrane electrofusion. This technique has several attractive features compared to other artificial methods. Because it is based entirely on the electrical properties of the cells, the detailed biochemical and biological properties of the cells have little influence on the fusion process. Therefore, the method is applicable to a majority of cells. The only

*Giant Vesicles* Edited by P. L. Luisi and P. Walde

important parameters are electrical ones, which can be controlled easily and precisely.

The principle of electrofusion is the simultaneous electric breakdown of plasma membranes at the contact points with adjacent cells, when a short pulse of supercritical voltage for electric breakdown is applied to the cell pair. A lumen is formed between the coaxial pores at the breakdown site, which eventually enlarges to connect the cytoplasm of the fusion partner cells. The force involved in making cell–cell contact before, during and after pulse application, and the force causing the simultaneous membrane breakdown, can be calculated from the conductivities and dielectric constants of the cytoplasm, the membrane, and the external medium. When all parameters are known precisely, the outcome of electrofusion can be predicted with certainty.

In the later part of the article, the effect of cell surface modification is considered. Namely, polyethylene glycol molecules are anchored to the membrane surface to provide a steric barrier to fusion. The purpose is to provide a means to reduce unwanted fusion (i.e. stealthiness). Giant liposomes provide an excellent model with which to test this theory, because their external, internal and membrane conductivities and dielectric constants can be controlled precisely, and their surfaces can be easily modified.

## 2. FORCES INVOLVED IN FUSION OF GIANT LIPOSOMES

A preliminary requirement of electrofusion is the prefusion alignment and attachment of cells or giant liposomes. There are several ways of aligning cells or giant liposomes prior to the application of the fusion-causing (fusing) electric pulse. One way is to compact the cells or giant liposomes into pellets by centrifugation [1]. Another method, which is what we considered here, is to use the dielectrophoresis force, by applying an AC electric field to a suspension of cells or giant liposomes in a medium, the electric permittivity (conductivity and dielectric constant) of which is different from those of the contents of these cells or giant liposomes. Because the instantaneous ionic concentrations within the pole regions (regions facing the electrodes) differ from those immediately outside these regions, due to different relaxation times of the ions, these cells or giant liposomes become instantaneous electric dipoles. This is depicted in Figure 16.1.

These dipoles are attracted to one another, and are attracted to or repelled from the field gradient, depending on the applied field frequency. The attractive force $F_D$ is given by [2]

$$F_D = 2\pi a^3 \varepsilon_e \, \mathrm{Re}[\mathbf{K}(\omega)] \nabla E^2 \tag{1}$$

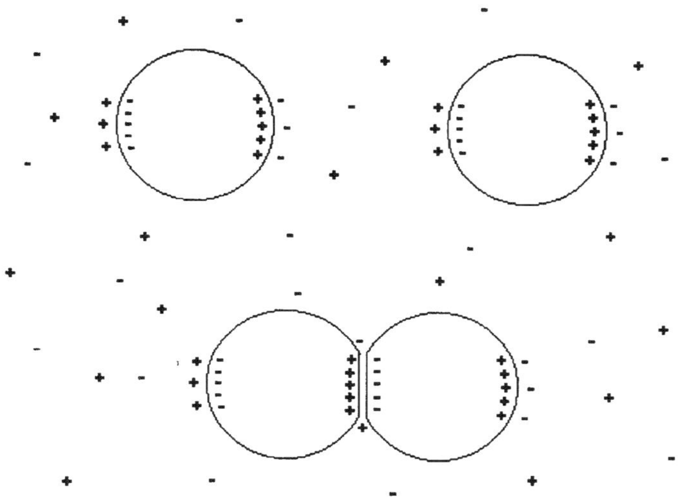

**Figure 16.1** A diagram representing the polarization of ions in an AC field inside and outside a pair of giant liposomes. The inside conductivity is depicted to be higher than that of the outside medium. The pair of liposomes are driven together by dipole attraction. An applied high-voltage pulse will break the intervesicular barrier and fuse the two liposomes together.

where $a$ is the radius of a spherical cell or giant liposome (hereto referred to only as a giant liposome), $E$ is the applied electric field, $\omega$ is the frequency of the applied field, and $\mathbf{K}$ $(\omega)$ is the complex permittivity of the system, defined as

$$\mathbf{K}(\omega) = (\boldsymbol{\epsilon}_i - \boldsymbol{\epsilon}_e)/(\boldsymbol{\epsilon}_i + 2\boldsymbol{\epsilon}_e) \tag{2}$$

The subscripts e and i indicate the parameters for external and internal fluid, respectively. The complex permittivity is defined as

$$\boldsymbol{\epsilon} = \varepsilon + j\sigma/\omega \tag{3}$$

where $\sigma$ and $\varepsilon$ are the conductivity and the dielectric constant of the fluid, respectively. The function Re represents the real part of a complex value. The above analysis is applicable only to a two-phase system. For giant liposome suspensions, the permittivity and the thickness of the vesicle wall must be taken into account. A detailed analysis for vesicle and cell suspension can by found in Kaler and Jones [3], with a modified expression for $\mathbf{K}(\omega)$.

From the above analysis, one can see that the dielectrophoresis force depends on a number of parameters, among them the conductivity of the internal and external fluid, as well as the amplitude, gradient, and frequency of the applied field. For a given giant liposome suspension, the dielectrophoretic spectrum can be deduced and measured. This has been reported by Stoicheva and Hui [4].

Now consider the fusion force after the giant liposomes are adequately aligned. The voltage generated across each membrane wall of the giant liposome of radius $a$ is [5]

$$V = 1/5aE_{\mathrm{eff}}/(1+\omega^2\tau_{\mathrm{m}}^2)^{1/2} \tag{4}$$

where the respective effective field $E_{\mathrm{eff}}$ and polarizability $P_{\mathrm{eff}}$ are:

$$E_{\mathrm{eff}} = E + P_{\mathrm{eff}}/(16\pi\varepsilon_0\varepsilon_{\mathrm{e}}a^3) \tag{5}$$

$$P_{\mathrm{eff}} = (4/3)\pi\varepsilon_0\varepsilon_{\mathrm{e}}a^3\mathrm{Re}[\mathbf{K}(\omega)]E \tag{6}$$

and the membrane relaxation time $\tau_{\mathrm{m}}$, defined as the time taken to charge the membrane at the poles of the giant liposome, is

$$\tau_{\mathrm{m}} = aC_{\mathrm{m}}(1/\sigma_{\mathrm{i}} + 0.5/\sigma_{\mathrm{e}}) \tag{7}$$

Here $C_{\mathrm{m}}$ is the membrane capacitance, and $\sigma_{\mathrm{i}}$ and $\sigma_{\mathrm{e}}$ are the internal and external conductivities. There are two pulse-generated pressures to be considered, the $\Pi_{\mathrm{m}}$ across each vesicle wall and the $\Pi_{\mathrm{d}}$ across the sandwich of these two walls and the thin layer of external medium between them:

$$\Pi_{\mathrm{m}} = 0.5\varepsilon_0\varepsilon_{\mathrm{e}}\{1.5aE_{\mathrm{eff}}/h(1+\omega^2\tau_{\mathrm{m}}^2)^{1/2}\}^2 \tag{8}$$

and

$$\Pi_{\mathrm{d}} = P_{\mathrm{eff}}^2/(8\varepsilon_0\varepsilon_{\mathrm{e}}a^2A^2) \tag{9}$$

where $A$ is the contact area between the two lipid bilayer walls of the vesicles [6] and $h$ is the thickness of the membrane. The total pressure is the sum of these pressures. The breakdown pressure is a complex function of the size of the giant liposome, the conductivities of the system, as well as the applied pulse field and the duration of the pulse. Because the pressures build up with time during the pulse, a pulse too short will not have enough time to build the critical pressure necessary to fuse a pair of giant liposomes of a given size. For a suspension of giant liposomes with a distribution of sizes, the fusion thresholds in terms of pulse strength or pulse width will be sigmoidal curves, the slopes of which represent integrals of the size distribution.

The fusion pressure is applied against the forces that normally keep the lipid bilayers apart. These forces are well known and have been measured by various methods [7,8]. Two adjacent bilayers are attracted to each other by the van der Waals force, but are repelled by the electrostatic force at a longer range, and the hydration–protrusion force at a shorter distance. Modification of the bilayer surfaces by incorporating molecules that constitute steric bafflers has been shown to increase the interbilayer repulsion force [9,10]. Such an increase is manifested as an increase of the fusion pressure threshold, when all other parameters are held constant.

## 3. EXPERIMENTAL METHODS

Giant liposomes were prepared by the floatation method [4]. Chloroform solutions of 3 mg of egg phosphatidylcholine (PC), or bovine brain phosphatidylserine (PS) (Avanti, Alabaster, AL), or a 1 : 1 mixture of both, were vacuum dried on the bottom of a round glass flask. For surface-modified giant liposomes, 5 mol % of poly-(ethylene glycol) (MW 5000) conjugated distearoylphosphatidylethanolamine (PEG-DSPE) was added to the lipids before drying. After the organic solvent was thoroughly removed, 10 ml of 0.5–0.75 mM NaCl was gently added to the lipid film, and the flask with contents was heated to 70 °C for 3 h. Floated giant liposomes were collected, concentrated by centrifugation at 10 000 rpm, and resuspended in 2.5–3.75% (w/v) of poly(ethylene glycol) (MW 8000). The PEG8000 solution was to serve as an iso-osmotic but low-conductivity external medium. It also provided a high-contrast background, due to its sufficiently different refractive index, for visualizing the giant liposomes in phase contrast microscopy.

About 50–100 μl of giant liposome suspension was introduced into a microscope chamber, the bottom of which supports a pattern of interdigital gold electrodes [11]. The distance between the electrodes was 200 μm. A cover slip was placed over the giant liposome suspension. Individual giant liposomes were clearly visible through the glass chamber.

A sinusoidal AC electric field of 370 kHz and 345 V $cm^{-1}$ peak-to-peak field strength was applied across the electrodes. Pearl chains of giant liposomes formed under the dielectrophoresis force. Single electric pulses at a field strength of 1.6–1.7 kV $cm^{-1}$, at various widths of up to 1.4 ms, were applied while the AC field was instantaneously shut-off. After the pulse application, the AC field was resumed for at least 30 s, and then turned-off. The microscopic image of the entire process was recorded on video tape. The images were processed and analyzed digitally, using a Datacube (Peabody, MA) Maxvision image analyzer. The numbers of fused and unfused giant liposomes before and after pulse application were counted. Fusion threshold values in terms of pulse field strength and width were determined. The experiments were repeated for both modified and unmodified giant liposomes.

## 4. RESULTS AND DISCUSSION

### 4.1 Fusion thresholds for unmodified giant liposomes

According to the above analysis, fusion will occur when the applied pulse field is sufficiently high and long enough to cause the build-up of transbilayer potential, reaching the bilayer breakdown potential of approximately 1 V within the pulsing time, and raising the pressure to be sufficiently high to overcome the interbilayer repulsion force. From eq. (4), for a given applied pulse amplitude, only liposomes exceeding a given size may reach this bilayer breakdown potential of their walls. For instance, with an applied pulse of 1.7 kV cm$^{-1}$, only liposomes greater than 4 μm in radius may attain the breakdown transbilayer potential. In our system of giant liposome suspensions, $\sigma_i = 58\,\mu\text{S cm}^{-1}$ and $\sigma_e = 9\,\mu\text{S cm}^{-1}$; therefore, $\tau_m$ is 20 μs for a giant liposome of 4 μm radius. Thus, the minimum fusing pulse length is expected to be about 20 μs. Larger liposomes require a longer time to reach the bilayer breakdown potential. A plot of the fusion yield will represent the size profile of the giant liposome suspension.

Both neutral PC and charged PS giant liposomes were used in the experiment. Pearl chains of giant liposomes were formed in the AC field. For larger liposomes, the dipole force was sufficient to stretch the initially spherical liposomes into ellipsoidal shapes. This was more evident after the pulse, when even larger liposomes were formed as the result of fusion. Some were stretched into a tubular shape across the gaps between electrodes. They resumed a more relaxed shape after the AC field was turned off, assuming a spherical or a star shape, depending on the volume-to-surface ratio. A sequence of video image is presented in Figure 16.2.

Figure 16.3 is the plot of fusion yield as a function of the applied pulse width. The smallest liposomes undergoing fusion are about 4 μm in radius. The minimum pulse width when fusion can be observed is 19 μs, as predicted from the calculation. Liposomes of charged lipids, such as PS, require a longer pulse to attain a similar level of fusion as that of liposomes made of neutral lipids, such as PC. Although the surface charge can modify the effective bilayer potential, the main influence of the surface charge is the introduction of electrostatic repulsion between apposing bilayers. Modification of the external conductivity shifts the fusion pulse width threshold as predicted [4].

## 4.2 Modification of lipid bilayer surfaces

Incorporation of PEG-DSPE to the bilayer extends the hydration–protrusion boundary further from the bilayer surface, thus posing a higher energy barrier to the fusion. The amount of PEG-DSPE required for a complete coverage, thus forming a steric shield over the bilayer surface, may be calculated from the molecular weight and the configuration of the PEG moiety. At low surface concentrations, PEG assumes a mushroom configuration, the half diameter of which is approximately equal to the Flory diameter of the polymer in solution [12]. when the surface concentration reaches the molecular crowding limit, PEG assumes the brush configuration by extending further into the aqueous medium. A diagram showing the polymer shielding effect on surface forces is given in Figure 16.4.

The extended repulsion boundary not only obstructs bilayer–bilayer approach, but also inhibits molecular adsorption to the surface. Measurements of protein adsorption and cell adhesion versus the surface concentration of PEG-DSPE on the phospholipid surface verify the inhibition effect of the surface-grafted PEG (Figure 16.5 [13]). The continuous line in each panel is calculated from the

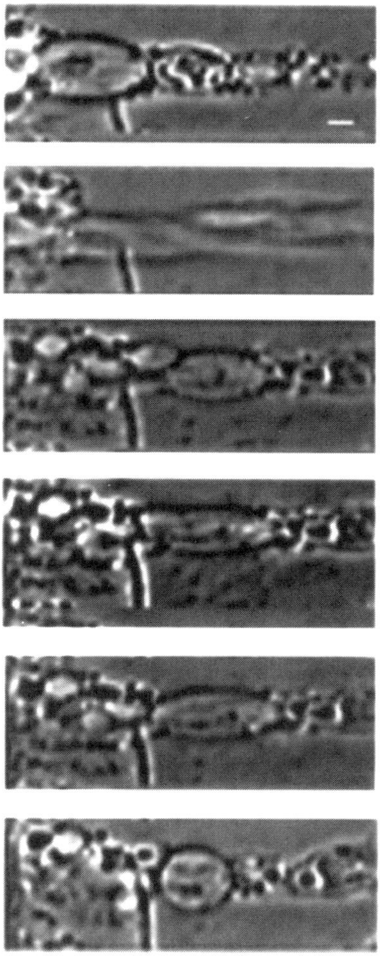

**Figure 16.2** A series of phase contrast micrographs depicting the time sequence of electrofusion of giant liposomes of mixed PC:PS (1:1). The recording times of the micrographs with respect to the 1.7 kV cm$^{-1}$, 250 μs pulse application are, from top to bottom, −1 s, 0 s, 1 s, 5 s, 30 s and after the 150 V cm$^{-1}$, 50 kHz AC field is turned off, (bar = 10 μm).

percentage of surface coverage by the polymer, assuming the mushroom configuration below the confluence limit and the onset for brush configuration to appear after the limit. The agreement is excellent. The polymer shielding effect has been utilized to reduce unwanted liposome loss due to phagocytosis by the mononuclear phagocytic system, in order to prolong the circulating time of drug-carrying liposomes. Hence, the name stealth liposome was concocted [14].

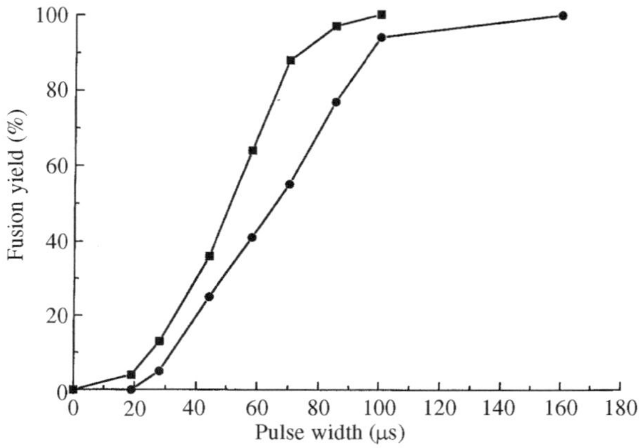


**Figure 16.3** The fusion yield of giant liposomes of (■) PC and (●) PS as functions of the applied pulse width. The applied pulse field is 1.7 kV cm$^{-1}$.

## 4.3 Fusion thresholds for modified giant liposomes

Giant liposomes with 5 mol % of (PEG-DSPE) were used to test whether the fusion pulse width thresholds were shifted by the surface shielding polymers. Figure 16.6 shows the fusion threshold of liposomes with added PEG-DSPE. The surface shielding offered by the PEG moiety increases the energy barrier for electrofusion. As a result, the pulse widths required to fuse these surface-modified giant liposomes

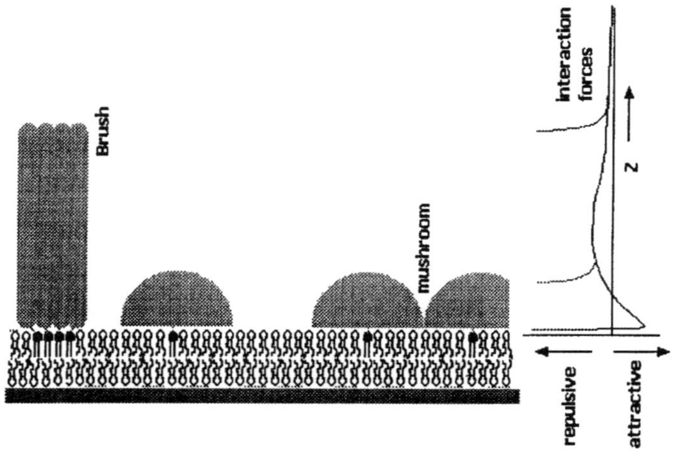


**Figure 16.4** Diagram of the steric shielding effect of surface-anchored polymers on the force field of the bilayer surface. The lipids with the solid head groups are polymer-bearing lipids PEG-DSPE. The curve represents qualitatively the combination of hydration–protrusion, van der Waals and electrostatic forces.

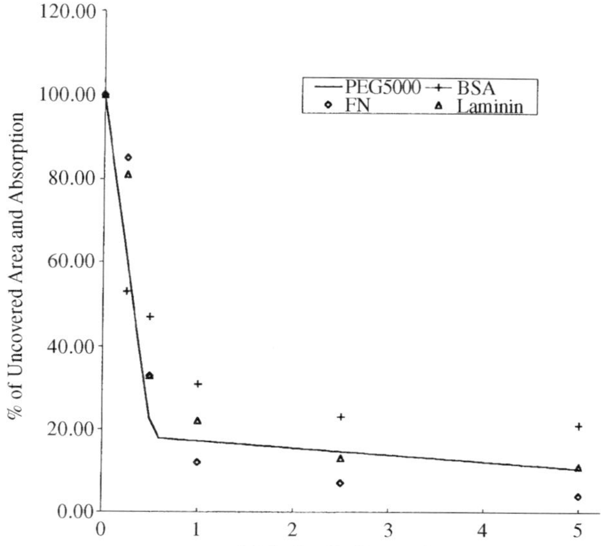


(a)

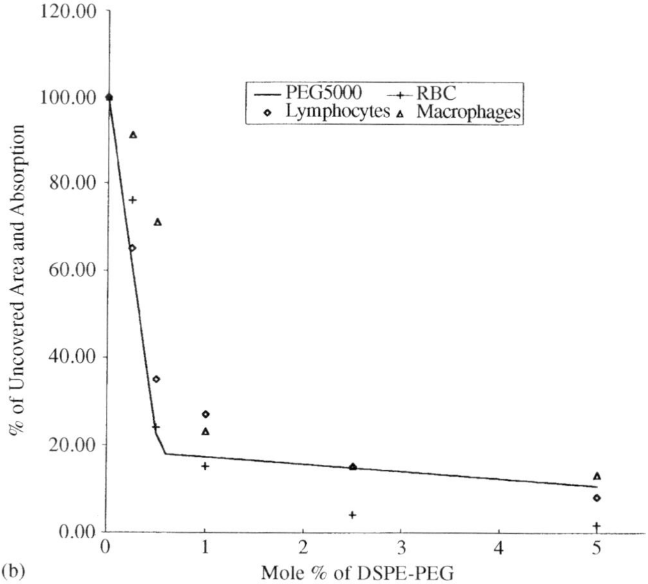


(b)

**Figure 16.5** (a) The relative adsorption of bovine serum albumin (BSA), fibronectin (FN) and laminin on a phospholipid surface containing various mole percentages of anchored DSPE-PEG5000. (b) The relative adhesion of erythrocytes (RBC), lymphocytes, and macrophages on the same surface (From [13], Courtesy of Elsevier Science).

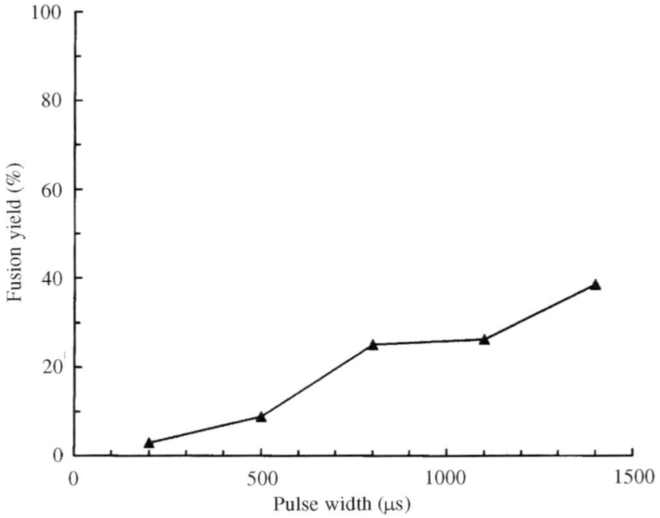


**Figure 16.6** The fusion yield of giant liposomes of PC : PS (1 : 1) mixture, containing 5 mol % of surface-anchored PE-PEG5000, as functions of the applied pulse width. The applied pulse field is 1.6 kV $cm^{-1}$.

are longer than those for unmodified ones of similar sizes, as shown in Figure 16.3, and the fusion is not as complete. In order to determine if the longer pulse width threshold and the incomplete fusion are entirely attributable to overcoming the repulsive forces of the surface-anchored polymers [9, 10], not only the surface coverage, but also the compressibility and lateral mobility of the surface-anchored PEG needs to be taken into consideration. The number of unknown parameters do not allow an unambiguous interpretation at this stage.

## 5. CONCLUSION

Giant liposomes provide an opportunity to precisely measure and calculate the forces involved in membrane fusion. The precise understanding of these forces not only helps to design better liposomes for various biotechnological applications, such as drug and gene delivery, but also provide a physical basis for understanding the many processes of biological membrane fusion. In the case of electrofusion, the giant liposome model supports most of the predictions of physical theory.

## 6. ACKNOWLEDGMENT

This work is supported by grant GM 30969 from the US National Institutes of Health.

## 7. REFERENCES

1. I. G. Abidor, L-H. Li and S. W. Hui, *Biophys. J.*, **67**, 427–435 (1994).
2. H. A. Pohl, *Dielectrophoresis*, Cambridge University Press, Cambridge, 1978.
3. K. V. I. S. Kaler and T. B. Jones, *Biophys. J.*, **57**, 173–182 (1990).
4. N. G. Stoicheva and S. W. Hui, *Biochim. Biophys. Acta.*, **1195**, 31–35 (1994).
5. C. Holzapfel, J. Vienken and U. Zimmermann, *J. Membrane Biol.*, **67**, 13–26 (1982).
6. D. A. Stenger, K. V. I. S. Kaler, and S. W. Hui, *Biophys. J.*, **59**, 1074–84 (1991).
7. R. P. Rand and V. A. Parsegian, *Biochim. Biophys. Acta*, **988**, 351–376 (1989).
8. J. Isrealachvili and H. Wennerstrom, *J. Phys. Chem.*, **96**, 520–531 (1991).
9. T. L. Kuhl, D. E. Leckband, D. D. Lasic and J. N. Israelachvili, *Biophys. J.*, **66**, 1479–1488 (1994).
10. A. K. Kenworthy, S. A. Simon and T. J. McIntosh, *Biophys. J.*, **68**, 1903–20 (1995).
11. D. A. Stenger, R. T. Kubiniec, W. J. Purucker, H. Liang and S. W. Hui, *Hybridoma.*, **7**, 505–518 (1988).
12. P. G. de Gennes, in: *Physical Basis of Cell–Cell Adhesion* (ed.) P. Bongrand, CRC Press, Boca Raton, FL, 1988, pp. 39–60.
13. H. Du, P. Chandaroy and S. W. Hui, *Biochim. Biophys. Acta*, **1356**, 236–248 (1997).
14. M. C. Woodle and D. D. Lasic, *Biochim. Biophys. Acta*, **1113**, 171–199 (1992).

# Chapter 17

# Membrane Roughness and Dispersive Phase as Effects of Higher-order Bending in Fluid Membranes

*BEATE KLÖSGEN AND WOLFGANG HELFRICH*
*Freie Universität, Berlin, Germany*

## 1. INTRODUCTION

Amphiphilic molecules such as lipids have an extremely low solubility in water and tend to self-associate and form lyotropic liquid crystals [1] if their monomer concentration in aqueous surroundings exceeds a critical value (the critical monomer concentration, (CMC), [2]). This self-organization is due to the hydrophobic effect [2]. It leads to the formation of micro- and mesoscopic[i] structures, among them micellar, hexagonal, lamellar and cubic phases [3] and the respective transitions between them.

[i] Mesoscopic here is thought to signify structures between the macroscopic (optically visible, above about 100 μm) and the microscopic or nanometer scale.
The contributions of the normal bending elasticity prevent the formation of permanent highly curved pieces of fluid membranes. The typical length scale of cubic structures for example is around 250 Å. Also, no stable unilamellar vesicles are found below a radius of about 20 nm [29] because the high curvature requires a lot of energy and results in high membrane tension and an affinity for fusion to end up with a bigger, and less curved, new vesicle.

*Giant Vesicles* Edited by P. L. Luisi and P. Walde

In particular, the planar bilayer phase manifests itself in several thermotropic subphases ($L_\beta$, $L'_\beta$, $P'_\beta$ and $L_\alpha$)[ii] that differ as to the orientational order and mobility of the lipid molecules. The common characteristic of this polymorphism is that the molecules arrange themselves into structures with a hydrophobic core by orienting their hydrophilic portions towards the water environment.

Fluid almost planar membranes ($L_\alpha$ phase) are of special biological interest because they are the basic organizational structure of every living cell (fluid mosaic model of biomembranes [5]). In the $L_\alpha$ phase, the molecules can freely diffuse within their monolayer and the membranes are very flexible. One first consequence is that the bilayers tend to form closed compartments (so-called vesicles) because open hydrophobic edges are energetically unfavorable. Lipid vesicles often serve as the simplest model of biological cells. Real biomembranes consist of the fluid lipid bilayer and macromolecules (such as proteins) that are associated with or incorporated into it. The lipid matrix provides a soft and adjustable interface for these solvated molecules; it is essential for the proper functioning of a cell or organelle [6].

The elastic properties of fluid lipid membranes comprise only contributions of stretching and bending deformations (as there is no shear deformation in a fluid). Expanding the bilayer requires a lot of energy so that membranes usually do not change their stretching state. However, bending elasticity is in the range of thermal energy and thus plays an important role:

- Soft membranes can bend spontaneously to form closed compartments[iii] with fluid walls.
- The shapes of these vesicles are given by minima of the bending elastic energy.
- Thermally induced fluctuations contribute to the interaction between membranes.

The theoretical approach to describe all these effects was formulated long ago [7], at the time purely on the basis of Hookean deformations that contribute to the bending elastic energy density of fluid membranes:

$$g \approx \tfrac{1}{2}\kappa(c_1 + c_2 - c_0)^2 + \bar{\kappa}_1(c_1 c_2) \tag{1}$$

[ii] This description was introduced to specify crystals that are ordered in less than three dimensions. The Latin capitals signify the lattice type of the molecular crystals. Thus L and P correspond to one-dimensional lamellar and two-dimensional oblique lattices, respectively. The Greek subscripts describe the detailed organization of the fatty acid chains and headgroups: $\alpha$ corresponds to the molten high-temperature state of the chains, whereas $\beta$ and $\beta'$ indicate that the chains are ordered, either parallel or tilted with respect to the director. In all these cases the headgroups are free to rotate. At even lower temperatures, a final solid state $\gamma$ is reached when headgroups also become ordered [4].

In the molten state, molecules may diffuse within their monolayer; therein the membrane resembles a two-dimensional liquid. Fluid membranes are soft and therefore easily deformable, whereas solid membranes are stiff and rather break on tearing than to adjust into a new shape. Nevertheless bending of fluid bilayers requires energy. The final equilibrium shape of an object that is confined by a fluid membrane is given by the minima of the bending elastic energy [8,9].

[iii] The free enthalpy of the system is reduced by the formation of a closed vesicular object because the energy that is necessary to bend a soft membrane is smaller than the amount required to maintain an interface between an otherwise open hydrophobic core and the water surroundings [29].

Herein, $g$ is the energy density ($\mathrm{J\,m^{-2}}$) of bending, and $c_1$ and $c_2$ are the principal and $c_0$ the spontaneous curvature. The bending moduli $\kappa$ and $\bar{\kappa}_1$ are associated with the mean $(c_1 + c_2)/2$ and the Gaussian $(c_1 c_2)$ curvature, respectively. The term $\bar{\kappa}_1(c_1 c_2)$ contributes an amount of $\Delta g = -4\pi\bar{\kappa}_1 < 0$ (for $\bar{\kappa}_1 > 0$) for every new passage between membranes that is formed but remains constant otherwise. It is therefore related to the topology of the membrane. The experimentally observed shapes of giant vesicles and their calculated forms, as found by minimization of eq. (1), correspond to each other [8–11]. The bending modulus $\kappa$ was measured to be of the order of $10^{-19}$ J which is about 10–20 kT [12–17].

However, usually by the use of phase contrast light microscopy techniques, membrane aspects have been observed that hint towards submicroscopic structures of the fluid membranes. Examples are localized stable wiggles in tubular vesicles [12] and an abundance of tethers in many samples. Conflicting results in the context of the induced adhesion [18,19] suggested a fine superstructure of fluid membranes with an additional hidden area [20,21], superimposed on the well-known undulations [22]. The existence of so-called dark bodies [23] in almost any of the lipid systems investigated and their formation processes (see below) implied the assumption that they are developed by the association of very small bilayer particles dispersed in the aqueous volume. Typical examples of dark bodies are shown in Figure 17.1 for DGDG.

## 2. DARK BODIES AND THE SUPERSTRUCTURE OF FLUID MEMBRANES

Dark bodies have sizes up to about 70 μm. They can be precipitated within minutes [23] in water chambers of highly swollen well-aligned multilayer systems. They

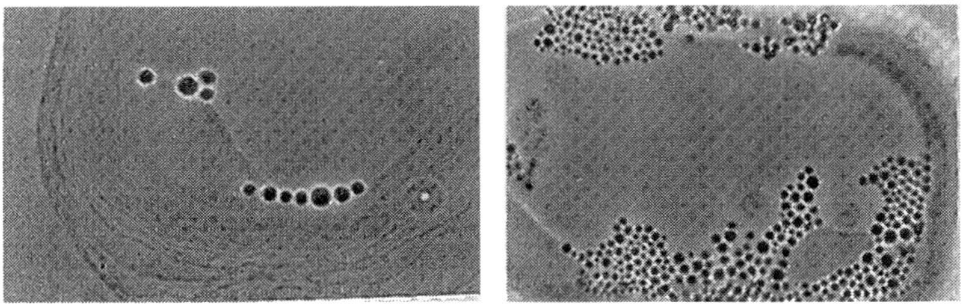

**Figure 17.1** Phase contrast optical micrographs showing dark bodies of digalactosoyl-diacyl-glycerol (DGDG; Lipid Products, Nutfield, Great Britain) in approximately 40 μm thick glass chambers[iv]. The sample was prepared in 50 mM NaCl solution. The left picture, taken during the formation process, exhibits dark bodies in coexistence with fluid membranes, whereas the right picture shows them numerously in a later phase. (Micrographs by courtesy of J. Thimmel).

[iv] DGDG is a mixture of lipids extracted from wheat flour. Dark bodies were abundant in many of the samples investigated, as well in systems of mixed lipids that had been extracted from natural tissue (egg PC, DGDG, cardiolipin, sphingomyelin) and also for several synthetic and pure PC systems (DOPC, POPC).

appear initially as rapidly moving tiny flakes, then grow steadily to become spherical dark bodies. The structural details are below the limit of optical resolution but from their properties in polarization microscopy it was concluded that their structure is isotropic, possibly of cubic symmetry. They seem to be made of connected bilayers but their reversible formation makes them different from the inert passages and their lattices [24]. The neighboring membranes are not discernibly involved in the assembling or the disassembling process of the dark bodies and their precursors. They can not be the immediate source of the material that constitutes the dark bodies. Apparently, there has to be a dispersive phase of optically unresolvable small bilayer particles in the aqueous medium. The elements of the dispersive phase should carry reactive sites that support the formation of clusters and the build-up of dark bodies and that also ease their disassembly back into the aqueous medium. However, direct evidence for the existence of such bilayer arrangements can only be given by use of high-resolution methods.

The situation was similar in the context of a submicroscopic superstructure of fluid membranes that was only indirectly hinted at in numerous observations [25]. The most striking implication came from inconsistent results concerning the adhesion energy of membranes [18,19] that turned out to differ by about two orders of magnitude, dependent on the approach used to analyze the experimental results.

The appearances of superstructure and dispersed membrane particles were investigated by two different transmission electron microscopy techniques, namely cryotransmission electron microscopy and freeze etching microscopy. Experimental details are given at the end of the article.

Using cryotransmission electron microscopy (cryo-TEM), abundant and reproducible evidence for a superstructure of 1,2-dioleoyl-*sn*-glycero-3-phosphocholine (DOPC) bilayers was obtained [26]. The micrographs suggest that the small fraction of vesicles enclosing one another develop passages (connections) between the bilayers. In contrast, the superstructure is basically a feature of disconnected single membranes. Typical examples are compiled in Figure 17.2. Among its modifications are isolated membrane bends or folds (Figure 17.2, left column) and a grainy membrane texture with a minimal grain spacing of 4–6 nm (Figure 17.2, right column). The two-dimensional graininess was observed with several densities. As the density decreases, the impression of short stretches of continuous lines rather than sequences of dark spots arises, and at the lowest densities the texture is reminiscent of an irregular network of lines. The spacing of the lines varies from 6 to about 20 nm. The graininess seems to avoid regions of high curvature. The object in the last micrograph of Figure 17.2 exhibits a tendency for phase separation into smooth and grainy membrane portions, even in the absence of curvature. Bounding lines constrict the vesicle; this suggests a line tension and supports the idea of two coexisting physical phases.

Transmission electron microscopy investigations performed on freeze-etching replicas of lipid vesicles of 1,2-dimyristoyl-*sn*-glycero-3-phosphocholine (DMPC)

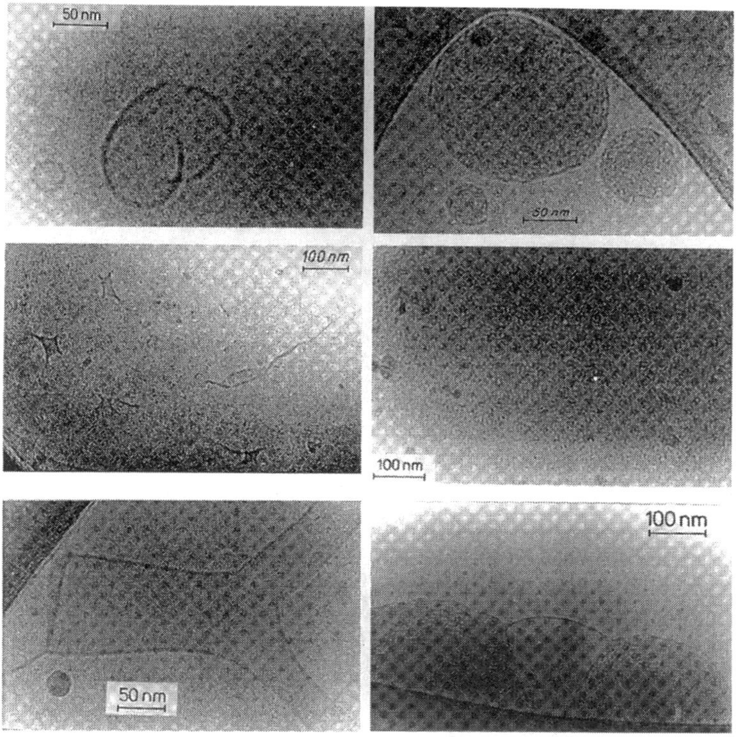


**Figure 17.2** Manifestations of superstructural aspects of DOPC bilayers, observed by cryo-TEM on surfaces of unilamellar vesicles that were prepared in pure water [26]. Left column: linear isolated membrane bends or folds. They are found as deep notches as in heart-shaped vesicles and also as extended overhangs on flat membranes. (The cushion-like contours indicate the presence of passages (overpasses) on the membrane.) The constrictions on the branched tube may be envisaged as lines of saddles. Right column: two-dimensional grainy membrane textures. The graininess was observed on small vesicles as well as on planar membranes and may even coexist with smooth bilayer portions.

and 1-palmitoyl-2-oleoyl-*sn*-glycero-3-phosphocholine (POPC) in aqueous suspensions revealed the presence of highly curved small particles of lipid bilayers [27]. Some are shown in Figure 17.3. These particles are not unique in shape and resemble thin cylinders with frequent narrow neck-like modulations of their diameters. They seem to consist of an aggregation of small and not really resolvable basic elements. In addition to a straight alignment, the aggregated dispersed particles may also form Y-shaped or T-shaped branches. They coexist with normal lipid vesicles and, occasionally, were seen as if emanating from the vesicle membranes. The observation is interpreted in terms of a dispersive bilayer phase. Such a phase of

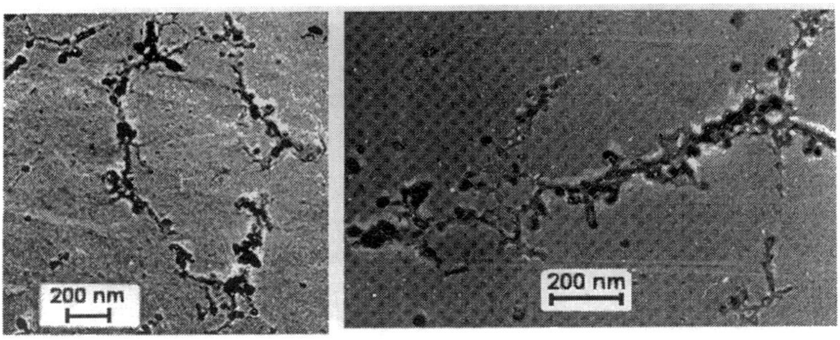


**Figure 17.3** Micrographs from electron microscopy obtained by investigations of freeze-etching replicas. The pictures exhibit traces of lipid particles, observed dispersed in an aqueous surroundings and in coexistence with lipid vesicles. The objects consist of elongated aggregations of knotted sticks, often with branches. Left picture: preparation of POPC, extruded in a salt solution stock suspension (NaCl, 5 mmol $l^{-1}$) and exposed to an osmotic gradient of $\Delta c \approx 10$ mosmol $l^{-1}$ prior to freezing. Right picture: preparation of DMPC under the same conditions. In all cases, and independent of the presence of salt in the suspension, the dispersed bilayer particles mostly consisted of connected structures. The orientation of the particles flat onto the ice surface is probably an artifact that results from the collapse of thin objects under their own weight when they are too far exposed from the surface during deep etching.

dispersed small bilayer particles was postulated as the material source of the so-called dark bodies because these objects were seen to form fast and reversibly from the aqueous volume between membranes.

The normal quadratic expression for the bending energy cannot explain the development either of the fine membrane texture and the highly localized folds or of the small and highly curved bilayer particles of the dispersive phase. The forma-tion of these structures was therefore attributed to additional bending terms, namely, those caused by Gaussian curvature contributions of higher than quadratic order. A modified bending energy density function is therefore proposed:

$$g \propto \tfrac{1}{2}\kappa(c_1 + c_2 - c_0)^2 + \bar{\kappa}_1(c_1 \times c_2) + \bar{\kappa}_2(c_1 \times c_2)^2 + \bar{\kappa}_4(c_1 \times c_2)^4 \qquad (2)$$

This function comprises higher-order terms of $(c_1 \times c_2)$ and two new moduli, $\bar{\kappa}_2$ and $\bar{\kappa}_4$, for the second- and fourth-order Gaussian bending, respectively. Values of $\bar{\kappa}_2 < 0$ and $\bar{\kappa}_4 > 0$ were estimated on the basis of the known module $\kappa$ and a simplified bilayer stress profile [28]. The higher-order terms will work to warp the membranes and support the formation of highly curved saddle structures whenever the net contribution of the new terms is negative, i.e. as long as $\kappa_2 < -\kappa_4(c_1 \times c_2)^2$. This holds for spheres ($c_1 = c_2$) and for saddles ($c_1 = -c_2$). Differing from the development of saddles (e.g. graininess, necks), the formation of spheres additionally requires normal bending energy. Therefore, the small spheres are energetically less favorable than the saddles. They are unstable as isolated objects and, if formed at all, they will probably decay quickly (such as by fusion) unless the possibly negative contribution of the higher order Gaussian terms $\bar{\kappa}_2(c_1 \times c_2)^2$ and $\bar{\kappa}_4(c_1 \times c_2)^4$ compensate for that of the normal bending term $\frac{1}{2}\kappa(c_1 + c_2 - c_0)^2$. The absence of spontaneous vesiculation in the systems suggests that this is not the case. However, a neck – although being promising with regard to bending energy – cannot exist as a singular structure because of its open borderline which is energetically very expensive [29]. So there is a need for a cap. The combination of necks and knots seems to be favorable.

The basic element of the dispersive phase possibly is a highly curved knot with a (half-) neck. Such an object most probably is only a temporary configuration with short lifetime. It may nevertheless act as an active site with a tendency to fuse with another basic element. The resulting object will have a reduced energy as compared to the two essentially spherical basic elements. It forms because the open borderline is avoided and the complete neck consists exclusively of saddles, so it requires no bending energy. That kind of fusion will gradually lead to the assembly of several elements into lengthy knotted sticks. The ends of the sticks must be spherical caps. Individual knotted sticks can join and develop T- or Y-shaped branches by the formation of a new neck between two knots. The energy of the system can be gradually reduced by a stepwise transformation of spherical curvature into saddle curvature and the reduction of spherical portions on the knots. Obviously, the dispersive phase remains active as long as highly curved knots are present and accessible. The branching or the association of branches would finally result in the buildup of extended three-dimensional structures like the connected structures that are presumably present in the dark bodies. Observations of dark body formation indeed suggested a high reactivity of the underlying structures that allowed for a quick assembly or disassembly of the dark bodies. The nonquadratic contributions to the bending energy density could thus in time change a lipid molecule arrangement which initially was present as (slightly curved) bilayers into another, but highly curved, bilayer structure such as the elements and composed bigger particles of the dispersive phase.

The situation is somehow less complicated in the case of the membrane roughness found by cryo-TEM. All the observed variations in the normal smoothness are due to saddle formations. The new Hamiltonian of eq. (2) favors such shapes. Preliminary results from Monte Carlo simulations [30] at least reveal the development of a fine saddle structure on an initially flat membrane. They make use of the modified bending Hamiltonian and put in reasonable values for the moduli. The new surface appearance resembles an egg carton and has a lower free energy density than the planar membrane. It was found as a highly ordered structure at low temperatures and seen to experience a phase transition into a melted state. Herein, the simulations show deviations from the strictly ordered pattern of the low-temperature regime: the pattern was seen to change in time, and neighboring extremes could aggregate. The experimentally observed graininess as seen by cryo-TEM is probably a melted egg-carton structure.

## 3. EXPERIMENTAL DETAILS

### 3.1 Electron microscopy

In cryo-TEM, the specimen consists of a thin vitrified suspension of the lipid structures to be investigated. The cryo-microscope (CM12, Philips) was operated at 100 kV ($\lambda \approx 3.6$ pm) with a primary magnification of 60 000 and at a resolution below 10 Å. A nitrogen-cooled specimen holder (Gatan) set the sample temperature to about 100 K. Image contrast was achieved by phase contrast applying a defocus $\Delta z = -1.2$ μm in order to enhance the contrast of structures around the size of the membrane bilayer.

The freeze-etching technique is indirect in that it is based on the observation of replicas made from frozen suspensions of the respective structures. Two 80 kV ($\lambda = 4.3$ pm) electron microscopes (Elmiskop, Siemens; EM300, Philips) were operated in the normal transmission mode, adjusting to Gauss focus at moderate magnification (up to 13 000). The resolution is limited by the granularity of the coating film (up to about 5 nm).

Data were recorded on electron-sensitive film material (Agfa Scientia).

### 3.2 Cryopreservation

Vitrification of the sample suspensions for cryo-TEM is intended as a tool to optimally conserve the *in vitro* lipid structures in their aqueous surroundings. Special microscope grids for the specimen preparation were used as commercially available (Ted Pella Inc., Redding, CA). They consisted of copper grids (300 mesh) that were covered with a perforated plastic foil (lacey carbon foil on formvar). Thin fluid films of the lipid suspensions were spanned in the holes of the foil to establish most efficient cooling conditions for vitrification [31]. A CEVC (controlled environment vitrification chamber, [32]) device was used to establish a defined starting temperature of the freezing process and to maintain an atmosphere of high humidity in order to prevent evaporation of water from the thin film. In the CEVC, the grid was held by a tweezers coupled to a plunger mechanism to catapult it into liquid-nitrogen-cooled liquid ethane (about 80 K). At freezing rates greater than $10^4\,K\,s^{-1}$ this provides an instantaneous vitrification of the samples. A cryotransfer system kept the frozen sample at the temperature of liquid nitrogen until it was mounted into the cooled specimen holder and investigated.

### 3.3 Replica formation

The sample first had to be frozen as a small droplet (diameter about 1.5 mm) in liquid propane (boiling point 228.6 K, melting point 84.4 K) that was kept around its melting temperature by indirect contact with liquid nitrogen (boiling point 77.3 K).

Then it was transferred into a cooled cutter/coater device (Ba 360M, BALTEC, Liechtenstein). A flat sample surface was prepared by a series of increasingly fine cutting steps. From this surface, the objects to be investigated were exposed by etching. Immediate deposition of an intimately coating thin contrasting film (Pt/C film, about 2 nm) and subsequent deposition of a mechanically stabilizing carbon layer resulted in the formation of a replica that was then transferred onto a standard copper grid for investigation in the transmission microscope.

More details of the experimental procedures are given elsewhere [26,27,33].

## 4. ACKNOWLEDGMENT

Part of this work was supported by Deutsche Forschungsgemeinschaft (DFG) through SFB 312.

## 5. REFERENCES

1. V. Luzzatti and P. A. Spegt, *Nature*, **215**, 701 (1967).
2. C. Tanford, *The hydrophobic effect*, Wiley, New York, 1991.
3. J. M. Seddon and R. Templer, in *Structure and Dynamics of Membranes, From Cells to Vesicles, Handbook of Biological Physics*, vol. 1A. (eds) R. Lipowsky and E. Sackmann, Elsevier Science, 1995, p. 97.
4. J. Charvolin and A. Tardieu, *Sol. State Phys., suppl.*, **14**, 209 (1978).
5. S. J. Singer and G. L. Nicholson, *Science*, **175**, 720 (1972).
6. P. K. J. Kinnunen, *Chem. Phys. Lipids*, **57**, 375 (1991).
7. W. Helfrich, Z. *Naturforschung*, **28**c, 1/12, 693 (1973).
8. H. J. Deuling and W. Helfrich, *J. Physique*, **37**, 1335 (1976).
9. U. Seifert, K. Berndl and R. Lipowsky, *Phys. Rev. A*, **44**, 1181 (1991).
10. X. Michalet and D. Bensimon, *J. Phys. II France*, **5**, 263 (1995).
11. H.-G. Döbereiner, E. Evans, M. Kraus, U. Seifert and M. Wortis, *Phys. Rev. E*, **55** (4), 4558 (1997).
12. G. Beblik, R. M. Servuss and W. Helfrich, *J. Physique*, **46**, 1773 (1985).
13. E. Sackmann, H.-P. Duwe and H. Engelhardt, *Faraday. Disc. Chem. Soc.*, **81**, 281 (1986).
14. M. Mutz and W. Helfrich, *J. Phys. France*, **51**, 991 (1990).
15. M. I. Angelova, S. Soléau, P. Méléard, J. F. Faucon and P. Bothorel, in *The Structure and Conformation of Amphiphilic Membranes*, (eds) R. Lipowsky, D. Richter and K. Kremer, Springer, Berlin, 1991, p. 178.
16. G. Niggemann, M. Kummrow and W. Helfrich, *J. Phys. II France*, **5**, 413 (1995).
17. P. Méléard, C. Gerbeaud, T. Pott, L. Fernandez-Puente, I. Bivas, M. D. Mitov, J. Dufourcq and P. Bothorel, *Biophys. J.*, **72**, 2616 (1997).
18. R. M. Servuss and W. Helfrich, *J. Phys. France*, **50**, 809 (1989).
19. W. Harbich and W. Helfrich, *J. Phys. France*, **51**, 1027 (1990).
20. W. Helfrich, *Liquid Cryst.*, **5**, 1647 (1989).
21. W. Helfrich and B. Klösgen, in *Dynamical Phenomena at Interfaces, Surfaces and Membranes*, (eds) D. Beysens, N. Boccara and G. Forgacs, Nova Science, New York, 1993, p. 11.

22. W. Helfrich and R. M. Servuss, *Il Nuovo Cimento*, **D3**, 137 (1984).
23. W. Harbich and W. Helfrich, *Chem. Phys. Lipids*, **55**, 191 (1990).
24. W. Harbich, R. M. Servuss and W. Helfrich, *Z. Naturforsch.*, **33a**, 1013 (1978).
25. W. Helfrich, in *Structure and Dynamics of Membranes*. Handbook of Biological Physics, vol. 1B, (eds) R. Lipowsky and E. Sackmann, Elsevier, Amsterdam, 1995, p. 691.
26. B. Klösgen and W. Helfrich, *Biophys. J.*, **73**, 3016 (1997).
27. A. Dunger, B. Klösgen and W. Helfrich, *Chem. Phys. Lipids*, in press.
28. R. Goetz and W. Helfrich, *J. Phys. France*, **6**, 215 (1996).
29. P. Fromherz, C. Roecker and D. Rüppel, *Faraday Discuss. Chem. Soc.*, **81**, 39 (1986).
30. A. Jud, PhD Thesis, Freie Universität Berlin, Germany, 1999; A. Jud and W. Helfrich, unpublished results.
31. Y. Talmon, *Ber. Bunsenges. Phys. Chem.*, **100**, 364 (1996).
32. J. R. Bellare, H. T. Davis, L. E. Scriven and Y. Talmon, *J. Electron Microsc. Technique*, **10**, 87 (1988).
33. B. Klösgen and W. Helfrich, *Eur. Biophys. J.*, **22**, 329 (1993).

# Chapter 18

# Swelling and Separation of DOPC Multilayer Systems

*JOHANNES THIMMEL, BEATE KLÖSGEN AND WOLFGANG HELFRICH*
*Freie Universität, Berlin, Germany*

*GERT RAPP*
*DESY, Hamburg, Germany*

## 1. INTRODUCTION

Phosphatidylcholines (PCs) above their main transition temperature form in excess pure water a multilamellar system that is characterized by a repeat distance [1]. At 30°C, it was measured to be 63.1 Å for DOPC [2]. A stack of membranes, however, can also be prepared in such a way that even single membranes detach and are observable under the light microscope. This property was originally described as a method to prepare giant vesicles [3]. The results of a first X-ray diffraction study of the underlying membrane disassembly [4] proved that an initial period of water uptake, as revealed by the respective lamellar Bragg pattern signals, was often followed in a second stage by a continuous and complete decay of the Bragg peak intensities at constant peak position. The decay time of the peaks varied from sample to sample and in a few cases, the lamellar order remained unchanged over weeks.

The following discussion reports on new studies of multilayer swelling and membrane separation. In all samples, X-ray scattering reproducibly exhibits a disassembly of the multilamellar system which begins when its repeat distance is close to the full hydration value. Both the time for the system to reach this distance and the decay rate of the respective X-ray scattering signal depend on temperature. In simultaneous light microscopy investigations the lipid structures emanating from

*Giant Vesicles* Edited by P. L. Luisi and P. Walde

an initial lipid ring were observed and found to differ between the two stages of swelling.

## 2. MATERIALS AND METHODS

1,2-dioleoyl-*sn*-glycero-3-phosphocholine (DOPC) with a purity above 99% was used as purchased (Avanti Polar Lipids, Alabaster, AL or Sigma, Deisenhofen, FRG). For both the optical and the X-ray investigations the lipid was deposited from a chloroform solution (10 or 20 mg ml$^{-1}$) into an X-ray glass capillary (Hilgenberg, Malsberg, FRG, diameter 0.5 or 0.8 mm, thickness $\approx 10\,\mu m$) as a well confined open DOPC ring (Figure 18.1). Storage of the sample at low pressure (below 100 Pa) for more than 12 h removed solvent remnants. Swelling was initiated by filling the capillary to a lipid : water ratio of 1 wt % with distilled, degassed water (Seradest, Millipore, Eschborn, FRG). A paraffin seal kept the water concentration constant, up to a temperature of 65°C. The capillaries were mounted into a temperature stage and heated to the final swelling temperature within less than 3 min.

The diffraction measurements started 7–25 min after water addition. Small angle X-ray scattering (SAXS) measurements were performed with a rotating anode setup ($CuK_\alpha$ line, $\lambda = 1.54$ Å) in the laboratory and at the beamline XR13 of the European Molecular Biology Laboratory (EMBL) outstation at Deutsches Elektronen Synchrotron (DESY), Hamburg ($\lambda = 1.5$ Å). Spectra of the thermostated samples were recorded at a typical time resolution of 1 min with a linear position sensitive detector (PSD). The synchrotron measurements were complemented by additionally recording the two-dimensional diffraction pattern on an image plate system (FUJI, BAS 2000) at the beginning and at the end of each experiment.

The morphology of the swelling structures was optically observed. In a few cases, a low resolution eyepiece served to record images of the sample simultaneously with the measurement of the corresponding X-ray diffraction patterns. Alternatively, the

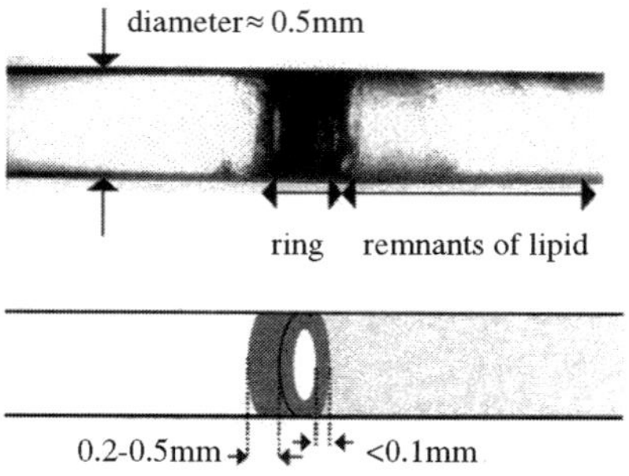


**Figure 18.1** Above: bright field micrograph of an X-ray glass capillary with a dry lipid ring. Below: diagram of the micrograph.

whole swelling process was monitored for several samples using a high-resolution microscope (Leitz Ortholux II, Pol BK) in phase contrast and polarization mode.

## 3. RESULTS

### 3.1 X-Ray diffraction

Usually, the lipid material for X-ray investigations is organized in a packed volume of multilamellar liposomes. The deposition of the lipid as a thin ring is advantageous because the steric hindrance is low: the structures emanating during swelling from the ring are provided with enough free space to develop. Additionally, the thickness of the ring is sufficient to yield strong X-ray signals.

From the beginning, the diffraction pattern reveals first- and second-order Bragg reflexes, characteristic of the multilamellar system. A considerable part of the image plate (IP) signal shown in Figure 18.2 originates from membranes aligned parallel to the capillary wall. The intensity integration over the oriented spot within an arc of about $5^\circ$ $I$ ($5^\circ$) yielded a relative orientation $\alpha = I(5^\circ)/I(180^\circ)$ of 18%. For an ensemble of seven similarly prepared samples, $\alpha$ was found to be between 17% and 97%. The swelling behaviour did not depend on the degree of orientation $\alpha$.

The change within one sample heated to 35°C is shown in Figure 18.3 for a series of original PSD spectra that show the first- and second-order Bragg reflex intensities to decrease in the course of time. The evaluation of these spectra (Figure 18.4) exhibits two stages of swelling: during the initial asymptotic increase of the lamellar

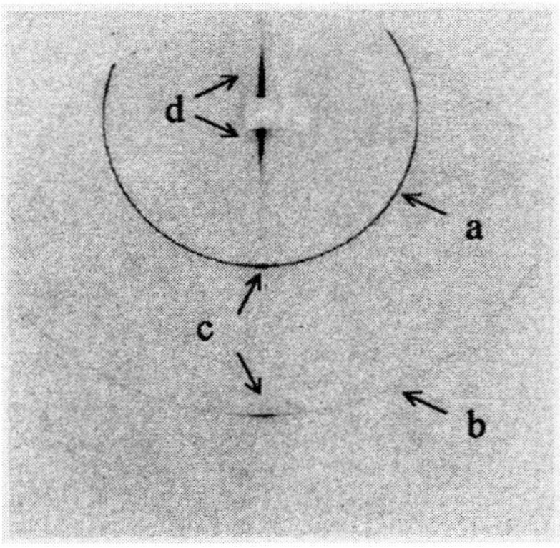


**Figure 18.2** Bragg pattern recorded on an image plate at the beginning of swelling: (a) first-order and (b) second-order reflections are visible. Both arcs exhibit (c) a pronounced spot that originates from the orientation of the lamellae parallel to the capillary surface. (d) The central elongated stripes close to the beam stop are an artefact that results from the total reflection of the primary beam at the surface of the glass capillary.

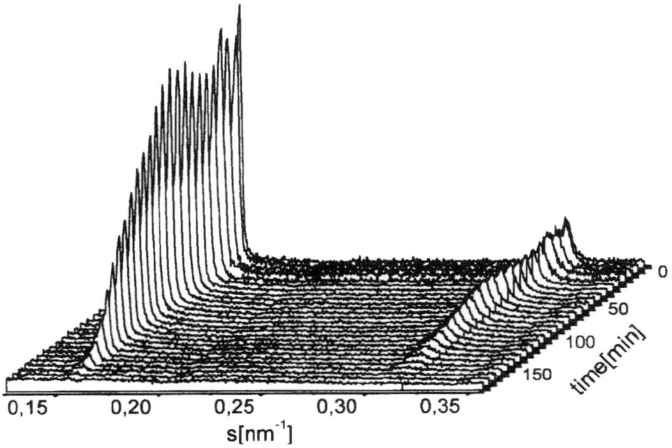


**Figure 18.3** Series of diffraction spectra $I$(s) of a DOPC sample that was heated at $t = 20$ min within 2 min from 20°C to the final swelling temperature of 35°C. $I$: scattering intensity, reciprocal spacing: $s = 1/d = 2\sin(\theta/\lambda)$; DOPC:water = 0.8 : 100 (by weight).

repeat distance $d$ to a value of about 64.5 Å the intensities of the Bragg reflexes remain nearly constant (stage I). Subsequently, the peak intensities drop continuously at constant membrane spacing (stage II). This almost linear decrease of the intensity with time proceeds until hardly any lamellar order is discernible by X-ray scattering. The scattering background increases slightly but no new Bragg peaks

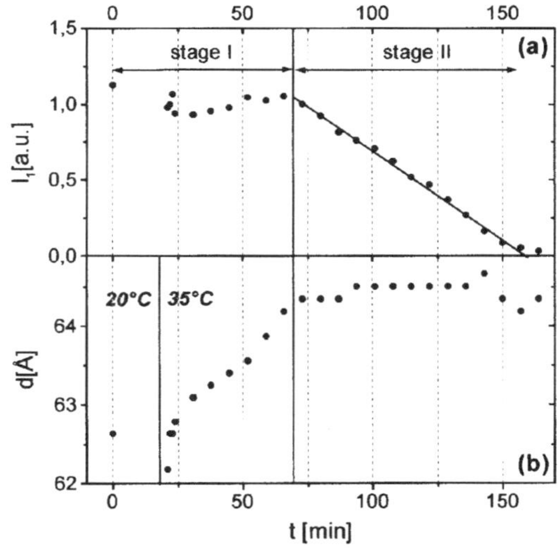


**Figure 18.4** Analysis of the time dependence of the spectra of Figure 18.3: (a) maximal intensities of the first-order $I_1$ versus $t$; (b) repeat distance $d$ of the multilayer system versus $t$.

develop. The new structure either has no long-range order or its characteristic length must be larger than the low-angle limit of the set-up (650 Å).

Swelling experiments performed at several temperatures between 32°C and 62°C on 44 samples yielded the same results as described above. The time interval for the maximum repeat distance $d$ to be reached during stage I decreases almost linearly with increasing temperature. Simultaneously, the rate of decomposition of the lamellar order increases strongly. At elevated temperatures ($t > 50°C$), the onset of the decrease of scattering intensity often starts slightly before the system has reached its maximal repeat distance.

### 3.2 Light microscopy

Domains with a pronounced birefringence, apparently liposomes and myelin cylinders [5] (MCs) (Figure 18.5(a)), are characteristic structures emerging from the ring soon after water uptake (stage I). When the sample is heated to reach the final swelling temperature the MCs may grow considerably in length. During this period, no distinct membrane structures are discernible in the vicinity of the ring where lipid remnants from the chloroform deposition are left (Figure 18.1, top). In stage II, beginning with the onset of the intensity decrease as monitored by X-ray diffraction (Figure 18.4(a)), small membrane objects such as tubules with a rather low lamellarity arise more or less simultaneously all over the sample. They are especially well discernible in the sample regions apart from the ring. In the course of time most of the strongly birefringent domains in the sample vanish but new MCs (Figure 18.5(b)) emanate from the lipid ring. As judged from their polarization properties, the concentric lamellae of these MCs are less densely packed than those of the initial multilayer system (Figure 18.5(a)) that gave rise to the Bragg reflections. A crude analysis of their respective phase shifts yields an estimated mean repeat distance $d_{est}$ greater than 650 Å for the new MCs. However, the ($d_{est}$)-values scattered a lot within the same sample if this analysis was applied to several optically homogeneously packed objects. The unbound state of membranes in the late swelling phase cannot be characterized by a uniform repeat distance as in stage I.

## 4. DISCUSSION AND CONCLUSION

By combining optical and X-ray investigations, new features of the system (DOPC in excess pure water) were discovered during the process of swelling. One of the most striking results is that in the whole sample paucilamellar objects coincidentally become visible around the time when the repeat distance of the bulk lipid of the ring is close to the full hydration value. The ubiquitous occurrence at the beginning of stage II demonstrates the independence of the swelling behavior from the number of membranes present in the stack. Including the results obtained on the repeat

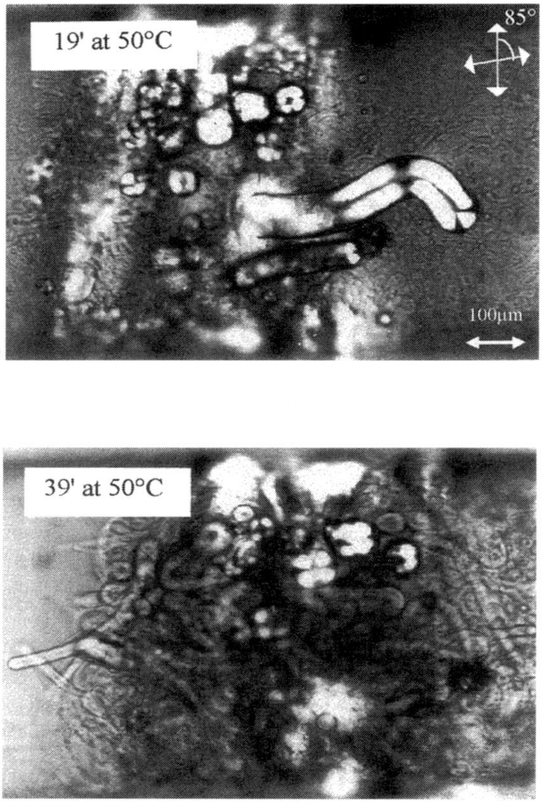


**Figure 18.5** Polarization micrograph (almost crossed polarizers) of a lipid ring in a capillary. (a) After 19 minutes at 50°C (at the beginning of stage II) faint membrane objects start to grow and the bright birefringent structures characteristic of stage I (e.g. MCs) regress. (b) The same sample detail 20 min later, during stage II. New but loosely packed multilamellar structures are visible. Several bright objects have already completely disappeared.

distances measured with diffraction, this suggests that a temperature-dependent threshold distance (and amplitude of fluctuations) initiates the transition into the released state.

In other words, the almost planar fully hydrated multilayer system is not the state of lowest energy density of fluid uncharged membranes like the PCs. Instead there seems to be a trapped state in all systems wherein the membranes are prepared to form closed shells (as in liposomes). The lamellar equilibrium distance thus signifies an intermediate lamellar ordered membrane state that precedes the disassembly of the multilayers. The final state of lowest energy cannot yet be specified.

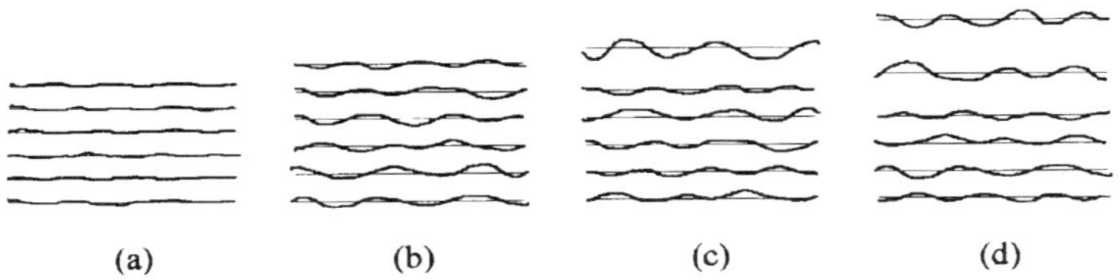


**Figure 18.6** Schematics of the process of swelling. The six outermost membranes of a bilayer stack embedded in water are shown at four successive states. (a) Water just added, and the membranes are still less separated than the full hydration repeat distance $d_{\mathrm{hyd}}$. (b) As the membranes reach $d_{\mathrm{hyd}}$ their fluctuations increase in comparison with state (a) [2]. (c) The outermost membrane starts to separate farther from the rest while the inner ones still maintain their uniform distance. (d) Step by step, the membranes successively peel away from the stack.

Normal membrane undulations [2,6] even at low lateral tensions are probably not sufficient to explain this loss of lamellar order. Instead, a model is proposed [7] that includes higher-order contributions to the bending elastic energy. This promotes the excitation of local, highly curved hats and an associated drop of the effective bending rigidity. As a consequence, enhanced undulations could increase the repulsive forces between adjacent membranes. The bilayer release from the stack should start with the outermost membranes that are not hindered to form hats and to increase their distance (Figure 18.6). As the separation continues, the membrane multilayer transforms into an unbounded state. Defects [8] are probably necessary to provide free water access and membrane flow. Their presence can thus be regarded as a technical precondition for the observation of membrane separation.

## 5. ACKNOWLEDGMENT

Part of the work was supported by the Deutsche Forschungsgemeinschaft through SFB 312.

## 6. REFERENCES

1. R. P. Rand and V. A. Parsegian, *Biochem. Biophys. Acta*, **988**, 351 (1989).
2. S. Tristam-Nagle, H. I. Petrache and J. F. Nagle, *Biophys. J.*, **75**, 917 (1998).
3. J. Reeves and R. Dowben, *J. Cell Physiol.*, **73**, 49 (1969).
4. J. Hartung, W. Helfrich and B. Klösgen, *Biophys. Chem.*, **49**, 77 (1994).
5. J. Nageotte, Morphologie des Gels Lipoïdes, *Actualités Scientifiques et Industrielles*, (ed. Hermann & C$^{ie}$), **432 & 434**, 1936, p. 1.
6. W. Helfrich and R.-M. Servuss, *Il Nuovo Cimento*, **3D**, 137 (1984).
7. W. Helfrich, unpublished results.
8. S. Chandrasekhar, *Liquid Crystals*, 2nd edn, Cambridge Press, 1992, p. 237.

Chapter 19

# Dynamic Aspects of Fatty Acid Vesicles: pH-induced Vesicle–Micelle Transition and Dilution-induced Formation of Giant Vesicles

*AYAKO GOTO, AKIHIRO SUZUKI, HISASHI YOSHIOKA AND RENSUKE GOTO*
*University of Shizuoka, Shizuoka-shi, Japan*

*TOYOKO IMAE*
*Nagoya University, Nagoya-shi, Japan*

*KEIJI YAMAZAKI*
*Otuka Electronic Co. Ltd, Hachioji-shi, Japan*

*PETER WALDE*
*ETH, Zürich, Switzerland*

## 1. INTRODUCTION

Linear single-chain surfactants are generally known as micelle-forming amphiphiles which, above a critical concentration, in aqueous solution aggregate into micellar structures [1]. In certain cases, it has been shown that a transformation of micelles composed of such linear single-chain amphiphiles into (mixed) vesicles can be

*Giant Vesicles* Edited by P. L. Luisi and P. Walde

achieved by co-addition of a second single-chain amphiphile. These mixed vesicles have been characterized, for example, in the case of sodium dodecylbenzenesulfonate–cetyl trimethylammonium tosylate [2], sodium *N*-lauroyl-sarcosinate–*N*-dodecylpyridinium chloride [3], sodium dodecyl sulfate–dodecanol [4,5], or sodium (or potassium) oleate–oleic acid [4,6–11]. The chemical structure of oleic acid is shown in Figure 19.1. The latter system is particularly interesting because the two amphiphiles building the bilayer of the vesicles – oleate and oleic acid – are chemically very similar. Vesicle formation has been observed under conditions where about half of the molecules are protonated (oleic acid) and half of the molecules are deprotonated (oleate) [4,6–11].

Generally, the preparation of mixed fatty acid–soap vesicles, in the case of oleic acid–oleate, always produces small (and large) as well as giant vesicles [10]. If vesicles composed of oleic acid and oleate at a total lipid concentration of 20 mM are prepared at pH 8.5, electron and light microscopy analysis indicate that the resulting suspension contains vesicles that are very polydisperse with diameters ranging from below 1 μm to much greater than 0.1 μm; the vesicles are often oligo- or multilamellar [10].

In contrast with phospholipid vesicles, the oleic acid–oleate vesicle system is characterized by a relatively high monomer solubility. The critical sodium oleate concentration for micelle formation (cmc) at about pH 10.5 is in the range 0.7–1.4 mM [10]; it has been reported that the monomer solubility at pH 7.4 is around 10–20 μM [11]. (For comparison, the monomer concentration of 1,2-dipalmitoyl-*sn*-glycero-3-phosphocholine (DPPC) is known to be around $10^{-10}$ M [12].) Due to the relatively high monomer concentration, several physicochemical properties of oleic acid–oleate vesicles are different from vesicles made from phospholipids, such as phosphatidylcholines. The kinetics of lipid exchange is expected to be faster in the

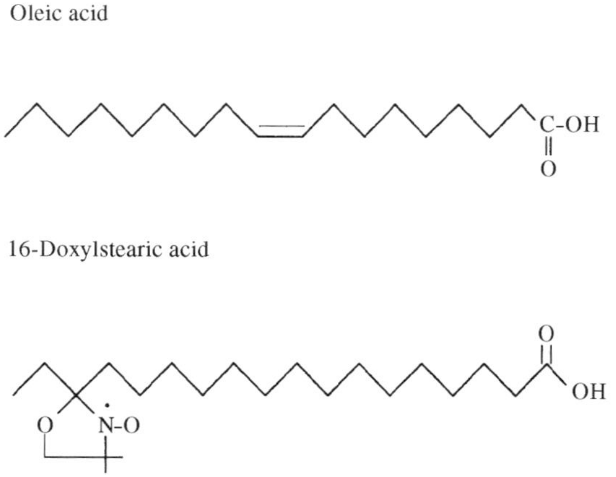


**Figure 19.1** Chemical structures of oleic acid (*cis*-9-octadecenoic acid) and 16-doxylstearic acid (16-DS).

case of single-chain amphiphiles in comparison with double-chain phospholipids [13]. In an earlier investigation [14] it was noted that a 10-fold dilution of a concentrated oleic acid–oleate vesicle suspension led to the formation of an increased amount of giant oleic acid–oleate vesicles, possibly through a vesicle reorganization mechanism that most likely involved fusion of smaller vesicles. This has never been observed with vesicles from phosphatidylcholine, such as POPC (1-palmitoyl-2-oleoyl-*sn*-glycero-3-phosphocholine).

In the present study, electron spin resonance (ESR) spectra of a fatty-acid spin probe (16-doxylstearic acid, see Figure 19.1) incorporated into an oleic acid–oleate system were measured in order to get insight into the pH-dependent aggregation properties of sodium oleate and oleic acid. Furthermore, the dilution-induced transformation of submicrometer-sized particles (micelles and/or vesicles) into giant oleic acid–oleate vesicles was investigated by electrophoretic light scattering measurements.

## 2. EXPERIMENTAL

Sodium oleate was purchased from Sigma (USA). ESR spectra were recorded on a JEOL JESRE3X spectrometer between 10 and 50°C using a flat quartz cell. 16-Doxylstearic acid (16-DS), which is stearic acid spinlabeled at carbon 16, was used as a spin probe and was obtained from Aldrich (USA). Sodium oleate solutions at a concentration of 25 mM were prepared by dissolving sodium oleate in pure water or in 0.2 M bicine (*N*,*N*-bis-(2-hydroxyethyl)glycine) buffer. 16-DS was dissolved in a sodium oleate micellar solution (16-DS : sodium oleate = 1 : 200) and the pH was adjusted to the desired value by the addition of 1 M HCl or 1 M NaOH. The ESR spectra of 16-DS were recorded in the presence of various amounts of sodium oleate (0–50 mM), and $Mn^{2+}$ in MgO was used as the external reference in order to get the corresponding peak positions precisely. The size distribution of the oleic acid–oleate aggregates was measured by an electrophoretic light-scattering photometer (ELS800 from Otsuka Electronics Co., Japan). Giant vesicles were observed by difference contrast light microscopy (using an Olympus 1 × 70, Japan, or an Axioplan microscope from Zeiss, Germany) and laser confocal scanning microscopy (an ACAS 575 UVC from Meridian Instruments, Inc., USA), in which 1-acyl-2-[6]-[(7-nitro-2-1,3-benzoxadiazol-4-yl)amino]caproyl]-*sn*-glycero-3-phosphocholine (NBD-PC) was used as a fluorescent probe.

## 3. RESULTS AND DISCUSSION

### 3.1 Dynamic properties of oleic acid–oleate vesicles studied by ESR

Figure 19.2 shows the ESR spectra of 16-DS in the presence of 25 mM sodium oleate–oleic acid at a pH 12.1, pH 8.7, and pH 6.4 in the absence of any buffers. The

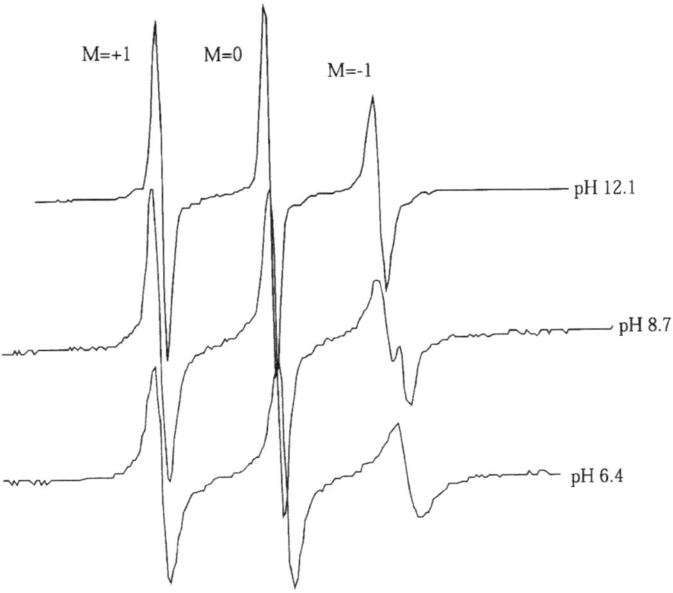


**Figure 19.2** ESR spectra of 16-DS in 25 mM oleic acid–oleate suspensions at a pH of 12.1, 8.7, and 6.4.

pH of the solutions/suspensions was adjusted by NaOH or HCl (see experimental part). The ESR spectra recorded at pH 12.1 or at pH 6.4 were highly symmetric, suggesting that each 16-DS molecule under these conditions exists in a single state. At pH 12.1, where oleate is known to form micelles above the cmc, the ESR spectra were measured as a function of the oleate concentration, and it was found that the shapes of the high-field line (abbreviated as the $M = +1$ line) were symmetric over the whole concentration range. The peak position of the line, however, shifted to lower magnetic field and the width increased with increasing concentration of sodium oleate, reaching a constant value above 15 mM (data not shown). This observation indicates that 16-DS is distributed between the micellar and aqueous domains, and exchanges very quickly between them; furthermore, almost all 16-DS is incorporated into micelles above 15 mM oleate.

In the case of pH 6.4, two lines were obviously overlapping at oleic acid–oleate concentrations below 15 mM. Based on the peak positions and shapes it is evident that the sharp peak corresponds to 16-DS in the aqueous medium, whereas the broader one most probably corresponds to 16-DS in bilayers (vesicles). Based on the observations that two peaks exist at pH 6.4, it is suggested that the exchange rate of the surfactant between the vesicles and the aqueous medium is slow with respect to the ESR time-scale. By increasing the concentration of oleic acid–oleate, the signal corresponding to the spin probe in the aqueous domain was reduced and the spectrum became symmetric above a total oleic acid–oleate concentration of 25 mM. Therefore, almost all 16-DS can be assumed to be incorporated into bilayers at 25 mM at a pH of 6.4. It has been argued before that oleic acid–oleate below pH 8.0 exists as an oil [8,11]. However, under the experimental conditions used in the present work, ESR measurements indicate that below a pH of 8, the oleic acid–oleate aggregates are much more ordered than oil droplets and they are most likely composed of closely packed bilayers.

From the ESR spectra recorded of 25 mM oleic acid–oleate above a pH 10.4 and below pH 6.5, the rotational correlation time, $\tau_c$, was calculated [15] to be $3.9 \times 10^{-10}$ s in the case of the micelles (pH 10.4), and $6.1 \times 10^{-10}$ s in the case of the bilayers (pH 6.4), respectively. The $\tau_c$ of 16-DS in pure water (below a cmc of approximately 10 μM) was $9.3 \times 10^{-11}$ s. Taking into account that $\tau_c$ generally depends on the microviscosity and that the microviscosity of water is nearly 1 cp, the corresponding microviscosity values are estimated to be 4 cp and 6 cp in the micelles and in the bilayers, respectively. Furthermore, as the hyperfine coupling constants ($aN$) of 16-DS were 15.4 G and 14.5 G above pH = 10.4 and below pH = 6.5, respectively, the micropolarity in the micelles is larger than that in the vesicles (Figure 19.3).

In the intermediate range of pH 10.4–7.0, the ESR lines of $M = +1$ became asymmetric as a result of the overlapping of two lines, as shown in Figure 19.2. At a first glance, this peak asymmetry seems to be due to the coexistence of micelles and vesicles with the probe molecules distributed between them. Alternative explanations, however, would be that (i) two kinds of vesicles exist or (ii) that the carboxylic acid of 16-DS incorporated in the vesicles is partly dissociated and the ESR spectrum of the ionic probe (the carboxylate) differs from that of the nonionic one (the carboxylic acid). However, this latter possibility is unlikely as the line

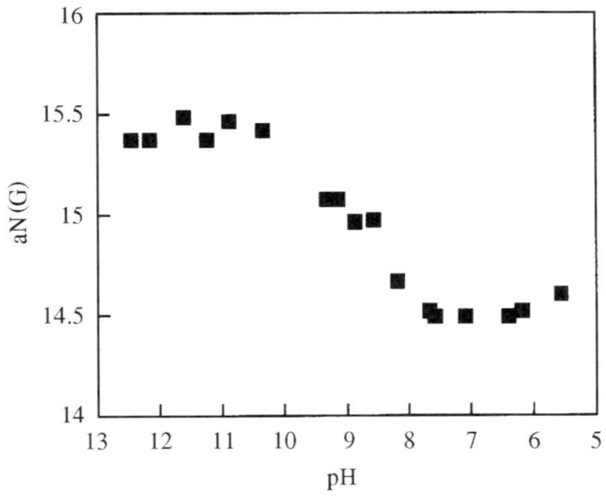


**Figure 19.3** ESR spectra of 16-DS in 25 mM oleic acid–oleate suspensions. The hyperfine coupling constant, $aN$, is plotted as a function of pH.

shapes of the ESR spectra of 16-DS in water hardly depended upon pH. furthermore, the ESR spectra of 16-DS incorporated into egg phosphatidylcholine vesicles were also independent of pH (data not shown). Concerning possibility (i), a series of ESR measurements were performed on a suspension containing 16-DS and a total concentration of 25 mM oleic acid–oleate at pH 8.5. The suspensions were extruded several times through polycarbonate filters with 200 nm and 50 nm pore diameters, respectively, in order to reduce the size and lamellarity of the vesicles as described before [10]. By doing so, the ESR signal remained essentially unchanged as compared with the control suspension prepared without extrusion. Furthermore, the presence of micelles at high pH and the now proposed coexistence of micelles and vesicles at intermediate pH represent both equilibrated states because the ESR line shapes of 16-DS in all the 25 mM oleic acid–oleate systems between pH 12.1 and pH 7.7 at 20°C were not changed after heating to 50°C and recooling to 20°C, or after cooling to 10°C and rewarming to 20°C.

The $M = -1$ line of the ESR spectra was also measured as a function of the total oleic acid–oleate concentration at pH 8.5. Two overlapping lines were always present in the whole concentration range. In the case of low concentrations, the two contributions are assumed to stem from 16-DS localized in the vesicle phase, and from 16-DS dissolved in the aqueous domain. With increasing oleic acid–oleate concentration, the latter peak shifted to lower magnetic fields, indicating that 16-DS is now completely distributed into the micellar domain. Therefore, the overlapped spectra of $M = -1$ at pH 10.4–7.0 consist of 16-DS incorporated into the vesicles, the micelles and the aqueous domain, meaning accordingly that vesicles and micelles always coexist in this intermediate pH range. From the ESR spectra it can be estimated that the ratio of micelles to vesicles becomes higher with increasing pH. At pH 8.5 and 25 mM oleic acid–oleate, vesicles clearly dominate over micelles.

The size distributions of the aggregates measured by electrophoretic light scattering showed that, above pH 9, there were two kinds of aggregates of diameters 240 nm and 20 nm, respectively (Figure 19.4). This observation supports the proposition of the coexistence of micelles and vesicles at a pH of about 9. Although a detailed analysis of the ESR spectra is in progress in order to quantify the pH region in which micelles and vesicles coexist, the micelle : vesicle ratio is expected to change b y changing the degree of protonation of oleic acid (or 16-DS).

Based upon the above results, for the oleic acid–oleate system at intermediate pH (10.4–7.0) the scheme shown in Figure 19.5 is proposed, in which vesicles, micelles, and monomers always coexist, and the 16-DS probe molecules are distributed between vesicles, micelles, and the aqueous domain. Although the probe molecules can exchange position quickly (a rapid equilibrium between the monomers and the solubilized 16-DS), in the case of the bilayers (vesicles), the rate of exchange of 16-DS between the bilayers (vesicles) and the aqueous domain is considerably lower than for the micelles.

Using difference contrast light microscopy and laser confocal scanning microscopy, the presence of giant vesicles could clearly be observed in the oleic acid–

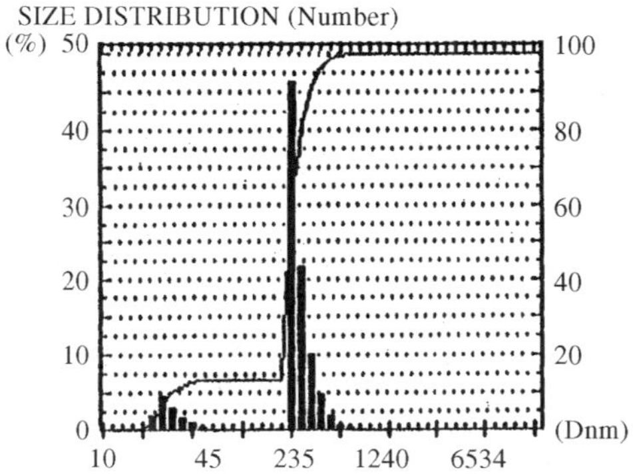


**Figure 19.4** Size distribution of oleic acid–oleate aggregates at pH 9.7 as determined by electrophoretic light scattering at a scattering angle of 90°.

oleate system, particularly at a pH of ≈ 8.5, near to the apparent p$K_a$ of aggregated oleic acid (Figure 19.6). This has been observed before [4,6,10] and suggests that under conditions where about half of the oleic acid molecules are deprotonated and half of the molecules are protonated, oleic acid–oleate dimers exist to a large extent in which one $H^+$ is shared by two carboxylate groups [9]. A high content of these dimers may favor the formation of giant vesicles that are composed of flat bilayers with a low radius of curvature.

### 3.2 Dilution-induced formation of oleic acid–oleate giant vesicles

As mentioned above, it has been found earlier [14] that extensive dilution of 80 mM oleic acid–oleate suspensions prepared in 0.2 M bicine buffer (pH 8.5) increased the

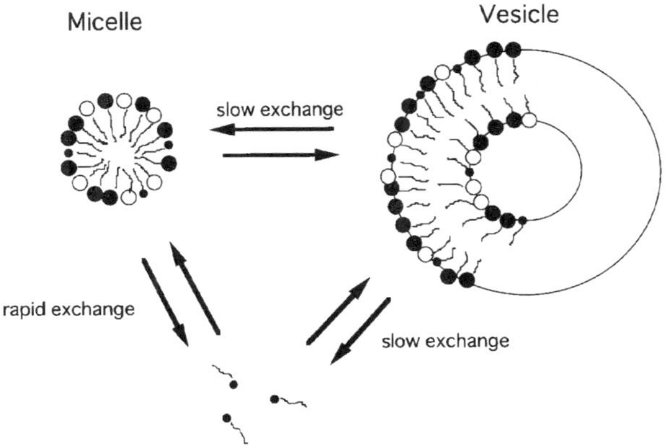


**Figure 19.5** Schematic representation of the oleic acid–oleate system at intermediate pH (10.4–7.0).

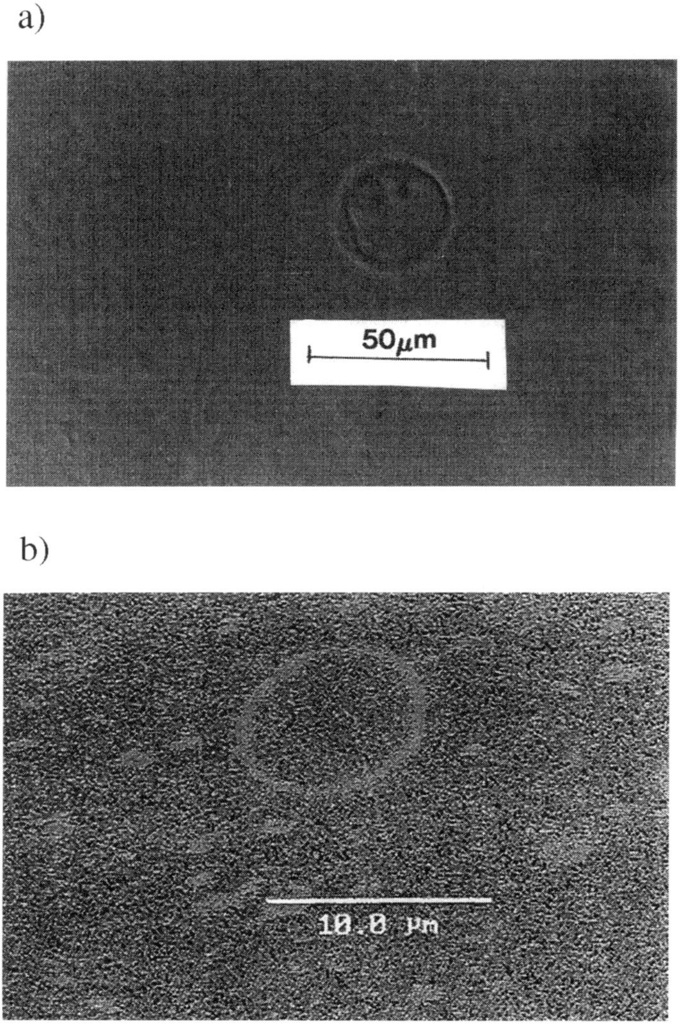


**Figure 19.6** (a) Difference contrast light micrograph and (b) laser scanning confocal micrograph of giant oleic acid–oleate vesicles at pH = 8.5.

relative amount of giant vesicles in the system, as determined by light microscopy. These observations have now been confirmed by electrophoretic light-scattering measurements. Dilution of 25 mM oleic acid–oleate in 0.2 M bicine buffer at a pH of 8.5 by a factor of 10 or 100 with 0.2 M bicine buffer of pH 8.5 led to a significant increase in the size distribution of the aggregates (vesicles) formed. Whereas with a

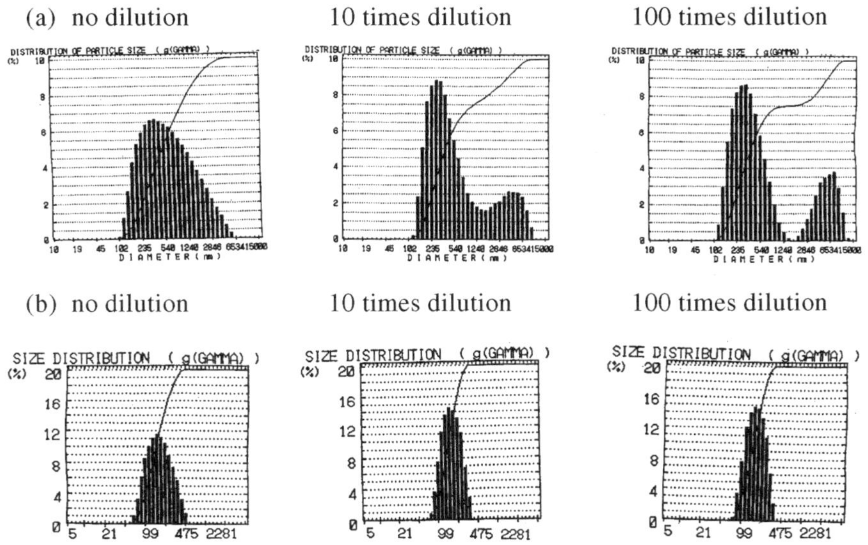


**Figure 19.7** Dilution-induced change in the size distribution of oleic acid–oleate vesicles at pH 8.5 (0.2 M bicine buffer): (a) nonextruded vesicles; (b) vesicles extruded through polycarbonate membranes with a mean pore diameter of 0.2 μm.

10-fold dilution, the size distribution shifted to a larger mean size, with a 100-fold dilution a new peak of larger size appeared (corresponding to giant vesicles), as shown in Figure 19.7(a). A similar dilution-induced formations of giant vesicles were observed in the whole pH range between pH 7.6 and 9.0 (data not shown).

As described above, the experimental evidence suggests that oleic acid–oleate vesicles always coexist with micelles and monomers, and one may propose that the formation of giant vesicles involves a mechanism in which micellar aggregates (and possibly nonaggregated monomers) play a crucial role. This type of dilution-induced reorganization of vesicles and formation of giant vesicles has not been observed for conventional phosphatidylcholine vesicles. It may, however, be a general property of mixed vesicle systems that are composed of single-chain amphiphiles in which at least one of the components can form micelles and which have a relatively high monomer solubility.

## 4. REFERENCES

1. J. N. Israelachvili, *Intermolecular and Surface Forces*, 2nd edn, Academic Press, London, 1992.
2. E. W. Kaler, K. L. Herrington, A. K. Murthy and J. A. N. Zasadinski, *J. Phys. Chem.*, **96**, 6698 (1992).

3. M. Ambühl, F. Bangerter, P. L. Luisi, P. Skrabal and H. J. Watzke, *Langmuir*, **9**, 36 (1993).
4. W. R. Hargreaves and D. W. Deamer, *Biochemistry*, **17**, 3759 (1978).
5. M. Bergström and J. C. Eriksson, *Langmuir*, **14**, 288 (1998).
6. J. M. Gebicki and M. Hicks, *Nature*, **243**, 232 (1973).
7. R. Bittman and L. Blau, *Biochim. Biophys. Acta*, **863**, 115 (1986).
8. D. P. Cistola, J. A. Hamilton, D. Jackson and D. M. Small, *Biochemistry*, **27**, 1881 (1988).
9. T. H. Haines, *Proc. Natl. Acad. Sci. USA*, **80**, 160 (1983).
10. P. Walde, R. Wick, M. Fresta, A. Mangone and P. L. Luisi, *J. Am. Chem. Soc.*, **116**, 11649 (1994).
11. K. Edwards, M. Silvander and G. Karlsson, *Langmuir*, **11**, 2429 (1995).
12. R. Smith and C. Tanford, *J. Mol. Biol.*, **67**, 75 (1972).
13. M. C. Phillips, W. J. Johnson and G. H. Rothblat, *Biochim. Biophys. Acta*, **906**, 223 (1987).
14. R. Wick, P. Walde and P. L. Luisi, *J. Am. Chem. Soc.*, **117**, 1435 (1995).
15. H. Yoshioka, *J. Colloid Interface Sci.*, **63**, 378 (1978).

# Part Five

# Chemical and Biological Aspects

# Chapter 20

# Giant Liposomes as Model Biomembranes for Roles of Lipids in Cellular Signalling

*PAAVO K. J. KINNUNEN AND JUHA M. HOLOPAINEN*
*University of Helsinki, Finland*

*MIGLENA I. ANGELOVA*
*Bulgarian Academy of Sciences, Sofia, Bulgaria*

## 1. INTRODUCTION

Biomembranes are best described as liquid-crystalline, cooperative, and adaptive supramolecular assemblies and the modern view emphasizes coupling between their organization and function [2 – 4]. In cells the organization of these supramolecular assemblies of proteins and lipids involves dynamic ordering representing both:

- spontaneously forming self-organizing assemblies due to intermolecular forces at thermodynamic equilibrium;
- dissipative nonequilibrium structures, maintained by energy input.

These systems are highly dynamic and apt to regulation by a number of membrane-binding ligands such as hormones and growth factors, metabolites, ions, pH, drugs, proteins, and changes in membrane lipid composition, as well as physical parameters such as membrane potential, osmotic forces and hydration, pressure, and temperature [2].

*Giant Vesicles* Edited by P. L. Luisi and P. Walde

Based on the intrinsic properties of lipids the functions of biomembranes can be controlled on the following hierarchy levels, which furthermore are in several cases connected and operate concomitantly:

- membrane lateral organization;
- the physical (phase) state of the membrane and the bulk properties of the membrane;
- three-dimensional organization of the membrane assembly.

The following sections briefly review the first two levels; the third level is then demonstrated experimentally.

## 2. MEMBRANE LATERAL ORGANIZATION

Fluorescence microscopy studies have provided perhaps the most compelling evidence for lateral ordering and domains in cell membranes [2]. Considerations arising from the material properties of lipids have demonstrated that, inherently to all many-body systems, membranes involving them will spontaneously organize on different lengths and timescales [3]. The most profound manifestation of this order is represented by the compositional fluctuations accompanying lipid-phase transitions [5,6]. Other mechanisms involve straightforward lipid–lipid, lipid–ligand/protein interactions due to electrostatics, hydrogen bonding, and hydrophobic matching condition. Accordingly, microdomain formation can be induced for instance by lipid–lipid interactions [7,8], $Ca^{++}$ [9], polyamines [10], peripheral proteins such as histone H1 [11], ribosomes [12], nucleic acids [13], dehydration [14], hydrophobic mismatch both between constituent lipids [15], as well as between lipids and integral membrane proteins [16–18]. The sensitivity of lateral order in biomembranes to impurities such as metabolites and drugs, as well as physical parameters such as membrane potential, temperature, and pressure, derives from the sensitivity of their lipid phase to these factors [2]. Modulation of lipid domain structure and dynamics by the membrane-partitioning drug cyclosporin A has been recently demonstrated [19].

Gel-state membrane domains have been concluded to exist in cell membranes and their formation appears to depend on sphingomyelin and cholesterol [20,21]. With the recognition of the significance of functional ordering in biomembranes [2,3] there has been a recent surge of interest amongst cell biologists in cellular membranes. Specific cell membrane regions called caveolae, have been attributed a central role in cellular signalling, for instance [22,23]. A new approach to these domains, also called rafts involves their isolation by detergent extraction of cell membranes at 4°C [24]. However, attempts to isolate detergent-resistant membranes at physiological temperatures have been unsuccessful. A paradigm or phase separation in lipid mixtures is represented by the region of gel–fluid coexistence. Accordingly, the detergent-resistant membrane domains could represent the formation of extensively segregated gel-like phases at reduced temperatures. However, the data presented here demonstrate that domains and capping can also occur at physiological temperatures and in fluid membranes (see below).

## 3. REGULATION OF CELLULAR FUNCTIONS BY BULK MEMBRANE PROPERTIES

Simple *in vitro* demonstration of a membrane responding to a physical change is the osmotic pressure gradient causing swelling (reduced lateral lipid packing) of a liposome, which subsequently leads to the accessibility of the lipids to the action of phospholipase A2 [25]. This concept of an enzyme responding directly to a change in physical state of the bilayer substrate is intriguing on a general level. As phospholipases are centrally involved in cellular signalling, it is obvious that cascades comprised solely of protein-mediated effects could be seriously defective. Instead, changes in the physical properties of a cell's membranes could be directly sensed by proteins, thereby initiating or conveying signals to other relevant cellular machineries. Another example is provided by the correlation between negative spontaneous curvature and the activity of diglucosyldiacylglycerol synthase from *Acholeplasma laidlawii* observed under *in vitro* conditions [26].

An inherent property of lipids as liquid crystalline materials is their ability to undergo phase changes, which are generally accompanied by drastic alterations in the physical properties of the membrane [27]. One concept is that life as a form of matter has adopted the phase changes of lipids for the regulation of both metabolism as well as replication of cells [4,28,29]. Accordingly, phase changes of membranes, coupled to intracellular polymers in eukaryotes, should further correspond to specific changes in the physiological states of the cell, such as those represented by apoptosis and the distinct phases of the cell cycle. In other words, there would be correlation between the *physiological* state of the cell (or a cell organelle in an eukaryote) and the *physical* state of the membrane. Biomembranes thus represent platforms that integrate and control entire metabolic pathways in a cooperative manner, the activities of the associated proteins being dependent on the physical (phase) state of the given membrane. The latter is determined by the constituent lipids.

The best example so far for the importance of a bulk membrane property in regulating cell functions is provided by the control of growth in prokaryotes and eukaryotes [4]. Both cell types contain in their membranes substantial amounts of lipids forming, in isolation and under the appropriate conditions (e.g. temperature, presence of ions, protons, proteins, and metabolites), inverted nonlamellar phases (INL), such as hexagonal phase $H_{II}$ [30–33]. When present in sufficiently small quantities in mixed membranes with lipids forming lamellar phases, INLs increase the tendency of the bilayer for adopting negative curvature, so that even the membrane remains lamellar. Such membranes are defined as frustrated and there is increasing packing strain towards the interior of the hydrocarbon phase of the bilayer. Instead, the packing closer to the bilayer surface becomes more loose. Compared to a relaxed lamellar bilayer the free volume distribution and lateral pressure profile within the bilayer change. The magnitudes of the pressure in the interior of the bilayer and particularly in the interfacial region can be considerable [34].

A unifying feature in rapidly growing cells appears to be an increase in the content of INLs. Membranes containing INL have been proposed to act as a general, ancient signal for cell growth, activating specific sets of rate-limiting proteins required for cell replication, i.e. enhancing the synthesis of constituents of the cellular default machinery, DNA, lipids, and proteins [4,27,28]. On a mechanistic level the activities of these membrane proteins would be controlled by the lateral pressure profile which, in addition to direct effects on the conformation of integral proteins, may also induce the attachment and activation of peripheral membrane proteins by the so-called extended lipid anchorage. To this end, it is important to notice that the INL content can be regulated to some extent independently in the different organelle membranes of eukaryote cells and can thus be used to control specific metabolic pathways independently. Yet, in the S-phase the content of INL should increase in all organelles harboring machineries needed for replication. Aberrant control of the content of INL will result in malignant transformation and cancer.

## 4. CHANGES IN THREE-DIMENSIONAL ORGANIZATION

Since their discovery by Bangham, liposomes have been widely used as a biomembrane model since the late 1970s [35]. The majority of this work has relied on the use of multilamellar vesicles, small unilamellar vesicles (30–100 nm in diameter) and, more recently, large unilamellar vesicles ($\approx$ 100–200 nm). However, developments in the understanding of the physical properties of lipid membranes have made it clear that the small size of the above liposomes limits their value as models for larger-diameter structures, such as the plasma membrane. Investigation of processes such as undulation, healing, and budding has only become possible with the use of the so-called giant unilamellar vesicles (GUV, $\approx$ 50–500 μm in diameter) and has provided unprecedented insight to the understanding of the mechanistic basis of changes in vesicle morphology [36]. These aspects are covered in this volume by some of the pioneering investigators in this field.

Changes in plasma membrane morphology are frequently connected to alterations in cell function, such as initiation of growth and triggering of programmed cell death. These physiological states also involve changes in lipid composition; however, the relationship between the latter and membrane morphology has received little attention. One of the hallmarks of apoptosis is the shedding of vesicles from the plasma membrane to the extracellular space [37]. In addition, apoptotic cells are identified by loss of normal cell contacts, cell shrinkage, and DNA fragmentation.

Due to the remarkable similarity of the characteristic morphological changes in all cell lines studied, similar mechanisms have been suggested to be responsible. The mechanisms of shedding of the fluid-filled, structureless, and membrane-enclosed vesicles in apoptosis is not understood, but are thought to arise from the disconnection of the cell membrane and cytoskeleton [37]. At an early stage in apoptosis, activation of SMase is observed [38]. The possible role of the apoptotic messenger, ceramide, in causing changes in vesicle shape was explored.

Giant liposomes composed of 1-stearoyl-2-oleoyl-*sn*-glycero-3-phosphocholine (SOPC) and *N*-palmitoyl-sphingomyelin (C16 : 0-SM; Figure 20.1 shows the chemical structures of the lipids used in the present study) readily formed in an AC field [39] and were visualized by fluorescence microscopy of the included fluorescent lipid tracer BODIPY-sphingomyelin (Bdp-SM; Figure 20.2). In brief, lipid stock solutions were mixed in chloroform to obtain the desired composition. The solvent was evaporated under a stream of $N_2$ and subsequent evacuation under reduced

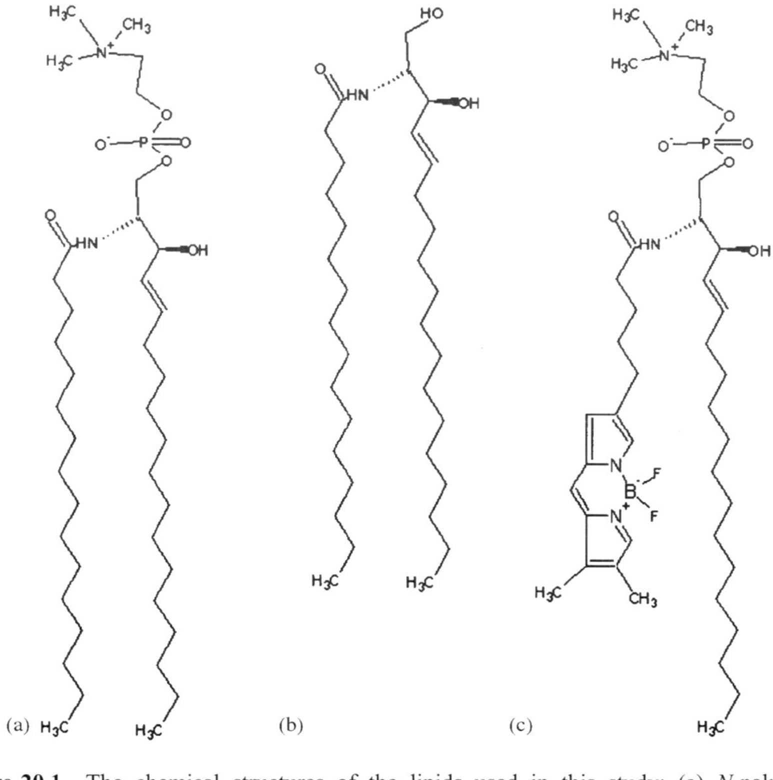

**Figure 20.1** The chemical structures of the lipids used in this study: (a) *N*-palmitoyl-sphingomyelin, (b) the corresponding *N*-palmitoyl-ceramide, and (c) the fluorescent probe BODIPY-sphingomyelin.

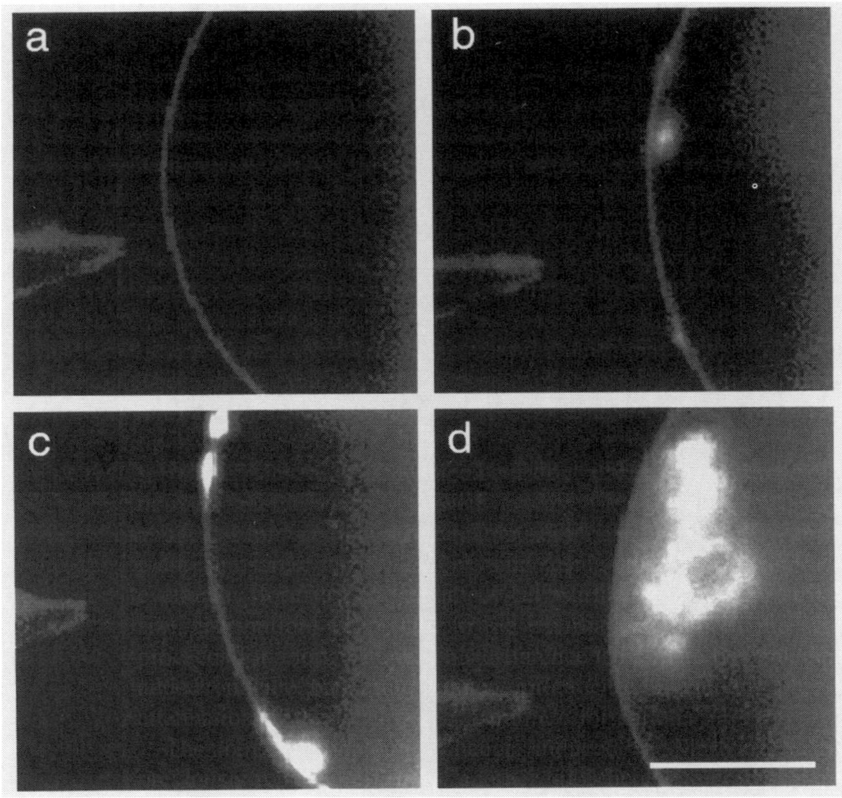


**Figure 20.2** Transformations of a single SOPC (Avanti)/C16 : 0-sphingomyelin (C16 : 0 SM, Northern Lipids, Vancouver, BC, Canada) giant vesicle, induced by a picoliter aliquot of SMase (SM phosphodiesterase, EC 3.1.4.12, from *B. cereus*, Sigma, one unit per ml, specific activity 100–300 units per mg protein in 9 mM $CaCl_2$, 1.8 mM $MgCl_2$ aqueous solution) applied in the vicinity of the vesicle outer surface. The mole fraction of BODIPY-labelled sphingomyelin (Bdp-SM, Molecular Probes) was $X = 0.05$. Fluorescent images were taken with a Peltier-cooled digital camera (a) before and after (b) 20 s, (c) 40 s, and (d) 2 min of enzyme administration. The scale bar corresponds to 100 μm. The excitation and emission wavelengths were selected with filters transmitting in the range 420–480 nm and > 500 nm, respectively. Micropipettes with inner tip diameters of 0.5–1 μm were drawn from borosilicate glass capillaries (outer diameter =1.2 mm). For easier handling only vesicles attached to the Pt electrode were employed.

pressure for at least 12 h. The dry residue was dissolved in diethylether : methanol 9 : 1 by volume to yield 1 mM final total lipid concentration. Approximately 1 μl of this solution was applied to the surface of a Pt electrode which was then dried with a stream of nitrogen and evacuated in vacuum for 1 h. The chamber with the electrodes and with a quartz window beneath was placed on the stage of an inverted fluorescence microscope. A 0.2 V AC voltage at 4 Hz was applied prior to adding 1.3 ml of 0.5 mM Hepes, pH 7.4 buffer. During the first minute of hydration the voltage was raised to 1 V. After 2 h the AC field was turned off.

The uniform distribution of the fluorescent SM is observed at the resolution of the optical microscope (Figure 20.2), in keeping with the miscibility of SM in PC membranes [40,41]. It has been demonstrated with large unilamellar liposomes that ceramide segregates into microdomains, in both fluid as well as in gel-state PC membranes [7,42]. To investigate the consequences of the asymmmetric enzymatic conversion of SM to ceramide in a fluid lipid membrane, the above giant PC/SM vesicles were subjected to the hydrolytic action of SMase. Within about 30 s from the external application of this enzyme to the vicinity of the outer membrane surface, diffuse domains with increased fluorescent intensity became evident (Figure 20.2). These membrane regions then increase in brightness and within about 1–2 min the domains are endocytosed as smaller vesicles into the internal cavity of the giant liposome.

When SMase was microinjected inside GUV, the appearance of diffuse fluorescent domains in the membrane were observed within about 30 s (Figure 20.3). After about 1–2 min small (4 μm in diameter) liposomes emerge on the outer surface of the giant liposome, their numbers increasing for several minutes. These vesicles eventually form a layer on the GUV surface and have a strikingly homogenous size distribution. Compared to the external addition of the enzyme, the reaction proceeds for a longer time. This difference is likely to be explained by the externally added enzyme becoming diluted into the bulk aqueous phase, resulting in the attenuation of the local rate of ceramide formation.

The above morphological changes induced by SMase are clearly distinct from the previously reported effects of another lipolytic enzyme, phospholipase A2. In brief, subjecting 1-palmitoyl-2-oleoyl-phosphatidylcholine (POPC) giant liposomes to either external or internal action of PLA2 causes these vesicles to burst and shrink, respectively [43]. These authors also showed that vesicle morphology is not altered by lysozyme, protein having no enzymatic activity towards the bilayer.

The physical determinants of vesicle shapes are reasonably well established and are explained in terms of excess area in one of the leaflets of a bilayer [44], area-to-volume ratio [45], bending elasticity, and spontaneous curvature [46–51]. Osmotically induced shape changes and fission of small vesicles have been shown previously [50]. Lateral domain formation, establishing domain boundaries, also contributes to vesicle budding [46,49,52]. Due to the coupling of curvature and local enrichment of lipids, introducing two or more components into the membrane results in a nonuniform membrane and leads to nonzero spontaneous curvature. The most simple two-component system is represented by the coexistence region in a membrane undergoing a phase transition between fluid and gel phases. To this end, upon increasing the temperature one can observe the pinching off of small vesicles (1 μm in diameter) from bovine brain SM giant vesicles [45]. An important conclusion from these data is that the dynamics required for these processes to occur

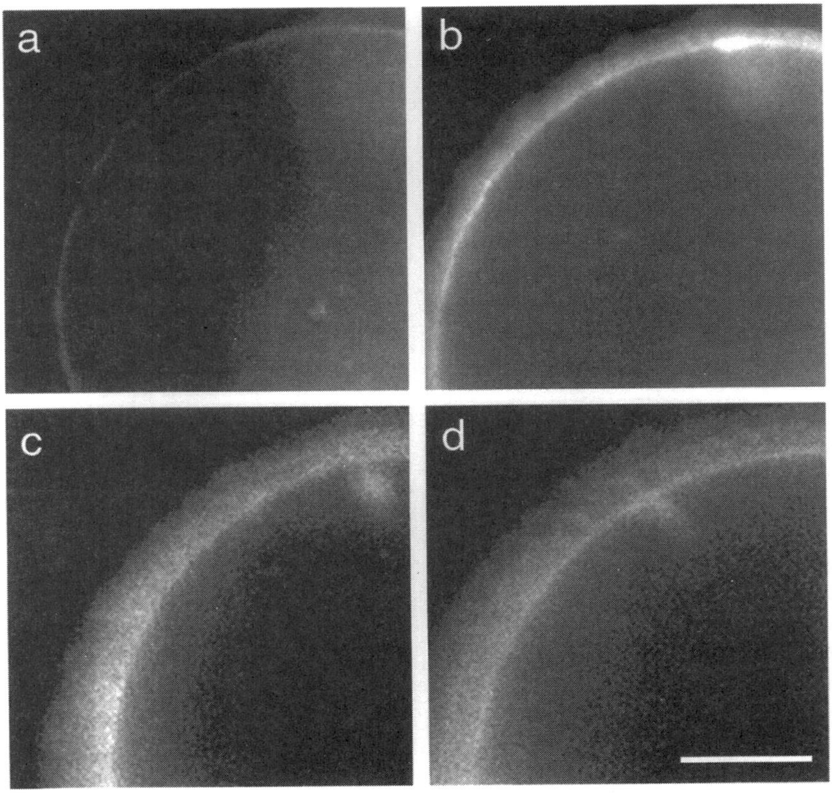


**Figure 20.3** Microinjection of SMase into a single PC/SM giant vesicle. Still fluorescence images were taken (a) prior to the enzyme addition, and (b) 1 min, (c) 3 min and (d) 10 min after the microinjection. The scale bar in panel D corresponds to 50 μm.

in living cells could be inherent in the shape phase diagram of the lipid membranes, thus relieving proteins from providing the driving force [45,50].

These results are depicted in Figure 20.4. The lipids in the SOPC/SM monolayer are initially randomly distributed, at least on the macroscopic length scales of the optical microscope (panel A). Cleavage of the phosphocholine headgroup from SM by SMase generates ceramide, a lipid with a small, weakly hydrated headgroup (panel B). Compared to POPC with a mean molecular area of 70 Å$^2$ [53], the area of ceramide in the monolayer is small, being about 40 Å$^2$ [54]. As a consequence, the ceramide-containing leaflet condenses, causing an area difference with respect to the adjacent leaflet of the bilayer. The ceramide formed then segregates into a domain (panel C), the likely driving force being intermolecular hydrogen bonding [42,55]. Due to its tendency to form inverted nonlamellar phases [56] an invagination is

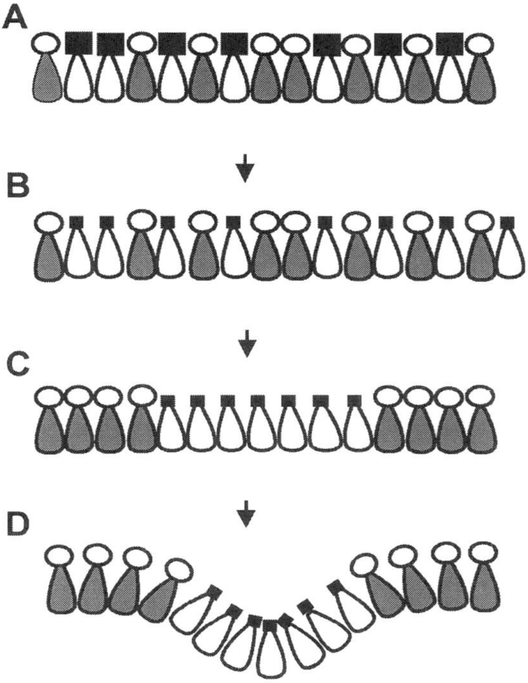


**Figure 20.4** A model for the mechanism of ceramide domain formation and invagination in a PC/SM membrane, induced by the asymmetric action of SMase. For the sake of clarity only the monolayer subject to the action of SMase is illustrated. The symbols used for the lipid headgroups are: (○) SOPC, (■) SM, (▪) ceramide. The scale bar corresponds to 50 μm.

formed by the ceramide-enriched membrane region (panel D). In addition to the area difference between the adjacent monolayers [44] and negative spontaneous curvature for the ceramide containing domain [56], augmented bending rigidity of the ceramide-enriched leaflet could be significant [46]. As a further progression (not shown) the domain grows in size and finally separates as a membrane enclosed cavity. Notably, although illustrated above as separate events, the formation of ceramide, its local enrichment, and membrane bending are likely to proceed simultaneously.

## 5. BIOLOGICAL IMPLICATIONS: CELLULAR SIGNALLING

In addition to cellular signalling mechanisms, vesicle budding and endocytosis are intimately involved in membrane trafficking and receptor-mediated internalization of ligands such as plasma low density lipoprotein, LDL. In the plasma membrane endocytosis proceeds via caveolae, 50–60 nm diameter invaginations [22]. These specialized membrane regions are enriched in ceramide and also contain G-protein-coupled receptors. A role for caveolae in signal transduction has been suggested [22,23]. Binding of interleukin-1$\beta$ to its receptor has been suggested to localize in an SM-rich plasma membrane domain with the characteristics of caveolae. Ligand binding was accompanied with the hydrolysis of SM to ceramide and the formation of the latter lipid was concluded to be highly compartmentalized in the cell surface [23]. It was recently shown that treatment of ATP-depleted macrophages and fibroblasts with exogenous SMase results within 10 min in the budding of a large number of vesicles from the plasma membrane into the cytoplasm of these cells [57]. These vesicles have a diameter of about 0.4 μm and seem to lack any protein coating. The data show that except for SMase no other proteins are necessary for endocytosis in giant liposomes.

In the light of the present results demonstrating budding of vesicles from giant liposomes after the microinjection of SMase into the vesicle interior, it seems plausible to propose that the enzymatic formation of ceramide may well cause the observed membrane blebbing in apoptosis, in a manner not requiring metabolic energy. On a general level these results suggest a crucial role for SMase in controlling both two- and three-dimensional ordering of cellular membranes. In connection with the localization of signalling proteins with sphingolipid-enriched domains in cellular membranes [22,24] the sphingomyelinase-controlled ceramide-driven endocytosis and shedding would thus provide a novel mechanism controlling cellular signalling by changes in the three-dimensional compartmentalization in the cell. Finally, it has been suggested that SMase and PC-specific phospholipase C of *Neisseria gonorrhoae*, *Staphylococcus aureus*, and species of mycobacteria would be involved in the mechanisms of their entry into human cells [58–60]. It is tempting to speculate that the bacterial SMase could induce endocytotic engulfing of these pathogenic bacteria by the eukaryote cells in a manner described here, with the bacterial cells attached to the surface of the invaginating ceramide-enriched domains. These issues are under investigation.

## 6. ACKNOWLEDGMENTS

This study was supported by the Finnish State Medical Research Council and Biocentrum Helsinki. J. M. H. is supported by the Finnish Medical foundation and M.D./Ph.D. programme of the University of Helsinki. M. I. A. is indebted to the French-Bulgarian Laboratory 'Vesicles and Membranes' supported by the CNRS and the Bulgarian Academy of Sciences.

## 7. REFERENCES

1. Y. A. Hannun, *Science*, **274**, 1855 (1996).
2. P. K. J. Kinnunen, *Chem. Phys. Lipids*, **57**, 375 (1991).

3. O. G. Mouritsen and P. K. J. Kinnunen, in *Biological Membranes. A Molecular Perspective from Computation and Experiment*, (eds) K. M. Merz and B. Roux, Birkhäuser, Boston, 1996, p. 465.
4. P. K. J. Kinnunen, in *Nonmedical Applications of Liposomes*, (eds) D. D. Lasic and Y. Barenholz, CRC Press, Boca Raton, FL, 1996, p. 153.
5. O. G. Mouritsen and K. Jørgensen, *Mol. Membr. Biol.*, **12**, 15 (1995).
6. A. Jutila and P. K. J. Kinnunen, *J. Phys. Chem.*, **101**, 7635 (1997).
7. J. M. Holopainen, J. Y. A. Lehtonen and P. K. J. Kinnunen, *Chem. Phys. Lipids*, **88**, 1 (1997).
8. P. Mustonen, J. Y. A. Lehtonen, A. Kõiv and P. K. J. Kinnunen, *Biochemistry*, **32**, 5373 (1993).
9. K. K. Eklund, J. Vuorinen, J. Mikkola, J. Virtanen and P. K. J. Kinnunen, *Biochemistry*, **27**, 3433 (1988).
10. K. K. Eklund and P. K. J. Kinnunen, *Chem. Phys. Lipids*, **39**, 109 (1986).
11. M. Rytömaa and P. K. J. Kinnunen, *Biochemistry*, **35**, 4529 (1996).
12. P. Kaihovaara, E. Raulo and P. K. J. Kinnunen, *Biochemistry*, **30**, 8380 (1991).
13. A. Kõiv, P. Mustonen and P. K. J. Kinnunen, *Chem. Phys. Lipids*, **70**, 1, (1994).
14. J. Y. A. Lehtonen and P. K. J. Kinnunen, *Biophys. J.*, **68**, 525 (1995).
15. J. Y. A. Lehtonen, J. M. Holopainen and P. K. J. Kinnunen, *Biophys. J.*, **70**, 1753 (1996).
16. J. Y. A. Lehtonen and P. K. J. Kinnunen, *Biophys. J.*, **72**, 1247 (1997).
17. F. Dumas, M. M. Sperotto, C. Lebrun, J.-F. Tocanne and O. G. Mouritsen, *Biophys. J.*, **73**, 1940 (1997).
18. O. G. Mouritsen, *Curr. Opin. Coll. Interface Sci.*, **3**, 78 (1998).
19. T. Söderlund, J. Y. A. Lehtonen and P. K. J. Kinnunen, *Mol. Pharmacol.*, **55**, 32 (1999).
20. F. Goodsaid-Zalduondo, D. A. Rintouil, J. C. Carlson and W. Hansel, *Proc. Natl. Acad. Sci. USA*, **79**, 4332 (1982).
21. R. L. Hoover, E. A. Dawidowicz, J. M. Robinson and M. J. Karnovsky, *J. Cell Biol.*, **97**, 73 (1983).
22. M. P. Lisanti, P. E. Scherer, Z. Tang and M. Sargiacomo, *Trends in Cell Biol.*, **4**, 231 (1994).
23. P. Liu and R. G. Anderson, *J. Biol. Chem.*, **270**, 27179 (1995).
24. D. A. Brown and E. London, *Annu. Rev. Dev. Cell Biol.*, **14**, 111 (1998).
25. J. Y. A. Lehtonen and P. K. J. Kinnunen, *Biophys. J.*, **68**, 1888 (1995).
26. O. Karlsson, M. Rytömaa, A. Dahlqvist, P. K. J. Kinnunen, and Å. Wieslander, *Biochemistry*, **35**, 10094 (1996).
27. P. K. J. Kinnunen and P. Laggner (eds), *Phospholipid Phase Transitions*, Special issue of *Chem. Phys. Lipids*, **57**, 109 (1991).
28. P. K. J. Kinnunen, A. Koiv, J. Y. A. Lehtonen, M. Rytömaa and P. Mustonen. *Chem. Phys. Lipids*, **73**, 181 (1994).
29. P. K. J. Kinnunen, *Chem. Phys. Lipids*, **81**, 151 (1996).
30. V. Luzzati, in *Biological Membranes*, (ed.) D. Chapman, Academic Press, London, 1968, p. 71.
31. G. Lindblom, and L. Rilfors, *Biochim. Biophys. Acta*, **988**, 222 (1989).
32. J. M. Seddon, *Biochim. Biophys. Acta*, **1031**, 1 (1990).
33. M. W. Tate, E. F. Eikenberry, D. C. Turner, E. Shyamsunder and S. M. Gruner, *Chem. Phys. Lipids*, **57**, 147 (1991).
34. R. S. Cantor, *J. Phys. Chem.*, **101**, 1723 (1997).
35. D. D. Lasic, in *Handbook of Nonmedical Applications of Liposomes*, (eds) D. D. Lasic and Y. Barenholz, CRC Press, Boca Raton, FL, 1996, p. 1.
36. F. M. Menger and J. S. Keiper, *Adv. Mat.*, **10**, 888 (1998).
37. G. Majno and I. Joris, *Am. J. Pathol.*, **146**, 3 (1995).

38. W. D. Jarvis, S. Grant and R. N. Kolesnick, *Clin. Cancer Res.*, **2**, 1 (1996).
39. M. I. Angelova and D. S. Dimitrov, *Faraday Discuss. Chem. Soc.*, **81**, 303 (1986).
40. W. I. Calhoun and G. G. Shipley, *Biochemistry*, **18**, 1717 (1979).
41. B. J. McKeone, H. J. Pownall and J. B. Massey, *Biochemistry*, **25**, 7711 (1986).
42. J. M. Holopainen, M. Subramanian and P. K. J. Kinnunen, *Biochemistry*, **37**, 17562 (1998).
43. R. Wick, M. I. Angelova, P. Walde and P. L. Luisi, *Chem. Biol.*, **3**, 105 (1996).
44. E. Farge and P. F. Devaux, *J. Phys. Chem.*, **97**, 2958 (1993).
45. H.-G. Döbereiner, J. Käs, D. Noppl, I. Sprenger and E. Sackmann, *Biophys. J.*, **65**, 1396 (1993).
46. W. Helfrich, *Z. Naturforsch.*, **28**, 693 (1973).
47. E. A. Evans, *Biophys. J.*, **14**, 923 (1974).
48. S. M. Gruner, *J. Phys. Chem.*, **93**, 7562 (1989).
49. R. Lipowsky, *J. Phys. II France*, **2**, 1825 (1992).
50. E. Sackmann, *FEBS Lett.*, **346**, 3 (1994).
51. U. Seifert and R. Lipowsky, in *Handbook of Nonmedical Applications of Liposomes*, (eds) D. D. Lasic and Y. Barenholz, CRC Press, Boca Raton, FL, 1996, p. 43.
52. F. Jülicher and R. Lipowsky, *Phys. Rev. Lett.*, **70**, 2964 (1993).
53. S.-S. Feng and R. C MacDonald, *Biophys. J.*, **69**, 460 (1995).
54. H. Löfgren and I. Pascher, *Chem. Phys. Lipids*, **20**, 273 (1977).
55. I. Pascher, *Biochim. Biophys. Acta*, **455**, 433 (1976).
56. M. B. Ruiz-Argüello, G. Basañez, F. M. Goñi and A. Alonso, *J. Biol. Chem.*, **271**, 26616 (1996).
57. X. Zha, L. M. Pierini. P. L. Leopold, P. J. Skiba, I. Tabas and F. Maxfield, *J. Cell Biol.*, **140**, 39 (1998).
58. I. Walev, U. Weller, S. Strauch, T. Foster and S. Bhakdi, *Infect. Immun.*, **64**, 2974 (1996).
59. K. A. Johansen, R. E. Gill and M. L. Vasil, *Infect. Immun.*, **64**, 3259 (1996).
60. H. Grassmé, E. Gulbins, B. Brenner, K. Ferlinz, K. Sandhoff, K. Harzer, F. Lang and T. F. Meyer, *Cell*, **91**, 605 (1997).

# Chapter 21

# Microinjection of Macromolecules in Giant Vesicles Prepared by Electroformation

*THOMAS OBERHOLZER AND ALINE FISCHER*
*ETH, Zürich, Switzerland*

## 1. INTRODUCTION

Since the late 1980s micrometer-sized giant vesicles (GVs) have attracted increasing interest and several methods have been described to produce them [1–4]. One such method permits the swelling of a lipid film in an alternating electric field and is, therefore, called electroformation ([5,6] see also Chapters 3 and 4). The characteristic feature of this method is that the giant vesicles remain attached to the platinum electrode and are, therefore, susceptible to micromanipulation.

Our interest is in using giant vesicles as microcompartments and as models for precursors to cells. To entrap solute molecules microinjection is used because this technique is mild and allows the controlled encapsulation of the solute molecules into a target giant vesicle [7].

One aim in preparing giant unilamellar vesicles is their use as microreactors [8]. For this application several requirements have to be fulfilled:

- The vesicle must be stable even after being touched by the microneedle [7,8].
- The selected target vesicle in which a solution has to be injected must endure a certain osmotic imbalance without bursting.
- For carrying out enzymatic reactions in a single giant vesicle it is important that (a) the injected volume of a single injection is not too small as highly concentrated

*Giant Vesicles* Edited by P. L. Luisi and P. Walde

protein solutions are often not available and can lead to precipitations in the microneedle and (b) that repetitive injections into the same vesicle are possible. (c) To avoid an enzymatic reaction occurring in the microneedle instead of the giant vesicle, it is important that the enzyme and the substrate solution are injected separately from one another, and for this, it is necessary that a target vesicle can be punctured several times with different solutions.

This chapter describes the conditions required for successful microinjections. Furthermore, examples of repetitive injections or multiple puncturing of a single giant vesicle are shown.

## 2. MATERIALS AND METHODS

### 2.1 Chemicals

1-Palmitoyl-2-oleoyl-*sn*-glycero-3-phosphocholine (POPC) was obtained from Avanti Polar Lipids (AL, USA). Didodecyldimethylammonium bromide, herring sperm DNA sodium salt, and yeast transfer RNA were purchased from Fluka (Buchs, Switzerland). The crude DNA was solubilized as a 2–4 mg $ml^{-1}$ solution and sonicated on ice for 30 min using a probe sonicator (Sonifier 250 from Branson, 20 W, duty cycle 50%). The mean length of the DNA molecules was shown by agarose electrophoresis to correspond to approximately 500 base pairs. YO-PRO-1 was purchased from Molecular Probes (OR, USA). Pancreatic DNase I (with a specific activity of 2000 U $mg^{-1}$) was obtained from Boehringer Mannheim (Mannheim, Germany).

### 2.2 Methods

#### *(a) Microscopic equipment*

All experiments were performed with an Axiovert 135TV inverted light microscope from Zeiss (Germany) using either differential interference contrast (DIC) or fluorescence as the detection mode. All observations were made with a $\times 20$ long working distance Achroplan lenses. Black and white images were recorded with a CCD video camera (XC-75CE from Sony) and processed by a personal computer equipped with a LYSP frame grabber and the software ImageAccess 1.5. Colour pictures were taken with a CCD C5810 video camera from Hamamatsu using a PC equipped with software also from Hamamatsu.

#### *(b) Investigation chamber, lipid film formation and vesicle formation*

The investigation chamber has been previously described [8]; 2 μl of a 0.2 mg $ml^{-1}$ phospholipid solution (in diethylether : methanol 9 : 1) were deposited on four regions of the two platinum wires. After the lipid film was dried by blowing a nitrogen stream over the wire for about 1 min, the investigation chamber was stored in a desiccator at about 10 Torr. The four regions containing phospholipid had a thickness of about 30–50 bilayers.

The investigation chamber was placed on the stage of a microscope and connected to a frequency generator (from Conrad Electronics, Germany), before 1 ml of aqueous solution was added. The applied AC electric field (1–5 V peak-to-peak value, frequency 2 Hz) was controlled by a cathode oscilloscope from Tektronix, model 5115.

### *(c) Preparation of the microneedles and microinjection experiments*

Microneedles were fabricated from borosilicate glass capillaries (outer diameter of 1.5 mm; inner diameter = 1.2 mm; containing a filament of 0.2 mm) purchased from Hilgenberg, Germany. The microneedles were prepared using a puller P30 from Sutter Instruments. The quality of the microneedles was controlled by measuring the threshold pressure at which nitrogen bubbles could be detected in ethanol following the method of Schnorf *et al.* [9]. For microinjection experiments, the microneedle was filled with an aqueous solution from a micropipette (from Eppendorf) and then fixed in a micromanipulator (from Saur, Germany). The movements of the microneedle towards the vesicles were controlled by a control stick (from Saur, Germany). The microinjections were performed with the help of a micromanipulator model 5242 (from Eppendorf) connected to an air compressor, type 6-J (from Jun-Air, Denmark).

## 3. RESULTS

### 3.1 Characteristic appearance of giant vesicles from POPC formed by electroformation

The first series of experiments investigated which morphological shapes were obtained when POPC giant vesicles were formed by electroformation. The resulting shapes were divided into four main groups as shown in Figure 21.1. This classification is very important for microinjection experiments because only certain shapes allow successful injection into the water pool of giant vesicles; other shapes burst or shrink when being punctured by the microneedle. The shapes were classified as (A) domes, (B) mushrooms, (C) spheres, and (D) cut-spheres. The domes are relatively flat structures which are definitively too small to be punctured. The mushrooms are also not suitable for microinjection experiments. Once they are punctured by the microneedle, they normally shrink immediately. Occasionally they seem to be relatively stable and survive the puncturing, but lose the injected material

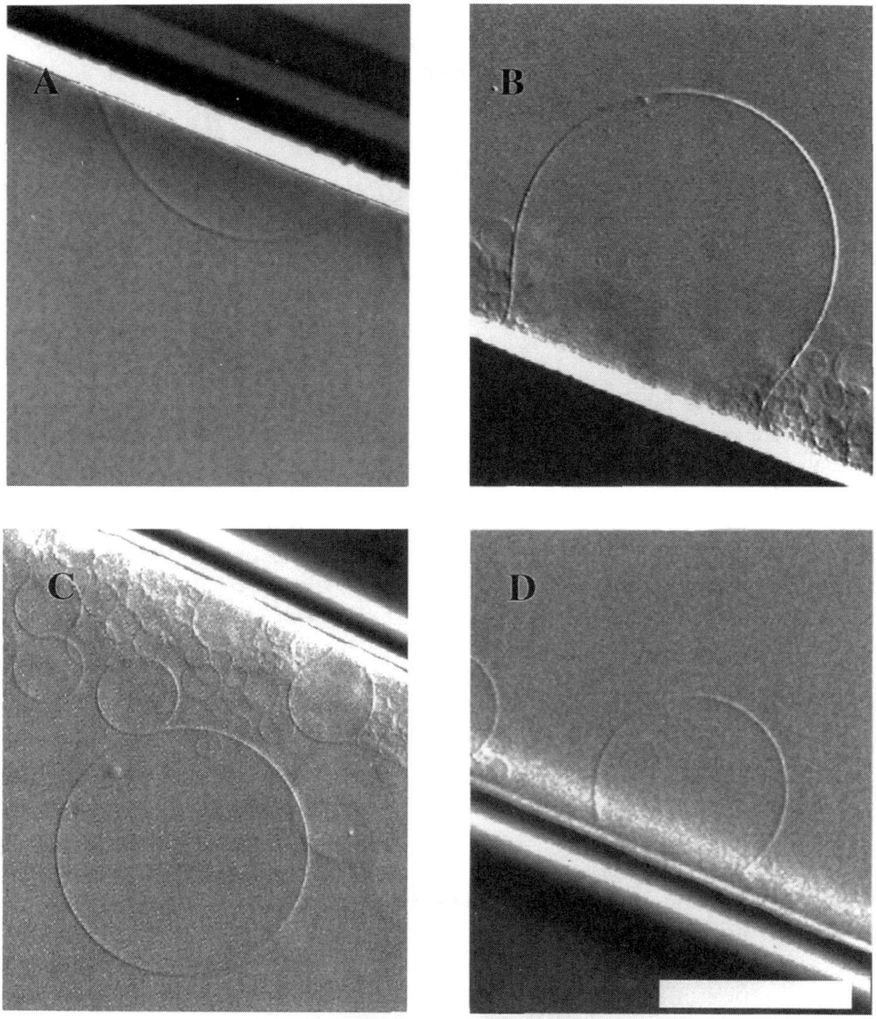


**Figure 21.1** Classification of giant POPC vesicles prepared by the electroformation method. The vesicles were grown at 2.5 V and 10 Hz for 2–4 h; light micrographs were taken in the differential interference contrast (DIC) mode: (A) domes, (B) mushrooms, (C) spheres, and (D) cut-spheres. Bar corresponds to 100 μm. (Figures B and C reproduced from *Langmuir*, **14**, 2712 (1998) by permission of the American Chemical Society).

later on. Spheres seem to be perfect candidates for such injection experiments. However, they are difficult to puncture. The reason is that the layer of vesicles between the sphere and the platinum wire is too thick. For this reason, the punctured sphere can avoid the mechanical stress applied by the microneedle and the vesicle to be punctured moves away. The best candidates for successful microinjections are the cut-spheres (Figure 21.1(D)). These are spherical vesicles that are in close proximity to the platinum wire and, therefore, cannot avoid the mechanical stress applied by the microneedle. The distinction between cut-spheres and mushrooms is often difficult; normally material has to be injected to enable a distinction to be made. Cut-spheres usually do not burst and, therefore, contain the injected material after several minutes, whereas mushrooms lose their injected contents within seconds. In addition, cut-spheres normally remain stable after being punctured, whereas mushrooms have the tendency to shrink.

## 3.2 Microinjections of nucleic acids into giant POPC vesicles

One of the main aims of microinjections into giant vesicles is their use as microreactor systems with dimensions comparable to eukaryotic cells. If enzymatic reactions are to be carried out in such a microreactor, the appropriate reagents have to be transported to the interior through the bilayer boundary. This transport can be done by passive permeation through the bilayer, by addition of the solute to be entrapped during the period of giant vesicle formation, or by active transport of the given solute into the water core of a giant vesicle. Of the several possibilities for active transport of hydrophilic solute molecules across the POPC bilayer, the method of microinjection as applied by cell biologists was selected, which is similar to the procedure of RNA injection in oocytes.

For the visualization of the injected nucleic acid material, a fluorescent dye molecule able to detect low amounts of nucleic acid but without a high background fluorescence was needed. All these requirements were fulfilled by the dye YO-PRO-1 [8,10]. Figure 21.2 shows such microinjection experiments of sonicated crude DNA. As can be seen, even concentrations of $10\,\mu g\,ml^{-1}$ DNA stained by YO-PRO-1 (dye concentration $1\,\mu M$) resulted in a detectable fluorescent vesicle. The time course of the fluorescent intensity was monitored and remained stable, indicating that the dye molecules present in the vesicle were sufficient to stain the DNA molecules (Figures 21.2(A) and 21.2(B)). In contrast to this finding, injections of a 10 times more concentrated DNA solution resulted in a fluorescent vesicle that became more fluorescent during a period of 1 h (Figures 21.2(C) and 21.2(D)). Obviously at a concentration of $1\,\mu M$ there are not enough dye molecules present to stain the injected DNA and, therefore, it takes a certain period for the dye molecules to permeate across the POPC bilayer. It can also be seen that the fluorescence intensity of the vesicle containing the higher amount of DNA seems to be more than 10 times higher.

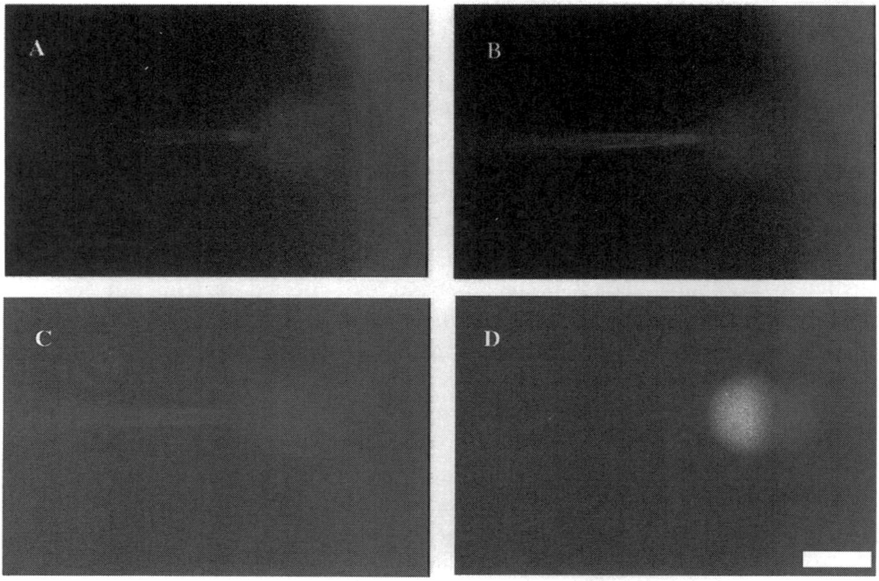


**Figure 21.2** Microinjection of DNA in POPC GVs, with staining of the DNA by YO-PRO-1. The vesicles were formed at 2.5 V and 10 Hz in the presence of 1 μM YO-PRO-1, then a crude DNA solution was injected (injection time of 1 s). (A) and (B): injection of 10 μg ml$^{-1}$ DNA, micrograph taken after (A) 1 min and (B) after 5 min. (C) and (D): injection of 100 μg ml$^{-1}$ DNA, 5 min and 1 h after injection. Micrographs were obtained in the fluorescence detection mode. Bar corresponds to 50 μm.

The injection time for these experiments was 1 s. To calculate the amount of nucleic acid that can be detected with this system in a single vesicle, one has to know the volume of liquid injected in one second. This is one of the most problematic points in this work; first, the microneedles have to be defined in a relatively simple way and, second, the amount of injected liquid has to be determined. For a routine determination of the microneedles the method of nitrogen bubble formation originally described by Schnorf *et al.* was used [9]. These microneedles can be used for subsequent experiments. To determine the injected volume two distinct experimental approaches were used, as follows. (1) Injection of water in oil and determination of the size of the water droplet by electron microscopy according to the method of McCanman *et al.* [11]. (2) Injection of [$^{35}$S]dATP into the investigation chamber and determination of the injected radioactivity by $\beta$-scintillation counting. Unfortunately, the results obtained by these two approaches were not consistent with each other: The first method resulted in volumes of 100–600 fl, whereas with the second approach apparent volumes of 1–20 pl (always determined for a single injection, for an injection period of 1 s with microneedles of comparable quality) were obtained. Therefore, it is still unclear how much nucleic acid was injected and what amount is required to be visualized by fluorescence microscopy. On the assumption that 2.5 pl were injected for the experiments in Figure 21.2(A, B), it can be assumed that the limit of detection for DNA molecules is in the range 10–50 fg. With RNAs as nucleic acids this limit of detection is about 10 times higher (data not shown).

### 3.3 Effect of repetitive microinjections on the size of the giant vesicle

Another question is the fate of the vesicles after injecting an additional volume. For example, could they burst? To answer this question, POPC GVs were formed as usual in the presence of nucleic acid dye YO-PRO-1. Then tRNA was injected several times into a selected GV. As can be seen from Figure 21.3(A), the vesicle had a diameter of about 130 μm before the vesicle was punctured. After 15 injections the vesicle diameter was slightly increased, and the injected fluorescent nucleic acid remained entrapped (Figures 21.3(C) and 21.3(D)). Even after 40 injections the vesicle was still stable and had a diameter of about 160 μm. An increase of 30 μm means that the vesicle volume has almost doubled (from an initial volume of 1.1 nl to a final volume of 2.1 nl). As described above it can be assumed that the mean injected volume is in the range of 500 fl to 5 pl and, therefore, it cannot be responsible for the total volume rise.

This experiment poses a further interesting question, namely where does all the additional lipid come from? It is unlikely that the volume increase can be explained with an increased elasticity of the POPC GVs [12]. It is more likely that additional POPC molecules are incorporated into the GV bilayer. But where do these delivered lipids come from? There are several possibilities. (1) The phospholipids may be delivered from the electrode surface. There is always a reservoir, as can be seen from Figure 21.1; the vesicles are in close contact and it is therefore possible that lipids can be incorporated from this reservoir. (2) The vesicles formed in the alternating electric field are not unilamellar or contain submicroscopic vesicles. In these cases the newly incorporated POPC molecules could come from such structures that are too small to be detected by microscopy.

### 3.4 Enzymatic reaction in GVs: The digestion of DNA by pancreatic DNase I

One of the main reasons that GVs are so attractive for studying is their use as microreactors. In such systems the progress of enzymatic reactions in a compartmentalized system can be studied *in situ* and in real time (see also chapter 22). An early example of such a system is the digestion of microinjected DNA by pancreatic DNase I in a GV. This system also shows the feasibility of this microinjection technique in GVs, because it demonstrates that multiple puncturing of a target GV is possible.

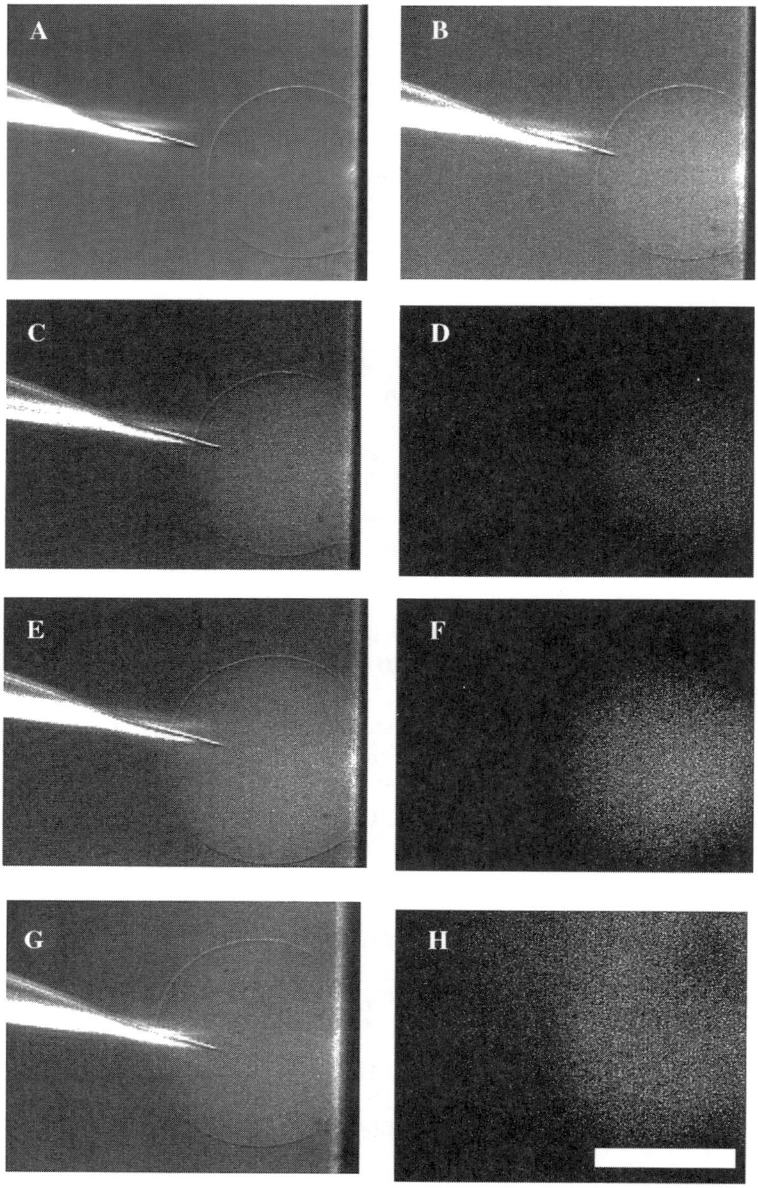


**Figure 21.3** Light micrographs showing the effect of repetitive microinjections into the same target GV. The POPC GVs were grown in an electric field (2 V, 10 Hz) in the presence of 50 μM YO-PRO-1. (A) and (B) show the GV before the injections. (C) and (D) show the same vesicle after 15, (E) and (F) 40, and (G) and (H) 50 injections. (A) (B) (C) (E) and (G) recorded in DIC mode; (D) (F) and (H) recorded in fluorescence detection mode. Bar corresponds to 100 μm. Reproduced from *Langmuir*, **14**, 2712 (1998) by permission of the American Chemical Society.

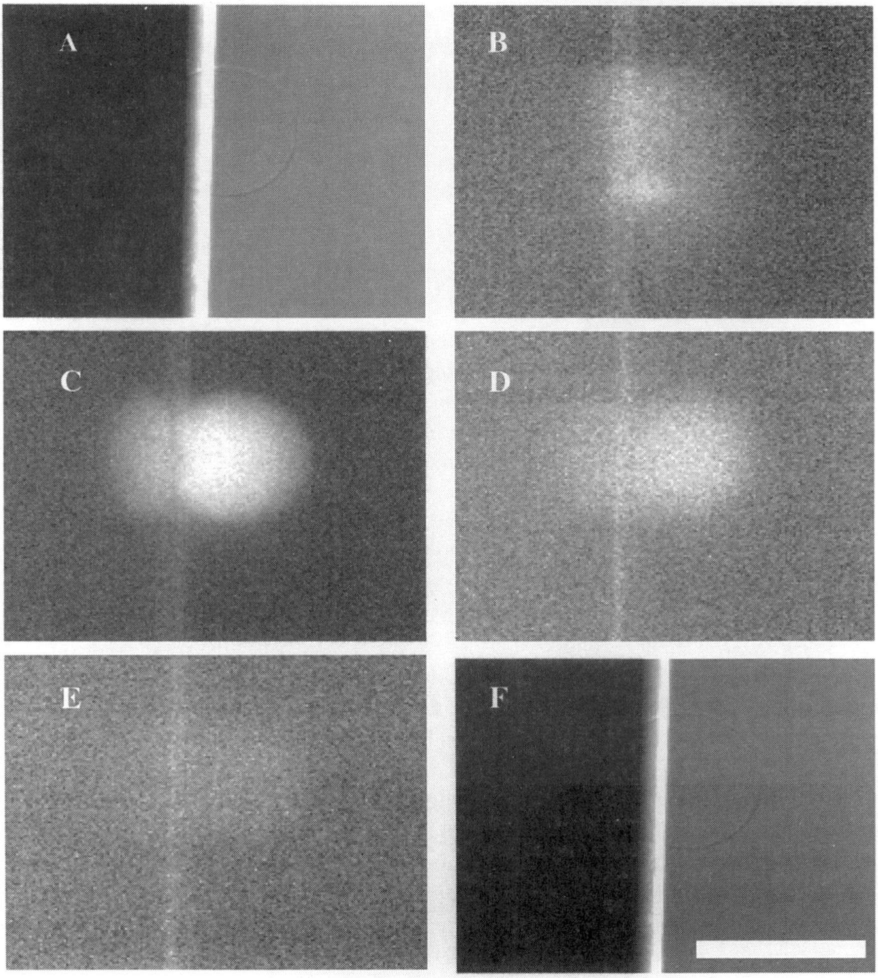


**Figure 21.4** Light micrographs demonstrating the entrapment of DNA into a target GV and the digestion of the entrapped DNA by pancreatic DNase I. (A) and (F) show the GV before and after the whole experiment (DIC mode). POPC GVs were formed at 2.5 V and 10 Hz for 3 h in the presence of 50 μM YO-PRO-1. (B) Immediately after the injection of a DNA solution (2 mg $ml^{-1}$) and (C) 15 min later. Then another puncturing with a pancreatic DNase I solution (10 U $ml^{-1}$) was carried out, and images were taken (D) 5 and (E) 15 min after the DNase injection. Bar corresponds to 100 μm.

The POPC GVs were grown in an alternating electric field in the presence of YO-PRO-1 as described above. Then DNA was injected from a 2 mg $ml^{-1}$ solution and the entrapped DNA was visualized (Figure 21.4(B) and (C)). After 15 min the selected vesicle was punctured again and this time a 5 $\mu g\,ml^{-1}$ DNase I solution was injected. The DNase now digests the entrapped DNA molecules and, after 15 min, almost no nucleic acid can be detected. The GV itself remained stable and unchanged during this process.

## 4. DISCUSSION

These investigations have demonstrated that cut-spherical POPC GVs formed in an alternating electric field for 2–4 h are stable enough for microinjection experiments. Other morphological structures cannot be used for such experiments, such as the mushroom structures. Spherical GVs – preferably the cut-spheres – can be touched and punctured by a microneedle; furthermore, after withdrawing the microneedle the GVs still retain their spherical shape.

Microinjections into such spherical GVs are possible and, in most cases, these punctured vesicles remain stable over a long period and do not lose their injected material. This work shows that macromolecules such as nucleic acids and proteins can be entrapped by microinjection and that the detection limit with the nucleic acid dye YO-PRO-1 is in the range 10–50 fg. With RNAs (mixture of tRNAs or midi variant RNA [13]) this limit is about 10 times higher (A. Fischer and T. Oberholzer, unpublished observation). A limitation of the present microinjection technique is the fact that the vesicles, in which a liquid has been injected, become less stable because of osmotic effects. Therefore, the injections of highly concentrated salt solutions have to be avoided. Such an injection would immediately lead to shrinkage (maybe it does grow, and then the membrane bursts) of the vesicle in such a way that it cannot be observed by this microscopy technique.

What is important for further applications of this approach (i.e. the microinjection of giant vesicles) is to what extent microinjections is possible with (i) single punctures and repetitive injections and (ii) several punctures with needles containing different substances. This report shows that both procedures are feasible. Repetitive injections are possible, in some cases up to 30–50 times (over a time period of about 10–30 min), before the vesicle becomes leaky (although this did not necessarily lead to destruction of the vesicle). It was also possible to inject various substances into the same GV by repetitive puncturing of the same vesicle (Figure 21.4).

## 5. REFERENCES

1. J. P. Reeves and R. M. Dowben, *J. Cell. Physiol.*, **73**, 49 (1969).
2. D. W. Deamer and A. D. Bangham, *Biochim. Biophys. Acta*, **443**, 629 (1976).
3. F. Szoka Jr. and D. Papahadjopoulos, *Annu. Rev. Biophys. Bioeng.*, **9**, 467 (1980).

4. N. Oku, J. F. Scheerer and R. C. McDonald, *Biochim. Biophys. Acta*, **692**, 384 (1982).
5. D. S. Dimitrov, J. Li, M. I. Angelova and R. K. Jain, *FEBS Lett.*, **176**, 398 (1984).
6. M. I. Angelova and D. S. Dimitrov, *Faraday Discuss. Chem. Soc.*, **81**, 303 (1986).
7. R. Wick, M. I. Angelova, P. Walde and P. L. Luisi, *Chem. Biol.*, **3**, 105 (1996).
8. P. Bucher, A. Fischer, P. L. Luisi, T. Oberholzer and P. Walde, *Langmuir*, **14**, 2712 (1998).
9. M. Schnorf, I. Potrykus and G. Neuhaus, *Exp. Cell. Res.*, **210**, 260 (1994).
10. R. P. Haugland, *Handbook of Fluorescent Probes and Research Chemicals*, 6th edn, Molecular Probes, Eugene, OR, 1996.
11. R. E. McCanman, D. G. McKenna and J. K. Ono, *Brain Res.*, **136**, 141 (1977).
12. B. Lerebours, E. Wehrli and H. Hauser, *Biochim. Biophys. Acta*, **1152**, 49 (1993).
13. C. K. Biebricher, S. Diekmann and R. Luce, *J. Mol. Biol.*, **154**, 629 (1982).

Chapter 22

# Enzymatic Reactions in Giant Vesicles

*PETER WALDE*
*ETH, Zürich, Switzerland*

## 1. INTRODUCTION

Although giant vesicles have been known for many years [1], most studies on giant vesicles so far have been devoted to the characterization of physicochemical properties, such as the bending elasticity [2–6] or shape transformations [7,8]. There are only a few known cases where giant vesicles have been studied in combination with biopolymers such as proteins (enzymes) [9–12]. The reason for this is not obvious, but it may be that a general scepticism concerning the preparation and handling of giant vesicles has hindered potential research in this field. Until recently, most researchers have probably been unaware of the fact that a convenient method has been available now for several years for preparing giant vesicles from phosphatidylcholines or phospholipid mixtures reproducibly. This method is known as the electroformation method [13–15] (see also Chapters 3, 4, and 21) and it has certain advantages over other known methods (see below).

This contribution is aimed at demonstrating with a few selected examples that giant vesicles prepared by electroformation are convenient systems with which to investigate the behaviour of enzymes. It is certainly a relatively new field of research where the action of enzymes can be directly studied by light microscopy in real time, without any particular need for indirect experiments or special sample preparations.

Two conceptually different cases are described:

- Giant vesicles composed of amphiphiles which are enzyme substrates; the action of the enzymes on these substrates will modify the chemical composition of the vesicles shell (the case of phospholipases D and $A_2$).

*Giant Vesicles* Edited by P. L. Luisi and P. Walde

- Giant vesicles as microreactor systems in which a water-soluble enzyme acts on water-soluble substrates within the aqueous interior of one single giant vesicle (the case of alkaline phosphatase).

An advantage of giant vesicles over conventional, submicrometer-size vesicles is that giant vesicles allow the continuous investigation of the action of enzymes on or inside an individual target vesicle.

## 2. EXPERIMENTAL

Unless otherwise mentioned, all giant vesicles were prepared from 1-palmitoyl-2-oleoyl-*sn*-glycero-3-phosphocholine (POPC) by electroformation [13–15], involving an open investigation chamber with two parallel platinum wires and a total volume of 1 ml [16–19]. The giant vesicles form on the wires from a POPC deposit, and the vesicles often remain adhered to the wire through small tethers [5]. This can be considered as a kind of immobilization as a result of which the vesicles remain at the site of formation, without moving away – an important aspect if the aim is to investigate one single focused target vesicle over a longer period of time. The vesicles prepared by electroformation are generally stable. If the investigation chamber is covered with a glass cover to minimize water evaporation, the vesicles do not change in size for at least one day.

Phospholipase D from *Streptomyces* sp. AA586 was obtained from Genzyme Diagnostics (West Malling, UK), *Naja naja* phospholipase $A_2$ was purchased from Sigma (USA), and fluorescein diphosphate from Molecular Probes (USA).

## 3. RESULTS AND DISCUSSION

### 3.1 Conventional versus giant vesicles

The main difference between conventional and giant vesicles is certainly the size and accordingly the bilayer curvature. Conventional vesicles typically have diameters between 0.05 and 0.5 μm, in comparison with 5–50 μm or more for giant vesicles. As a consequence of these differences, giant vesicles can be observed by light microscopy, whereas in the case of conventional vesicles one has to rely on more indirect methods, in particular electron microscopy or light scattering. Another consequence of the difference in size is the number of vesicles used in one set of experiments. In the case of conventional vesicles, one is usually dealing with several million vesicles in each experiment (Table 1), and each vesicle differs from the other, either by the size or lamellarity (vesicles are mainly kinetically trapped structures, characterized by a certain degree of polydispersity [20,21]). If one prepares conventional vesicles containing entrapped solute molecules (e.g. enzymes), one always has a solute distribution; that is, not all vesicles contain the same amount of solute.

**Table 1** Comparison of typical single phospholipid giant vesicles (diameters 5 or 50 µm) with typical conventional vesicle preparations (diameters 50 or 500 nm). The calculations are made for POPC as phospholipid, using a constant mean head group area of 0.72 $nm^2$ and a bilayer thickness of 3.7 nm [18]. (1 femtoliter = 1 fl = $1 \times 10^{-15}$ l)

| Parameter | Small | | Large | |
|---|---|---|---|---|
| *Conventional vesicles* | | | | |
| Vesicle diameter | 0.05 µm | | 0.5 µm | |
| Number of POPC molecules per vesicle | $1.88 \times 10^4$ | | $2.15 \times 10^6$ | |
| Internal aqueous volume of one single vesicle (fl) | $4.05 \times 10^{-5}$ | | $6.26 \times 10^{-2}$ | |
| Typical POPC concentrations used (mM) | 0.1 | 10 | 0.1 | 10 |
| Corresponding vesicle concentration (nM) | 5.31 | 531 | 0.0465 | 4.65 |
| Corresponding number of vesicles per ml | $3.19 \times 10^{12}$ | $3.19 \times 10^{14}$ | $2.80 \times 10^{10}$ | $2.80 \times 10^{12}$ |
| Corresponding aqueous volume trapped by the vesicles (%) | 0.0129 | 1.292 | 0.1753 | 17.53 |
| *Giant vesicle* | | | | |
| Vesicle diameter (µm) | 5 | | 50 | |
| Number of POPC molecules per vesicle | $2.18 \times 10^8$ | | $2.18 \times 10^{10}$ | |
| Internal aqueous volume of one single vesicle (fl) | 65.2 | | $6.54 \times 10^4$ | |

In contrast, when using giant vesicles, one single vesicle is used at a time (see Table 1). One of the consequences of this is that processes with conventional vesicles in which intervesicular interactions are involved may not occur in the case of a single giant vesicle (see the concluding remarks).

### 3.2 Studies with phospholipase D from *Streptomyces sp.* AA 586

Phospholipase D (EC 3.1.4.4) catalyzes the hydrolysis of the terminal phosphate ester bond of 1,2-diacyl-*sn*-glycero-3-phopholipids, leading in the case of phosphatidylcholines (such as POPC) to phosphatidic acid and choline (Figure 22.1) [22–25]. The enzyme appears to play an important role in a number of signal transduction pathways. It also catalyzes transalkylphosphatidylations, allowing the synthesis of a variety of glycerophospholipids with different polar head groups [26,27]; in this respect, phospholipase D from *Streptomyces* sp. AA586 has been successfully used to prepare a series of phosphatidyl nucleosides [28,29].

In a detailed $^1$H NMR study it has been shown that phospholipase D from *Streptomyces* sp. AA586 catalyzes the hydrolysis of POPC if added to conventional unilamellar vesicles, the yields always being about 50% [18]. Although not established, it is likely that only the POPC molecules originally present in the outer monolayers of the vesicles could be reached by the enzyme and therefore could be hydrolyzed to POPA and choline under the conditions of the experiments (e.g. in the absence of $Ca^{2+}$). The exchange of the phospholipids between the inner and the outer monolayer of the vesicles is expected to be rather slow, as the flip-flop time for phospholipids, as for phosphatidylcholines, is known to be of the order of several hours [30]. Furthermore, freeze-fracture electron microscopy has shown that the mean size of the vesicles remained the same during the course of the reaction [18].

In a comparative study, instead of conventional vesicles, giant POPC vesicles were used, and the effect of exo- and endo-vesicular addition of phospholipase D to single vesicles was investigated. In the case of exovesicular enzyme addition, the enzymes were added at a close distance to the target vesicle. In the case of endovesicular enzyme addition, the enzyme is directly injected into the target vesicle by the use of corresponding microneedles and a micromanipulation device [16,17,19]. Basically, two different sets of experiments were carried out, one using only POPC, and one using POPC vesicles containing fluorescently labelled phospholipase D substrate Texas Red-DOPE (POPC : Texas Red-DOPE = 50 : 1, by weight). The latter phospholipid was used to monitor the chemical transformation of the phospholipid upon the action of the enzyme. The vesicle membrane containing Texas Red-DOPE is fluorescent, and the release of the water-soluble fluorescent head group (Texas Red) upon the action of phospholipase D yields vesicles containing nonfluorescent DOPA. The initial conditions of the experiment are given in Table 2, and the light microscopic observations are illustrated in Figure 22.2. In good agreement with the results obtained from the NMR work on conventional vesicles, the reaction yields

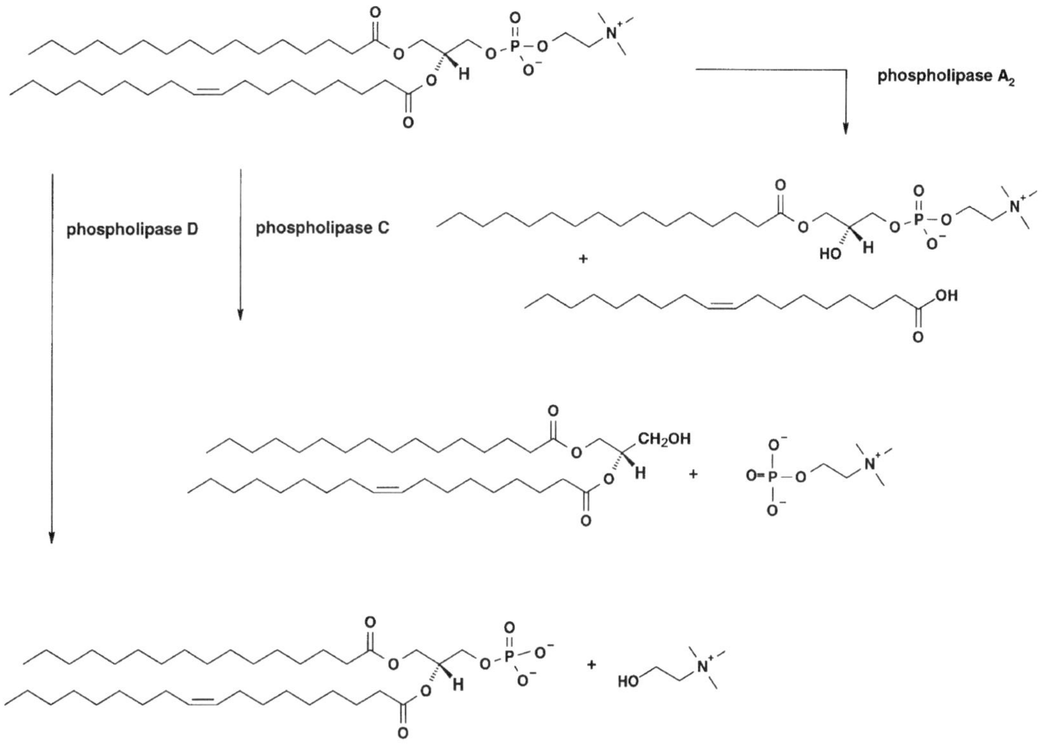


**Figure 22.1** POPC hydrolysis reactions catalyzed by the phospholipases $A_2$, C, and D [31,32].

**Table 2** Experimental conditions used for the experiments with giant POPC vesicles and exo- or endovesicular addition of phospholipases at room temperature

| Enzyme | Phospholipid (experimental conditions) | Mode of enzyme addition | Diameter of the target vesicle (μm) | Number of phospholipid molecules in the target vesicle | Ratio of phospholipid molecules to added enzyme molecules |
|---|---|---|---|---|---|
| Phospholipase D | POPC (10 mM Tris/HCl, pH 8.0) | exovesicular | ~60 | $\sim 3.1 \times 10^{10}$ | $8.5 \times 10^3$ |
| | POPC/Texas Red-DOPE (50 : 1, wt) (10 mM Tris/HCl, pH 7.6) | exovesicular | ~27 | $\sim 1.2 \times 10^6$ | $6.1 \times 10^3$ |
| Phospholipase $A_2$ | POPC (1 mM Tris/HCl, 5 mM $CaCl_2$, pH 8.0) | exovesicular | ~30 | $\sim 2.5 \times 10^7$ | ~320 |
| | POPC (1 mM Tris/HC1, 5 mM $CaCl_2$, pH 8.0) | endovesicular | ~18 | $\sim 7.7 \times 10^6$ | ~360 |

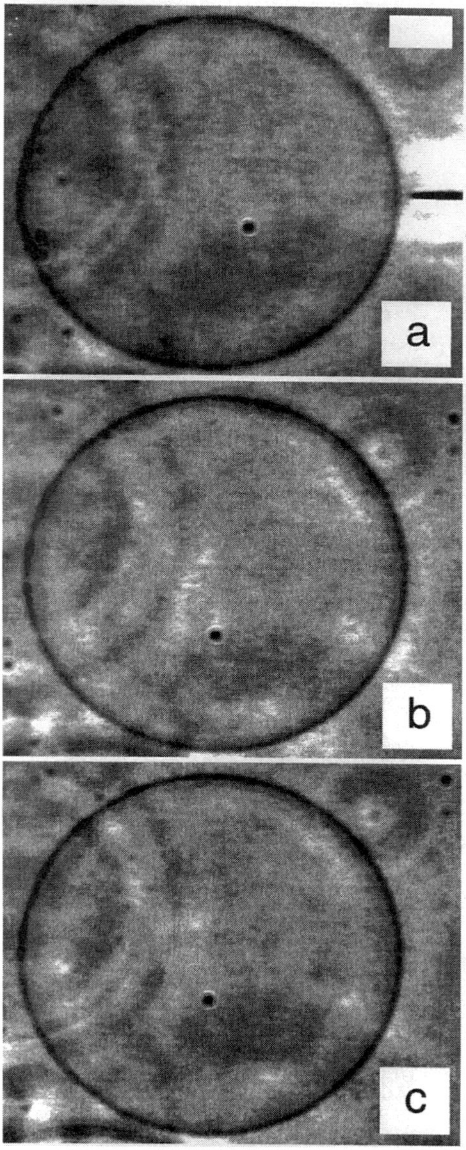


(A)

**Figure 22.2** The effect of exovesicular addition of phospholipase D to single giant vesicles. (A) Giant vesicle containing POPC. (B) and (C) Giant vesicle containing a mixture of POPC and fluorescent Texas Red-DOPE (POPC : Texas Red-DOPE = 50 : 1, by weight. (C) Changes in the fluorescent intensity: (○) no enzymes; (●) after addition of enzyme. Length of the bar 10 μm. Photographs were taken before enzyme addition (a), 20 minutes (b), and 60 minutes (c) after enzyme addition, respectively. Partly adapted from Ref. 18.

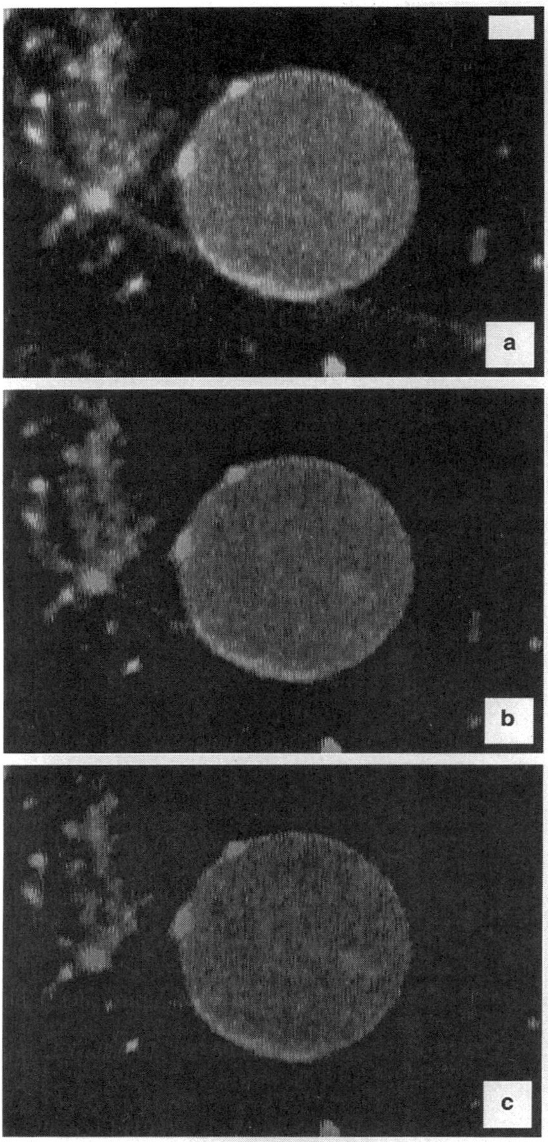


(B)

**Figure 22.2** (*continued*)

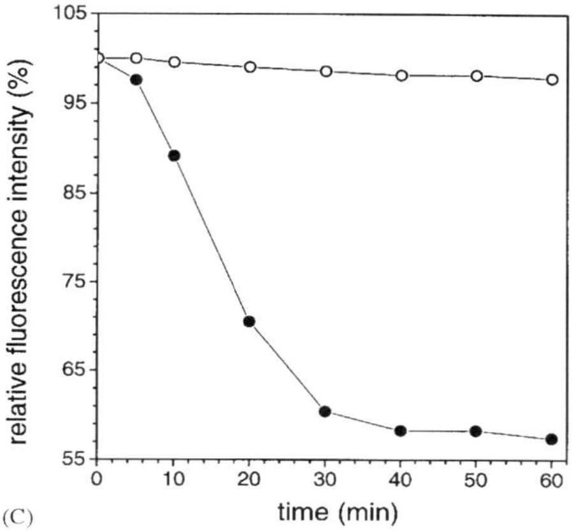

(C)

**Figure 22.2** (*continued*)

were again about 50%, as judged from Figure 22.2(C) (the fluorescent intensity of the vesicles prepared with fluorescently labelled phospholipids decreased to about 50% upon addition of phospholipase D). Furthermore, again in agreement with the study on conventional vesicles, the size of the giant vesicles did not change significantly (Figure 22.2(A) and (B)), indicating that the transformation of the zwitterionic POPC to the anionic POPA did not lead to a measurable membrane destabilization. Also, for the case of endovesicular phospholipase D addition, no change in size was observed.

### 3.3 Studies with phospholipase $A_2$ from *Naja naja*

Phospholipase $A_2$ (EC 3.1.1.4) from the venom of the cobra snake *Naja naja* is a $Ca^{2+}$-dependent enzyme that catalyzes the hydrolysis of the acyl ester bond at the *sn*-2 position of 1,2-diacyl-*sn*-glycero-3-phospholipids, such as phosphatidylcholines [23,31–33]. In the case of POPC, the products released are 1-palmitoyl-*sn*-glycero-3-phosphocholine (a lysoPC) and oleic acid (Figure 22.1). Although lysoPCs are known to be surface-active micelle-forming substances [34–36] with hemolytic activity [32], it has also been reported that under certain conditions lysoPC can form a stable bilayer type of organization in the presence of fatty acids [37]. Furthermore, it has been reported that the action of exovesicularly added phospholipases lead only to partial hydrolysis of phosphatidylcholines in small unilamellar vesicles – the inner monolayer of the vesicles apparently remaining intact [38,39]. Others have reported that the action of phospholipase $A_2$ on conventional vesicles leads to a complete phospholipid hydrolysis and vesicle destabilization if the reaction is carried out above the phase transition temperature of the phospholipid

[40]. Some of these discrepancies may possibly be clarified by using giant vesicles. With giant vesicles it is possible to monitor, in real time, morphological changes of the vesicles upon exo- or endovesicular enzyme addition. With the experimental conditions listed in Table 2, a destabilization of POPC vesicles has been observed, independent of whether phospholipase $A_2$ was added externally to a target vesicle, or whether it was microinjected inside the vesicle. In the case of endovesicular phospholipase $A_2$ addition, the target vesicle shrunk continuously after a certain lag phase, until the size of the vesicle became smaller than the resolution limit of the microscope, (Figure 22.3(A) and (B)). Exovesicular phospholipase $A_2$ addition led to a transformation of the target vesicle into smaller vesicles (or other lipid structures), again after a certain latency period (Figure 22.3(C)). The smaller structures partly reassembled to a few larger structures which then split again, until finally neither vesicles nor other lipid aggregates could be detected by light microscopy. Depending on the distance from which phospholipase $A_2$ was added to the target vesicle (at a fixed amount of added enzyme), the latency time changed; close addition shortened the lag phase.

Although there are differences between the results obtained from endo- or exovesicular enzyme additions, the target POPC vesicle was in both cases always completely destroyed. This is most likely caused by the formation of 1-palmitoyl-*sn*-glycero-3-phosphocholine and oleic acid which do not remain in a bilayered state together with unreacted POPC. In the case of endovesicular phospholipase $A_2$ addition, the aqueous space to which the reaction products may partition is rather limited – in contrast to the much larger bulk volume in the case of exovesicular enzyme addition – so one may understand, at least qualitatively, that the vesicle destruction process can be different in the two cases. The appearance of a lag phase may be related to the formation of a critical amount of reaction product which then initiates the vesicle destabilization., If all POPC molecules of one single vesicle (diameter 30 μm) were hydrolyzed to lysoPC and oleic acid, an overall lysoPC concentration of $1.3 \times 10^{-11}$ M would be obtained in the investigation chamber. The lysoPC molecules formed most likely partition into other vesicles present on the metal wire or, alternatively, if the local concentration is high enough, the lysoPC, at least transiently, may aggregate into micelles. The cmc of 1-palmitoyl-*sn*-glycero-3-phosphocholine is 7 μM [34].

### 3.4 Studies with alkaline phosphatase from calf intestine

In a series of experiments it has been shown that it is possible to prepare giant POPC vesicles by the electroformation method in the presence of (non-fluorescent) fluorescein diphosphate, a substrate for alkaline phosphatase. The vesicles thus formed contained in the interior aqueous space substrate molecules which, after

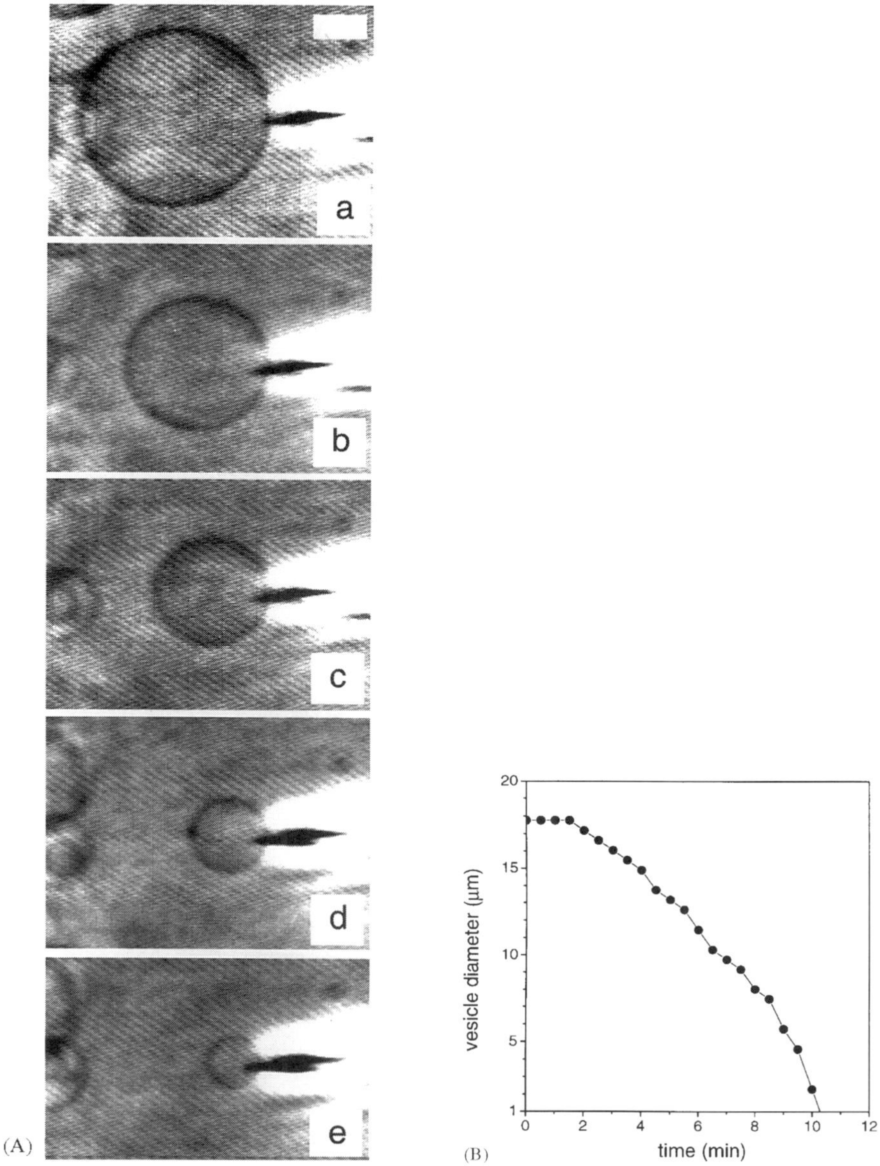


**Figure 22.3** The effect of adding phospholipase $A_2$ to single giant POPC vesicles. (A) and (B) Endovesicular enzyme addition. Photographs were taken before enzyme addition (a), 4 minutes (b), 6 minutes (c), 8 minutes (d), and 9 minutes (e) after enzyme addition, respectively.

(C)

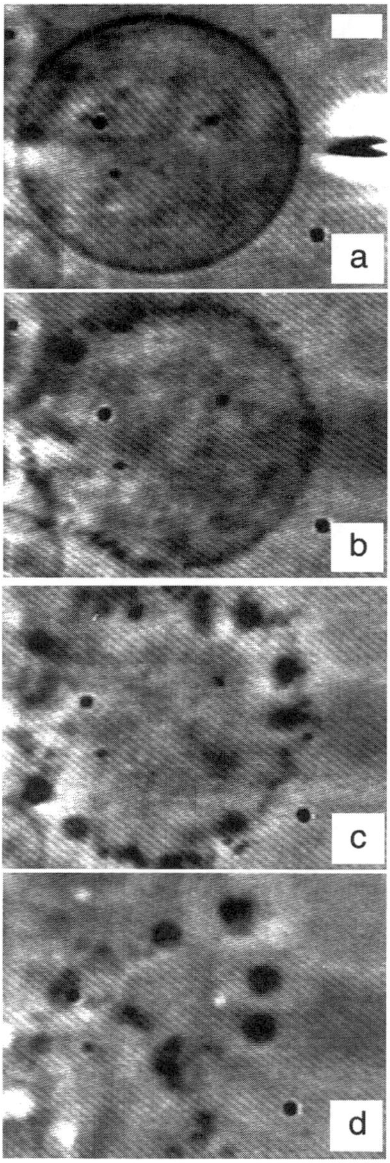


**Figure 22.3** (continued). (C) Exovesicular enzyme addition. Photographs were taken before enzyme addition (a), 12 seconds (b), 15 seconds (c), and 20 seconds (d) after enzyme addition, respectively. Length of bar 5 μm. Adapted from Ref. 16.

endovesicular addition of alkaline phosphatase, transformed into phosphate and fluorescein. The latter showed a strong fluorescence: $\lambda_{abs,max}$ at 490 nm and $\lambda_{em,max}$ at 514 nm (data not shown). In this way it was possible to carry out an enzymatic reaction with a water-soluble enzyme and water-soluble substrate molecules within a single giant vesicle. The same can be done with other enzyme–substrate systems.

## 4. CONCLUDING REMARKS

There is no doubt that giant vesicles, particularly if prepared by electroformation [13–15], are useful alternative systems with which to investigate the behavior of enzymes. As this is a relatively new field of research, many interesting experiments can be performed. First, one has to pay attention to the quantification of the chemical reaction which is catalyzed by the enzyme of interest. This has been done in the case of phospholipase D using a fluorescently labelled substrate (Section 3.2). Furthermore, with membrane-active enzymes, such as lipases or phospholipases, one has to consider that there may be a regulation of the enzyme activity through the packing density in the bilayer [41–43]. Therefore, studies on conventional vesicles may not be directly applicable to giant vesicles, due to the differences in the curvature and lipid packing. In addition, in the case of conventional vesicles, intervesicle interactions are possible through vesicle collisions as a result of Brownian motion. With processes such as vesicle fusion, different results may be obtained with giant vesicles, because in this case collisions do not occur, at least if the vesicles are prepared by electroformation.

In all the experiments with giant POPC vesicles and phospholipase C (which catalyzes the hydrolysis of POPC to 1-palmitoyl-2-oleoyl-glycerol and phosphocholine (Figure 22.1)) no vesicle transformation was detected, although it is known that phospholipase C can induce vesicle fusion, which is believed to arise from the formation of a critical amount of diacylglycerol. As a fusion process always involves at least two fusing objects, it is not surprising that no changes were seen with one single vesicle. However, this needs further investigation.

Concerning the results obtained with phospholipase $A_2$, a recent cryotransmission electron microscopy study using a venom phospholipase $A_2$ and conventional vesicles made from saturated phosphatidylcholines has shown that the vesicles are destabilized upon (exovesicular) enzyme addition, again after a starting lag phase [44]. This is in good agreement with the observations of Section 3.3 [16].

A particular advantage of giant vesicles prepared by electroformation is the possibility of solute microinjection, of either enzymes or other water-soluble substances. This defines a microreactor compartment composed of an aqueous interior and a bilayer shell, and is a possible precursor structure for contemporary cells.

## 5. REFERENCES


1. J. P. Reeves and R. M. Dowben, *J. Cell. Physiol.*, **73**, 49 (1969).
2. D. Needham and E. Evans, *Biochemistry*, **27**, 8261 (1988).
3. H. P. Duwe, J. Kaes and E. Sackmann, *J. Phys. France*, **51**, 945 (1990).
4. L. Bo and R. E. Waugh, *Biophys. J.*, **55**, 509 (1989).
5. P. Méléard, C. Gerbeaud, T. Pott, L. Fernandez-Puente, I. Bivas, M. D. Mitov, J. Dufourcq and P. Bothorel, *Biophys. J.*, **72**, 2616 (1997).
6. S. Svetina and B. Žekš, *Eur. Biophys. J.*, **17**, 101 (1989).
7. K. Berndl, J. Käs, R. Lipowsky, E. Sackmann and U. Seifert, *Europhys. Lett.*, **13**, 659 (1990).
8. J. Käs and E. Sackmann, *Biophys. J.*, **60**, 825 (1991).
9. A. Saitoh, K. Takiguchi, Y. Tanaka and H. Hotani, *Proc. Natl. Acad. Sci. USA*, **95**, 1026 (1998).
10. G. Decher, H. Ringsdorf, J. Venzmer, D. Bitter-Suermann and C. Weisgerber, *Biochim. Biophys. Acta*, **1023**, 357 (1990).
11. H. Miyata and H. Hotani, *Proc. Natl. Acad. Sci. USA*, **89**, 11547 (1992).
12. M. L. Longo, A. J. Waring, L. M. Gordon and D. A. Hammer, *Langmuir*, **14**, 2385 (1998).
13. M. I. Angelova and D. S. Dimitrov, *Faraday Discuss. Chem. Soc.*, **81**, 303 and 345 (1986).
14. D. S. Dimitrov and M. I. Angelova, *Prog. Colloid Polym. Sci.*, **73**, 48 (1987).
15. M. Angelova and D. S. Dimitrov, *Prog. Colloid Polym. Sci.*, **76**, 59 (1988).
16. R. Wick, M. I. Angelova, P. Walde and P. L. Luisi, *Chem. Biol.*, **3**, 105 (1996).
17. R. Wick and P. L. Luisi, *Chem. Biol*, **3**, 277 (1996).
18. V. Dorovska-Taran, R. Wick and P. Walde, *Anal. Biochem.*, **240**, 37 (1996).
19. P. Bucher, A. Fischer, P. L. Luisi, T. Oberholzer and P. Walde, *Langmuir*, **14**, 2712 (1998).
20. D. D. Lasic, *Biochem. J.*, **256**, 1 (1988).
21. D. D. Lasic, *J. Colloid Interface Sci.*, **140**, 302 (1990).
22. M. Heller, *Adv. Lipid Res.*, **16**, 267 (1978).
23. E. A. Dennis (ed), *Methods Enzymol.*, **197** (1991).
24. M. Waite, in *Biochemistry of Lipids, Lipoproteins and Membranes*, (eds) D. E. Vance and J. E. Vance, Elsevier, Amsterdam, 1996, p. 211.
25. M. F. Roberts, *FASEB J.*, **10**, 1159 (1996).
26. H. Eibl and S. Kovatchev, *Methods Enzymol.*, **72**, 632 (1981).
27. L. R. Juneja, T. Kazuoka, N. Goto, T. Yamane and S. Shimizu, *Biochim. Biophys. Acta*, **1003**, 277 (1989).
28. S. Shuto, S. Ueda, S. Imamura, K. Fukukawa, A. Matsuda and T. Ueda, *Tetrahedron Lett.*, **28**, 199 (1987).
29. S. Bonaccio, P. Walde and P. L. Luisi, *J. Phys. Chem.*, **98**, 6661; 10376 (1994).
30. E. Sackmann, *Ber. Bunsenges. Phys. Chem.*, **82**, 891 (1978).
31. E. A. Dennis, in *The Enzymes* (ed.) P. D. Boyer, 3rd edn, Vol. 16, Academic Press, New York, 1983, p. 307.
32. M. Waite, *The Phospholipases*, Plenum Press, New York, 1987.
33. E. A. Dennis, *J. Biol. Chem.*, **269**, 13057 (1994).
34. M. E. Haberland and J. A. Reynolds, *J. Biol. Chem.*, **250**, 6636 (1975).
35. G. Cevc and D. Marsh, *Phospholipid Bilayers: Physical Principles and Models*, John Wiley & Sons, New York 1987.
36. P. R. Cullis, D. B. Fenske and M. J. Hope, in *Biochemistry of Lipids, Lipoproteins and Membranes*, (eds.) D. E. Vance and J. E. Vance, Elsevier, Amsterdam, 1996, p. 1.
37. M. K. Jain, C. J. A. van Echtfeld, F. Ramirez, J. de Gier, G. H. de Haas and L. L. M. van Deenen, *Nature*, **284**, 486 (1980).

38. G. Scherphof, B. van Leeuwen, J. Wilschut and J. Damen, *Biochim. Biophys. Acta*, **732**, 595 (1983).
39. M. K. Jain and D. V. Jahagirda, *Biochim. Biophys. Acta*, **814**, 313 (1985).
40. C. R. Kensil and E. A. Dennis, *J. Biol. Chem.*, **254**, 5843 (1979).
41. R. Verger, M. C. E. Mieras and G. de Haas, *J. Biol. Chem.*, **248**, 4023 (1973).
42. R. A. Demel, W. S. M. Geurts van Kessel, R. F. A. Zwaal, B. Roelofsen and L. L. M. van Deenen, *Biochim. Biophys. Acta*, **406**, 97 (1975).
43. J. Y. A. Lehtonen and P. K. J. Kinnunen, *Biophys. J.*, **68**, 1888 (1995).
44. T. H. Callisen and Y. Talmon, *Biochemistry*, **37**, 10987 (1998).

# Chapter 23

# Giant Phospholipid Vesicles Entrapping Giant DNA

*SHIN-ICHIROU NOMURA AND KENICHI YOSHIKAWA*
*Kyoto University, Japan*

## 1. INTRODUCTION

All living organisms depend on closed membranes, where long DNA molecules confined narrow spaces of the order of several microns. The main constituents of these biomembranes are phospholipids. Thus, liposomes entrapping giant DNA molecules are expected to serve as a simple models of living cells. As typical natural DNA contains $10^6$–$10^9$ base pairs; conventional liposomes with a size of several hundreds Angstroms are too small to incorporate such giant molecules. In order to create a model of living cells, it is necessary to develop the experimental techniques with which to prepare giant liposomes of the order of several microns and to entrap giant DNAs under mild conditions, without any harmful chemical reagents or mechanical agitation. This chapter describes a method for the preparation of giant vesicles entrapping giant DNAs, by adapting the procedure of natural swelling [1–3] of phospholipid film with aqueous solution containing giant DNAs. Fluorescence microscopy, has been used to observe the individual DNAs entrapped within the liposome.

## 2. EXPERIMENT

Bacteriophage T4 dcDNA was purchased from Nippon Gene (Tokyo, Japan). 4′,6-Diamidino-2-phenylindole (DAPI) and 2-mercaptoethanol (ME) were purchased

*Giant Vesicles* Edited by P. L. Luisi and P. Walde

from Wako Pure Chemicals (Osaka, Japan). 1,2-Dioleoyl-*sn*-glycero-3-phosphocholine (DOPC) was obtained from Sigma Chemicals (St. Louis, MO). Magnesium chloride, and HEPES buffer were obtained from Nakarai Tesque (Kyoto, Japan). To obtain giant vesicles, the method of natural swelling has been slightly modified from that given in previous studies [4–6]. Using a lipid concentration one-tenth of that in the reported procedures gives giant unilamellar vesicles of spherical shape. DOPC film was formed from 1 mM solution of 1 : 2 (v/v) MeOH : $CHCl_3$, under an atmosphere of nitrogen gas. T4 DNA was dissolved in 10 mM HEPES buffer solution of pH 7.1 containing 10 mM $MgCl_2$. The final concentrations in the DNA solutions were 0.3 μM T4 DNA (in nucleotide units), 0.3 μM DAPI, and 2% (v/v) 2-ME. The film was then swollen with the T4 DNA solution in a test tube. The sample, was transferred to a glass slide and observed with a fluorescence microscope (Zeiss Axiovert 135 TV) equipped with a ×100 oil-immersed objective lens under the illumination of 358 nm light. The apparent conformation and the spatial position of the DNA molecules on the video frames were calibrated with an image processor (Argus 20, Hamamatsu Photonics, Hamamatsu, Japan). Special care was taken to clean the microscope slides and coverslips thoroughly before the observations.

## 3. RESULTS

The object (A) in Figure 23.1 exemplifies a single duplex T4 DNA molecule with a contour length of 57 μm [7], exhibiting translational and Brownian motion in an aqueous environment. As aqueous medium is generally a good solvent for highly charged DNA; in the figure the DNA chain takes the elongated conformation [8]. The object (B) shows a T4 DNA molecule in the presence of 10 mM $MgCl_2$ outside the liposomes, with a somewhat shrunken conformation compared to the elongated coil state (object (A)). The objects (C) show shrunken DNA molecules entrapped by a giant DOPC liposome. Careful inspection over time of changes in the thermal motion, as shown by fluorescence images, indicate that the DNA molecules, both inside and outside the liposomes, are in the intrachain segregated state; that is, elongated and collapsed parts coexist within a single DNA molecule. There exists considerable blurring, of the order of 0.3 μm, due to the resolution limit of the wavelength of the fluorescent light and, in addition, due to the high sensitivity of the Silicon Intensifier Tube (SIT) camera. The size of DNA molecules determined from the images is therefore increased by ca. 0.3 μm [9,10].

In order to estimate the actual size of the shrunken DNA in comparison with the elongated DNA coil ((A) in Figure 23.1), the Brownian motion has been analyzed Figure 23.2. From the slope of curve (b), the apparent hydrodynamic radius $R_H$ of the DNA outside the liposome is evaluated as $0.58 \pm 0.03$ μm (in the presence of

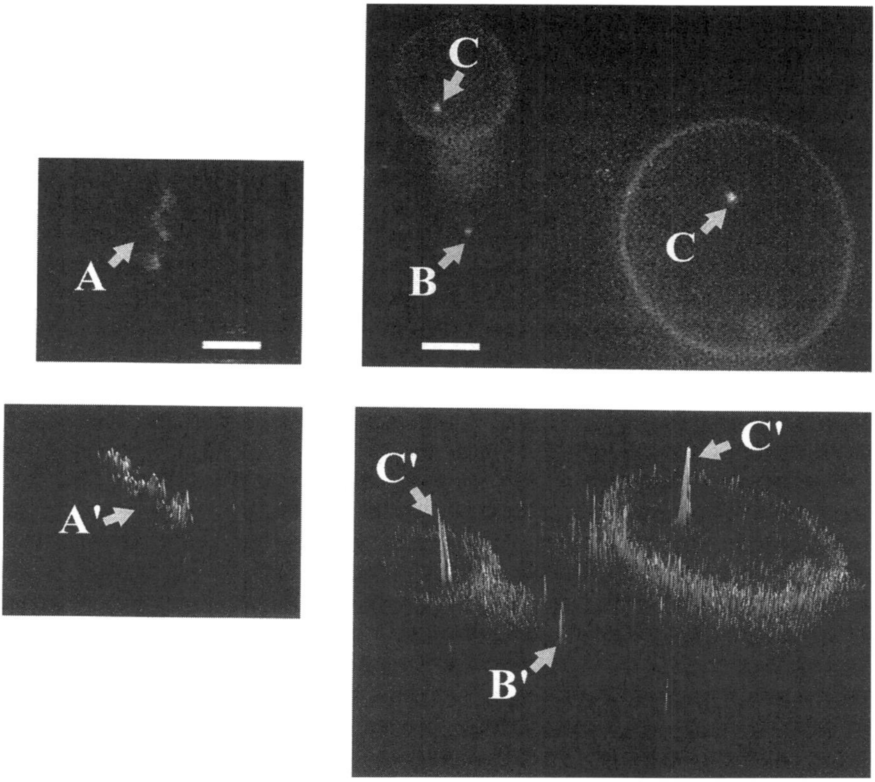


**Figure 23.1** (Upper) Fluorescence microscopic images of T4 dcDNA molecules: (A) in 10 mM $MgCl_2$ and 10 mM HEPES buffer solution (pH 7.1); (B) and (C) in the buffer solution with closed phospholipid liposomes outside and inside the liposomes, respectively. (Lower) Quasi three-dimensional pictures of the distribution of fluorescent intensity for the corresponding upper photographs. The bar is 10 μm. Careful inspection of the time-successive images, as in (B′) and (C′), reveals that the shrunken state corresponds to intrachain segregated state [10,11].

10 mM $MgCl_2$ after the correction on the convectional flow [11]). For the DNA trapped in the liposome, curve (c), the slope tends to decreases with time, indicating the effect of the finite space on the Brownian motion. From the initial slope of curve (c), the apparent value of $R_H$ of the encapsulated DNA is found to be $0.57 \pm 0.06$ μm, which is essentially the same as outside the liposome. However, from the plot of curve (a) and after correction for the effects convectional flow, the hydrodynamic radius for the coiled DNA is deduced to be $1.10 \pm 0.05$ μm.

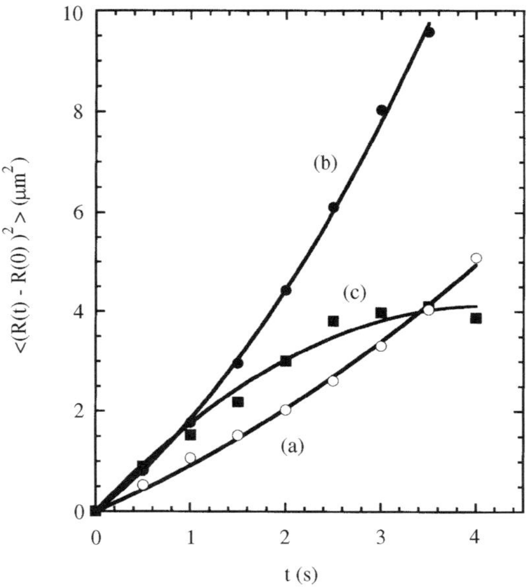


**Figure 23.2** Translational thermal motion of T4 DNA as represented by the time dependence of the mean-square displacement. Curve (a) is for the elongated T4 DNA molecule in bulk buffer solution. Curves (b) and (c) are for T4 DNA outside and inside a giant vesicle with the diameter of 20 μm, respectively. Curves (a), (b), and (c) correspond to (A), (B), and (C) in Figure 23.1, respectively.

## 4. DISCUSSION

Magnesium solution was used to generate giant liposomes with natural swelling. Magnesium ion has the potency to bind to the negatively charged phosphate group in phospholipid, suggesting that neutral phospholipid molecules behave as cationic lipid in the presence of magnesium ion. Thus, the osmotic pressure in the aqueous solution between the lipid bilayers increases, and this osmotic effect induces expansion of the interlayer distance. As a result, giant liposomes, several tens micrometers in size, are formed [6]. Magnesium ions are also expected to interact with the negatively charged DNA chain. From the systematic investigation of the higher-order structure of DNA in aqueous solution, it has been established that individual giant DNA molecules exhibit marked discrete transition, or first-order phase transition, from elongated coil states into collapsed globule states by the addition of multivalent cation [8,12]. It has been found that the threshold concentration of the multivalent cation, used to induce the transition, decreases with the increase of cation valency [13]. In other words, the discrete character of the transition induced by multivalent cations tends to be diminished with divalent cation. Under such conditions, the intrachain segregated state, or coexistence of elongated and collapsed parts, can appear as the intermediate conformation between fully elongated and fully collapsed states in the individual single chains, accompanied by a change in the chemical environment due to the presence of the phospholipid molecules. As for the fully collapsed state of T4 DNA, induced by trivalent cation $Fe^{3+}$, the hydrodynamic radius has been found to be $R_H = 0.065\ \mu m$ [11]. This means that the effective volume of the shrunken DNA with the $R_H$ around 0.5–0.6 μm in the presence of magnesium ions is almost 1000 times as large as that of the fully collapsed DNA, suggesting that the segregated state of DNA consists of a rather large portion of elongated coil. It is clear that phospholipid molecules have a significant effect on the conformation of giant DNA in the presence of magnesium ions. Further physicochemical studies on the effect of phospholipids are necessary.

## 5. CONCLUSIONS

This paper shows that giant liposomes, entrapping DNA molecules, can be produced under very mild conditions. It is expected that elongated DNA molecules would also be trapped within liposomes with a suitable choice of experimental conditions, including the adoption of the smaller concentration of magnesium ion for the swelling buffer solution. Further studies are required to establish a recipe for the model cell, together with the attempts for biological applications such as gene therapy and gene transfection [14].

## 6. REFERENCES

1. E. Boroske, M. Elwenspoek and W. Helfrich, *Biophys. J.*, **34**, 95 (1981).
2. H. Hotani, *J. Mol. Biol.*, **178**, 113 (1984).
3. G. Decher, H. Ringsdorf, J. Venzmer, D. Bitter-Suermann and C. Weisgerber, *Biochim. Biophys. Acta*, **1023**, 357 (1990).
4. S. M. Mel'nikov, V. G. Sergeyev, Y. S. Mel'nikova and K. Yoshikawa, *J. Chem. Soc., Faraday Trans.*, **93**, 283 (1997).
5. N. Kumazawa, Y. S. Mel'nikova and K. Yoshikawa, *Macromol. Symp.*, **106**, 219 (1996).
6. N. Magome, T. Takemura, and K. Yoshikawa, *Chem. Lett.*, **205** (1997).
7. R. E. Dickerson, H. R. Drew, B. N. Conner, R. M. Wing, A. V. Fratini and M. L. Kopka, *Science*, **216**, 475 (1982).
8. K. Yoshikawa, M. Takahashi, V. V. Vasilevskaya and A. R. Khokhlov, *Phys. Rev. Lett.*, **76**, 3029 (1996).
9. M. Matsumoto, T. Sakaguchi, H. Kimura, M. Doi, K. Minagawa, Y. Matsuzawa and K. Yoshikawa, *J. Polym. Sci.*, **30**, 779 (1992).
10. S. M. Mel'nikov, V. G. Sergeyev and K. Yoshikawa, *J. Am. Chem. Soc.*, **117**, 240 (1995).
11. Y. Yamasaki, and K. Yoshikawa, *J. Am. Chem. Soc.*, **119**, 10573 (1997).
12. V. A. Bloomfield, *Curr. Opin. Struct. Biol.*, **6**, 334 (1996).
13. M. Takahashi, K. Yoshikawa, V. V. Vasilevskaya and A. R. Khokholov, *J. Phys. Chem. B*, **101**, 9396 (1997).
14. D. D. Lasic, *Liposomes in Gene Delivery*, CRC Press Boca Raton, FL, 1997.

# Chapter 24

# Cell Deformation Mechanisms Studied with Actin-containing Giant Vesicles, a Cell-mimicking System

*HIDETAKE MIYATA AND KAZUO OHKI*
*Tohoku University, Sendai, Japan*

*GERARD MARRIOTT*
*Max Planck Institute for Biochemistry, München, Germany*

*SHUJI NISHIYAMA*
*TOTO Ltd, Kanagawa, Japan; Keio University, Yokohama, Japan*

*KEN-ICHIROU AKASHI, AND KAZUHIKO KINOSITA, JR.*
*Keio University, Yokohama, Japan; Core Research for Evolutional Science and Technology, Genetic Programming Team 13, Kawasaki, Japan*

## 1. INTRODUCTION

### 1.1 Cell locomotion

Cells can crawl on a substrate. The ability of cells to crawl supports many cellular activities such as invasion of cancer cells, wound repair of epithelial cells, or path-finding of neuronal cells. In general, the process of locomotion can be divided into several steps: protrusion of the front part of the cell, which is also called a leading

*Giant Vesicles* Edited by P. L. Luisi and P. Walde

edge; adhesion of the protrusion to the substrate; and traction of the cell body followed by a detachment of the trailing part of the cell [1,2]. These steps occur cyclically, thereby resulting in the forward movement of the cell. Protrusive formation is indispensable for the directed movement of the cell and, hence, elucidation of its molecular mechanism is highly important in understanding the crawling mechanism of the cell at the molecular level.

Previous electron microscopic studies have shown that a network of actin filaments exists immediately beneath the cell membrane at the leading edge. Within the network are intersecting actin filaments which form a gel with the aid of the cross-linking ability of the actin-binding protein (ABP). The barbed (fast-growing) end of the actin filaments in the network are directed toward the leading edge. Fluorescently labeled actin monomers, microinjected in the cell, were incorporated into the actin network at or near the barbed ends in the leading edge. Thus, the protrusive activity seems to be intimately associated with the addition of actin monomers to the barbed ends in the network [3].

A hypothesis that polymerizing actin pushes cell membranes against an external force has been put forward [4]. A recently proposed mechanism is that the growing filaments ratchet the thermally fluctuating membranes [5,6]. However, no experimental approach has been made to test these hypotheses in the cell, perhaps because many factors such as signaling molecules or various actin binding proteins are involved in the protrusive process [7,8] and the force-generating mechanism alone cannot be easily separated out for investigation.

## 1.2 Model systems to study the origin of protrusive force

To test the hypothesis that the growing actin filaments can deform the cell membrane, giant vesicles incorporating actin have been used as a model system. It is assumed if the growing actin filaments can exert a force on cell membranes and deform it, they can also deform the vesicle membranes. Several groups have independently developed and utilized these model systems and have shown that actin polymerization is necessary for vesicle shape change. The following discussion examines, the methods adopted by these groups for the encapsulation of actin monomers into giant vesicles, and their polymerizations are described in some detail. As mono- or divalent cations are necessary to polymerize actin *in vitro* [9], most model systems utilize ionophore to introduce mono- or divalent cations into the vesicles. As a consequence, the preparation of the vesicles was carried out in low-ionic-strength buffers, which do not induce actin polymerization.

In a system developed by Cortese *et al.* [10], vesicles encapsulating actin were prepared by injecting diethyl ether dissolving phosphatidylcholine (PC) and phosphatidylethanolamine together with a potassium ionophore, valinomycin, into a solution containing 48 μM monomeric actin. During the injection period (less than 5 min), the temperature was maintained at 52–55°C. The vesicle size was around 2–20 μm. After the dilution of actin outside the vesicles at least 10 times, the encapsulated actin was polymerized by adding 100 mM $K^+$ ions. Polymerization was confirmed by the absence of recovery of the fluorescence after photobleaching of the encapsulated actin which had been labeled with tetramethylrhodamine iodoacetoamide. Irregularly shaped vesicles were found to appear as a result of the actin polymerization: the vesicle shape change became apparent over 30 min after the start of the polymerization, although images of the vesicles during that period are not provided. It was also found that the actin filaments distributed nonuniformly within the vesicles. When gelsolin (an actin-capping protein) or filamin (an actin cross-linking protein) was also encapsulated, the vesicle shape became more spherical and the actin distribution became more uniform.

Janmey *et al.* [11] polymerized actin (47 μM) in vesicles prepared from PC (details are not given) by sequential addition of 0.1 μM A23187 and 5 mM $Ca^{2+}$ or $Mg^{2+}$. Based on observations made before and after the addition of the reagents, they concluded that filopodia-like protrusions extended from the vesicles. From dynamic light scattering measurements, they demonstrated that only in the presence of ionophore and divalent cations an increase of hydrodynamic radii of the vesicles occurred over 5–10 min, which is indicative of the shape change of the vesicles.

Barmann *et al.* [12] prepared actin-containing vesicles by swelling in a buffer solution containing 7.2 μM monomeric actin a dried lipid film made from a mixture of 95 mol% 1,2-dimyristoyl-*sn*-glycero-3-phosphocholine (DMPC) and 5 mol% ionophores, A23187, or valinomycin. The actin which was not incorporated was digested with the protease chymotrypsin, and was removed by ultrafiltration of the vesicle suspension over Amicon XM-300 membrane. With the actin labeled with the fluorophore nitrobenzoxadiazole, they found that after adding 100 mM $K^+$ or 2 mM $Mg^{2+}$ to the vesicle suspension the fluorescence increased severalfold over several hours, indicating that actin polymerized in the vesicles. They observed a shape change of the vesicles: at first, the vesicles became elongated, which is attributed to an osmotic effect caused by the addition of ions, rather than to the polymerization *per se*, but later in parallel with the polymerization, the vesicles became more spherical and finally assumed a stomatocyte-like shape. The study of this system has been extended [13,14].

## 2. TEMPERATURE-JUMP STRATEGY FOR POLYMERIZATION OF ACTIN ENCAPSULATED IN VESICLES

### 2.1 Temperature-jump strategy

The model systems described above demonstrate that the actin-containing vesicles change their shape as a result of actin polymerization. However, as giant vesicles can spontaneously change their shape due to a change in osmotic pressure [15,16] or shear force arising from a flow of the medium surrounding the vesicles [17], it is difficult to ascertain that the shape change resulted from the polymerization, unless the whole process of the deformation was observed. The temperature-jump strategy, a method in which the encapsulated actin is polymerized by raising the temperature, helps to resolve this issue. This method is less disturbing and enables continuous observation of individual vesicles. It has been successfully utilized by Hotani and Miyamoto to polymerize tubulin which had been encapsulated in giant vesicles [18].

## 2.2 Experimental methods

Actin was prepared from rabbit skeletal muscle by the method of Spudich and Watt [19] and was further purified with gel-filtration over Sephadex G-150 gel [20]. Monomeric actin, labeled with the fluorophore tetramethylrhodamine maleimide, was a generous gift of Dr Ewa Prochniewcz at the Department of Biochemistry, University of Minnesota. It was concluded from a preliminary experiment that Spudich–Watt actin contained minor factor(s) which greatly accelerated the polymerization of actin at 0.5 mM $Ca^{2+}$ and at 8°C: upon the addition of $Ca^{2+}$ ions, 100 µM Spudich–Watt actin rapidly polymerized, whereas the purified actin did not. $Ca^{2+}$ ions were chosen because they only promoted polymerization at higher temperatures (Miyata, unpublished observations).

Actin-containing vesicles were prepared by swelling for 1 h at 0°C a dried lipid film, made from a mixture of 70 µg DMPC and 70 µg cardiolipin (CL), in 25 µl of solution containing 100 µM monomeric actin and an internal buffer (2 mM tris(hydroxylmethyl)aminomethane, pH 8.0, 0.2 mM ATP, 0.5 mM $CaCl_2$, 0.3 mM $NaN_3$, 0.5 mM dithiothreitol). After the swelling, the concentrated vesicle suspension was diluted in the internal buffer 50 times to prevent actin polymerization outside the vesicles. A drop of the dilutent (about 3 µl) was sandwiched between a glass slide (24 × 60 mm) and a coverslip (22 × 22 mm), sealed with silicone grease and was subjected to observation under a dark-field microscope. To keep the temperature of the sample low before the observation, the glass slide was placed on a chilled (≈5°C) aluminum block. After selection of appropriate vesicles, the temperature of the sample was raised and was maintained at 29 ± 2°C by blowing warm air over the sample. Polymerization of the encapsulated actin was induced by this procedure, as shown from the development of birefringence in the liposomes; at 30°C polymerization was complete within about 30–40 min [21]. A separate ultracentrifugation assay demonstrated that about 85% of the actin had polymerized under these conditions [22].

Dark field microscopy identified many vesicles with diameters > 5 µm. Although many of them were multilamellar, as judged from the thick, rigid and extremely bright contours under the dark-field illumination, some seemed to possess lower lamellarity. This was suggested from the less bright and undulating contours; such vesicles were selected. No evaluation of the lamellarity of the selected vesicles was carried out.

## 2.3 Vesicle morphology and distribution of actin in vesicles

Figures 24.1(a–d) show differential interference contrast (DIC) micrographs of two typical vesicles found after the temperature jump. Panels a and b are two different views of an elongated vesicle with racket-like structure at one end (racket-like vesicle); sometimes the racket-like structure appeared at both ends. The flat feature of the end was revealed by occasional rotation due to the thermal motion of the vesicle around its long axis. During the rotation the flat feature of the end portion remained unchanged, indicating considerable rigidity of this portion. The elongated portion also exhibited rigidity which was evident from little bending motion. Panels c and d are two views of a vesicle having a disk-like morphology. This morphology was observed about 10 times more frequently than for the racket-like one. Despite vigorous tumbling motion, the morphological feature of these vesicles did not change, indicating the rigidity of the vesicles. In addition to the rigidity of the body these two types of vesicles possessed little fluctuating membranes. In contrast with this, vesicles encapsulating no actin possess highly undulating shape and contour, which is demonstrated in a series of dark-field micrographs of a vesicle prepared in the absence of actin (Figure 24.1(e)).

The deformation process of actin-encapsulating vesicles has been observed [21]. In case of the racket-like vesicle, a slightly elongated and flaccid vesicle grew longer. A disk-like vesicle was generated from a spherical vesicle as a result of flattening of the sphere. In both cases the shape change occurred over tens of minutes, which coincides well with the time for actin polymerization measured in a cuvette under the same conditions.

Figure 24.2 shows epifluorescence micrographs of racket- and disk-like vesicles containing the rhodamine-labeled actin polymerized under the conditions described above. As is clear from these micrographs, the fluorescent region is localized along the periphery of the deformed vesicles. Previous observations by polarization microscopy revealed that the periphery of the deformed vesicles was birefringent, indicating that actin filaments spontaneously aligned and formed a bundle structure running along the vesicle periphery [21]. The alignments deduced from these microscopic studies are schematically drawn in Figure 24.2(c).

## 2.4 Shape change mechanism

The above results demonstrate that when actin is polymerized in the vesicles, actin filaments spontaneously align to form a bundle structure along the liposome periphery and the vesicles change their shape. The qualitative explanation is based on several points. The solution of polymerized actin ($> 50\,\mu M$) has been demonstrated to develop birefringence, attributable to the appearance of domains containing aligned actin filaments [23]. Polarization microscopy confirmed that the birefringent domains, 10–20 μm wide and about 100 μm long appeared, when

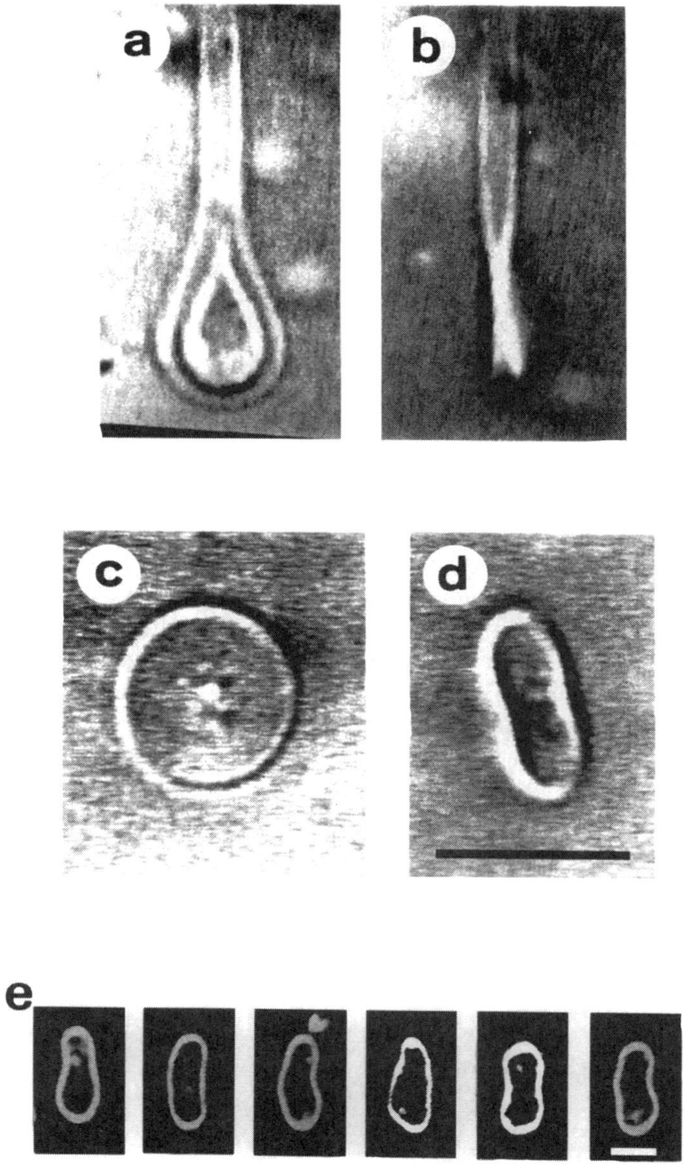


**Figure 24.1** Distinct morphologies of DMPC-CL vesicles containing polymerized actin; (a) and (b), show differential interference contrast (DIC) micrographs of a racket-like vesicle as viewed from two different angles; (c) and (d), are two different views of a disk-like vesicle; (e) shows a series of dark-field micrographs, acquired every few seconds, of a DMPC-CL vesicle prepared in the absence of actin, demonstrating flexibility of the lipid membrane and hence, the contour shape. Length of bar, 10 μm (a)–(d), 5 μm (e).

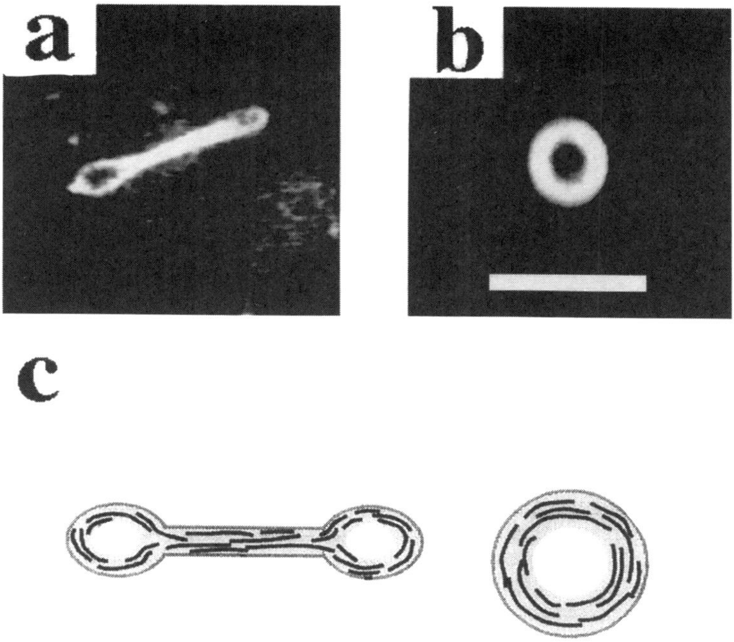


**Figure 24.2** Epi-fluorescence micrographs demonstrating the distribution of actin filaments in (a) racket-like and (b) disk-like DMPC-CL vesicles; only front views are shown. The actin encapsulated in these vesicles had been labeled with tetramethylrhodamine maleimide. The peripheral localization of actin within both types of the vesicles is evident (bar length 10 μm). (c) Schematic drawings of the mode of actin filaments inside the racket-like (left) and the disk-like (right) vesicles. Actin filaments reside in the shaded area; the lines in the area represent actin filaments, the alignment of which had been deduced from polarization microscopy.

100 μM actin was polymerized with 0.5 mM $Ca^{2+}$. A theory predicts phase separation of the mixture of rod-shaped polymers [24]. An elastic interaction between the actin filaments and the vesicle lipid membranes may be also involved, because the persistence length of the actin filaments (≈10 μm; [25]) is similar to the dimensions of the vesicles. The tentative scenario for the generation of the elongated vesicle is as follows. At an early stage of deformation a small number of actin filaments which appeared in a slightly elongated vesicle spontaneously aligned by a mechanism similar to the mechanism which promote the development of the birefringent domain. The phase-separation mechanism may have also participated. The aligned filaments then provided a scaffold for the formation of an actin bundle along the vesicle periphery. Further deformation of the vesicle results from an elongation of individual filaments in the bundle structure. The disk-like vesicles were possibly generated in a similar manner. In this case, however, the initial shape of the vesicle was spherical and a different vesicle shape resulted. The presumed

elastic interaction may be the driving force for the collapse of spherical vesicles observed during the generation of the disk-like vesicles. Evaluation of this scenario is a subject of future studies.

### 2.5 Comparison with other systems

Other groups have also reported that the shape change of the vesicles occurred as a result of actin polymerization, but without showing the shape of vesicles during the deformation, except for one study [12]. As pointed out above, lipid membrane is easily deformed, and one should continuously observe individual vesicles to ensure that the vesicle deformation is coupled to the actin polymerization.

Cortese *et al.* [10] demonstrated that the vesicles assumed irregular shapes and polymerized actin filaments exhibited a heterogeneous distribution within the vesicles, which is quite different from our results. The difference in the method of preparation of actin-encapsulating vesicles (diethyl ether injection method versus swelling method) or of the cation species and the concentration may be the source. The latter may have significantly affected the polymerization rate and hence the final vesicle shape.

Recent studies suggest binding of actin monomer to the lipid membranes [13,26,27]. Under our experimental conditions, no significant interaction between vesicles and actin filaments was detected [21]; however, it is always possible that during early stages of polymerization, actin monomers interacted with the membrane to form some ordered structure, which served as a scaffold for the alignment.

## 3. MANIPULATION OF DEFORMED VESICLES

### 3.1 Manipulation of encapsulated and polymerized actin with cytochalasin D

In the above tentative scenario on the vesicle shape change, the elastic interaction between the bundle of actin filaments and the lipid membrane was assumed to play a role. Therefore, it was of interest to see what kind of change occurs to the vesicle shape, if the bundle structure is altered by some means. Cytochalasin D (CD) was chosen for this purpose. CD is a fungal metabolite which enters the cell and disrupts the actin filament therein, when it is added to the medium surrounding the cell [28]. This drug has been widely used to investigate the role of actin in the cell [29]. The mechanism of the drug action is twofold: one is to bind to the barbed end of actin filaments and inhibit polymerization, the other is to sever (cut up) the filaments [30].

### 3.2 Experimental procedures

Vesicles of DMPC-CL encapsulating actin and $Ca^{2+}$ ions were prepared as described in Section 2.2, but with the internal buffer supplemented with 89 mM sucrose, and with dithiothreitol replaced with 5 mM 2-mercaptoethanol. For this experiment an open chamber and a micromanipulation system were employed to allow delivery of the drug to the targeted vesicles (Figure 24.3). The open chamber was constructed by adhering to a 24 × 36 mm coverslip a silicone rubber sheet (3 mm thick) with a 5 × 20 mm rectangular hole. A small aliquot of the solution of actin-containing vesicles was diluted 50-fold in a microfuge tube containing the internal buffer supplemented with 100 μM bovine serum albumin and 89 mM glucose (external buffer). Bovine serum albumin and glucose were included to balance the osmotic pressure. The dilution was kept at 30°C for over 1 h. Actin did not polymerize outside the vesicles. An appropriate amount of the dilution was transferred to the chamber and was observed at 20°C under a phase-contrast microscope. No significant depolymerization of the encapsulated actin occurred during the observation period [22]. CD was added either to the vesicle suspension in the microfuge tube or to the suspension in the observation chamber.

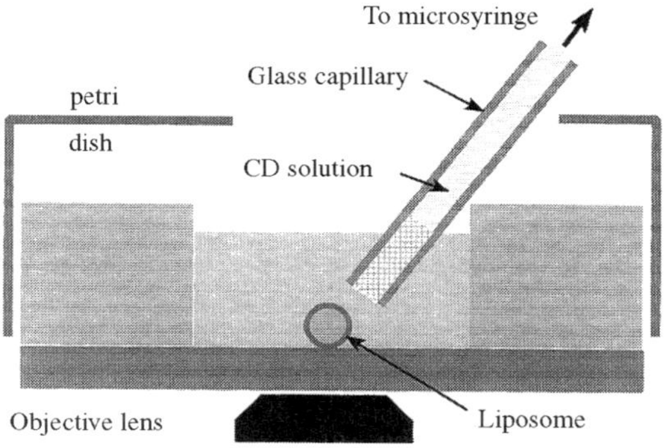


**Figure 24.3** Diagram of the method of delivery of cytochalasin D (CD) to a DMPC-CL vesicle in an open chamber on the microscope stage. The open chamber, containing the diluted vesicle suspension, was covered with a petri dish with a moistened tissue paper attached (not shown) to minimize evaporation. An appropriate vesicle was selected by phase-contrast microscopy, and a glass capillary (inner diameter = 1 mm) connected to a micro-syringe was placed near the vesicle with a micromanipulator (not shown). The figure is not drawn to scale. About 10–20 μl of the internal buffer containing CD (20–100 μM) had been withdrawn into the capillary. Owing to the difference in the refractive index between the internal and external buffer, it was possible to recognize the arrival of the buffer containing CD around the targeted vesicle. As the opening of the capillary was far larger than the dimension of the field of observation, the CD concentration around the vesicle should have been increased almost uniformly. This flow caused continuous displacement and tumbling of the targeted vesicle, sometimes causing blurred images but, owing to the flow, different views of the vesicles became available.

### 3.3 Vesicles with a new spindle-like morphology

Before the addition of CD the vesicles assumed disk- or racket-like shapes, but in the diluted vesicle suspension, to which 20–100 μM CD had been added, many spindle-shaped vesicles were found; the tumbling motion provided different views of individual vesicles. The axial ratio of the spindle-shaped vesicles increased somewhat with increasing CD concentration [22]. This result indicated that the spindle-shaped vesicles must have been generated from the disk- or racket-like vesicles. To confirm this, the drug was delivered to individual vesicles.

Figure 24.4 shows the phase contrast images of two vesicles obtained before (left panels) and after (right panels) the CD addition. Before the CD addition, both vesicles assumed the disk-like shape (only the front views are shown). Continuous observation confirmed that the spindle-shaped vesicles shown in the right panels did emerge from these vesicles. The shape change was complete within about 1 min of

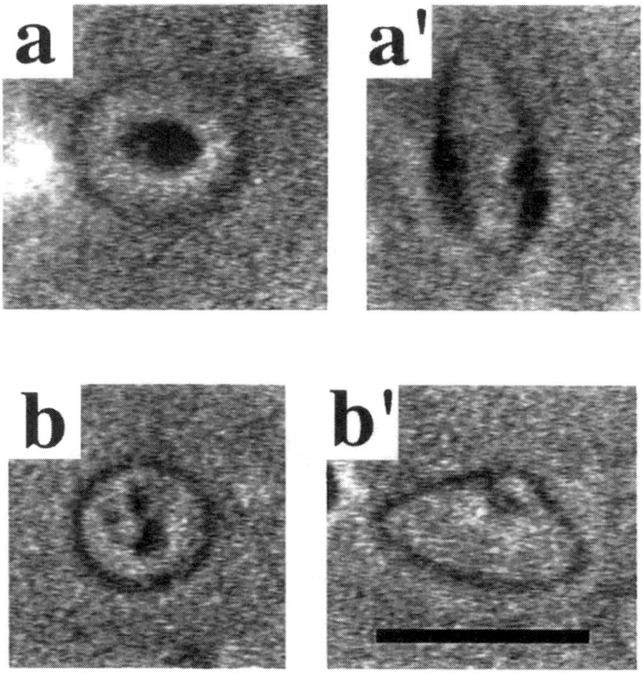


**Figure 24.4** Phase-contrast images of two DMPC-CL vesicles (a) and (b) before and (a′) and (b′) after the addition of cytochalasin D (CD). The pairs of panels (a) and (a′), and (b) and (b′) show the respective sets of vesicles. Before the addition of CD, both liposomes assumed the disk-like shape; after the addition, they transformed into spindle shapes. Continuous tumbling of the vesicles caused by the flow of CD solution blurred the vesicle images in (a) and (a′) (bar length 10 μm).

the change being noticeable under the microscope. The racket-like vesicles also transformed into the spindle-shaped vesicles. Some of the targeted vesicles became spherical without assuming the spindle-shape, whereas others transiently assumed the spindle shape and then became spherical. The final shape possibly depends on the amount of the drug entered the vesicles.

### 3.4 Mechanism of the shape change upon the addition of cytochalasin D

The following two experiments were to elucidate the changes that might occur to the actin filaments in the vesicles [22]. Observation of the polarized fluorescence in the spindle-shaped vesicles of actin, labeled with the fluorophore acrylodan [31], indicated that the alignment of the actin filaments became parallel to the long axis of the vesicle, which was quite different from that in the disk- or racket-like vesicles (Figure 24.2). Further, falling-ball viscometry and ultracentrifugation assay demonstrated that 10–100 μM CD caused instantaneous severing, but little depolymerization, of the 100 μM actin polymerized with 0.5 mM $Ca^{2+}$ ions. These experiments suggested that the primary effect of CD in the vesicles was the severing of the actin filaments. Presumably, the severed filaments became incapable of maintaining the bundle structure and the elastic interaction between the bundle and the vesicle membrane was altered, suggesting that the spindle-shape must be an energetically favorable arrangement of both actin filaments and the vesicle membrane.

### 3.5 Mechanical perturbation of actin-containing vesicles

The above argument assumes that an elastic interaction between the actin filaments and the vesicle membranes is an important determinant of the vesicle shape. Because manipulation of the actin bundle with CD resulted in the vesicle shape change, it would be interesting to examine if the other elastic entity, the vesicle membrane, is mechanically perturbed.

As illustrated in Figure 24.5, an aspiration pressure was applied with a micropipette [32]. A semiracket-like DMPC-CL vesicle containing 100 μM actin polymerized with 0.5 mM $Ca^{2+}$ (Figure 24.6, upper left panel) was aspirated with the pressure of a few cm $H_2O$, corresponding to a membrane tension of the order of $10^{-2}$ dyn cm$^{-1}$. This pressure was sufficient to make the vesicle immediately (within one video frame, 33 ms) assume a spherical shape (lower left panel). Upon the release of the aspirated vesicle (Figure 24.6, right panels), the vesicle gradually regained an elongated, but different, shape from the original. The in-plane membrane tension probably elastically deformed the actin bundle and, in the deformed bundle, the alignment of the actin filaments must have been disturbed; even after the release, the filaments did not restore the original alignment.

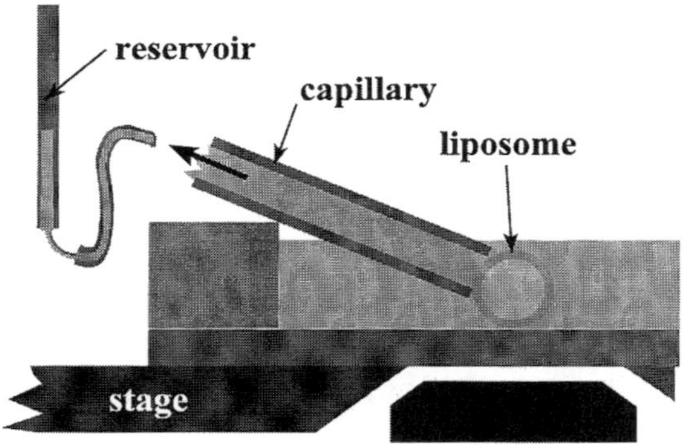


**Figure 24.5** Diagram of the method of microaspiration of a vesicle. The experiments were carried out in the open chamber described in the caption for Figure 24.3. With a micromanipulator a microcapillary (tip inner diameter $\approx 1\ \mu m$) was positioned near the targeted DMPC-CL vesicle containing polymerized actin. The aspiration pressure was adjusted by changing the height of the water column in the reservoir. The figure is not drawn to scale.

## 4. FUTURE PERSPECTIVES: EVOLVING MODEL SYSTEMS

### 4.1 Use of caged actin: a new polymerization strategy

In order to polymerize actin encapsulated in the giant vesicles, the temperature-jump strategy was used, which allowed continuous observation of the vesicles. With this method, however, it was not possible to employ ionic conditions closer to those of the physiological ones, i.e. higher ionic strengths, because at higher ionic strengths actin polymerizes even at 4°C. As a consequence, the speed of polymerization was considerably slower than would have been achievable at higher ionic strengths. This may be the reason for the difference in vesicle morphology as compared to those of Cortese *et al.* To develop this system further requires methods that allow the usage of higher ionic strengths.

Caged compounds, such as caged ATP, are now widely used to trigger various biological responses in both spatially and temporally controlled manners. If actin monomers can be reacted with a photolabile reagent that blocks the polymerization in the presence of physiological concentration of salts to form caged actin, this will be ideal. The agent 4,5-dimethoxy-2-nitrobenzyl chloroformate (NVOC-Cl), which labels several reactive lysine residues of an actin monomer, has been shown to effectively block actin polymerization under an ionic condition normally used for actin polymerization (for instance 0.1 M $K^+$ and 2 mM $Mg^{2+}$) [33]. In a preliminary experiment DMPC-CL vesicles were prepared containing 40 μM caged actin. Potassium ion concentrations between 20 and 30 mM and $Mg^{2+}$ concentrations in the range 0.2–0.4 mM were used to prepare DMPC-CL vesicles containing caged actin. Following dilution, the vesicle suspension was illuminated for 30–60 s by UV

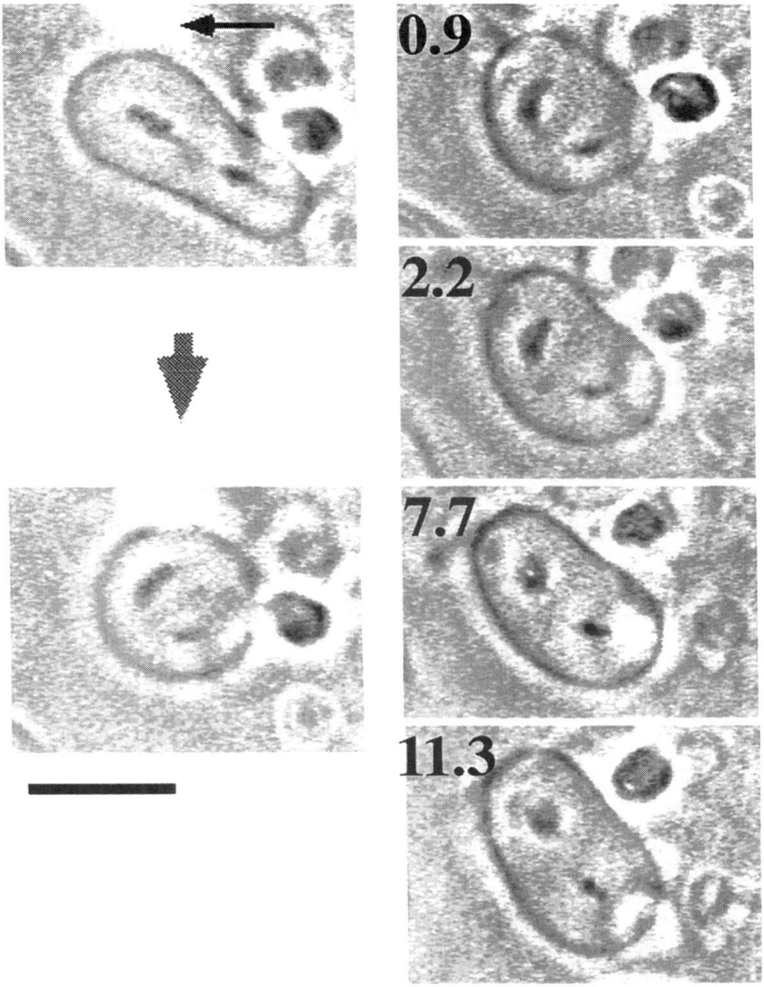


**Figure 24.6** Phase-contrast micrographs showing the result of microaspiration. Left panels: the tip of the microcapillary, indicated with an arrow, is situated near the top of the semiracket-like DMPC-CL vesicle containing polymerized actin. The direction of the capillary was vertical. Upon aspiration, the vesicle immediately became spherical due to the in-plane membrane tension. Right panels from top to bottom: following the release, the vesicle gradually regained an elongated shape, but different from the original one. The numbers in the right panels indicate the time in seconds elapsed from the moment of release (bar length 10 μm).

(300–400 nm) light from a 150 W mercury arc lamp, which caused gradual (over a few minutes) suppression of thermal motion of the entrapped small vesicles or lipid aggregates. This would be expected because the generated filaments sterically hinder the thermal motion of the vesicles or particles. This strategy now requires optimization of the conditions for encapsulation of both actin and cations, and the UV illumination time.

### 4.2 Model system for elucidation of the mechanical properties of protrusive cell structures

The actin-containing vesicles have demonstrated that polymerizing actin can deform the vesicle membranes. However, this qualitative observation does not immediately mean that the growing actin filaments can push the cell membranes, because the bending moduli of the cell membranes are 5–50 times larger than those of the vesicle membranes [34]. The in-plane tension of the cell membranes (of the order of $10^{-1}$–$10^{-2}$ dyn cm$^{-1}$, [35]) should also be considered. Hence, the magnitude of the force generated during the deformation of the vesicles should be measured to evaluate whether the actin polymerization alone develops enough force to deform the cell membranes. The microaspiration technique will be one way to realize this measurement.

The existence of various actin crosslinking or bundling proteins at the leading edge may be related to the mechanical strength of the network or the bundle structures of actin filaments at the leading edge. It may be that during the generation of the protrusive structure these actin-binding proteins play the role of stabilizing the structure against various forces, such as the cell membrane tension or the osmotic stress. This notion can be evaluated with the model system containing the actin bundling or crosslinking proteins in addition to actin. It will also be interesting to see what effect on the deformation process might be caused by introducing a connection between the actin filaments and the lipid membranes, because the existence of connection between actin filaments and membranes of *Dictyostelium* amoebae seems to be important in successful formation of pseudopods, another protrusive structure [36]. Incorporating factors that might govern the physical property of the leading edge will be one way to further evolve actin-encapsulating vesicles towards a system which better mimics the physical and mechanical aspect of the cell.

## 5. ACKNOWLEDGMENT

This work was supported by a grant from the Takeda Science Foundation, a grant from the Ministry of Education, Science, Sports and Culture of Japan.

## 6. REFERENCES


1. M. Abercrombie, *Proc. Royal Soc. London (Biol.)*, **207**, 129 (1980).
2. T. M. Mitchson and L. P. Cramer, *Cell*, **84**, 371 (1996).
3. J. V. Small, K. Rottner, I. Kaverina and K. I. Anderson, *Biochim. Biophys. Acta*, **1404**, 271 (1998).
4. T. L. Hill, *Proc. Natl. Acad. Sci. USA*, **78**, 5613 (1981).
5. C. S. Peskin, G. M. Odell and G. F. Oster, *Biophys. J.*, **65**, 316 (1993).
6. A. Mogliner and G. Oster, *Biophys. J.*, **71**, 3030 (1996).
7. T. P. Stossel, *Science*, **260**, 1086 (1993).
8. S. H. Zigmond, *Curr. Opin. Cell Biol.*, **8**, 66 (1998).
9. E. D. Korn, *Physiol. Rev.*, **62**, 672 (1982).
10. J. D. Cortese, B. Schwab III, C. Frieden and E. L. Elson, *Proc. Natl. Acad. Sci. USA*, **86**, 5773 (1989).
11. P. A. Janmey, C. C. Cunningham, G. F. Oster and T. P. Stossel, in *Mechanics of Swelling*, (ed.) T. K. Karalis, NATO ASI Series Vol. 64, Springer-Verlag, Berlin, 1992, p. 333.
12. M. Barmann, J. Kas, H. Kurzmeier and E. Sackmann, in *The Structure and Conformation of Amphiphilic Membranes*, (eds.) R. Lipowsky, D. Richter and K. Kremer, Springer-Verlag, Berlin Heidelberg, 1992, p. 137.
13. R. Grimm, M. Barmann, W. Hackl, D. Typke, E. Sackmann and W. Baumeister, *Biophys. J.*, **72**, 482 (1997).
14. W. Häckl, M. Barmann and E. Sackmann, *Phys. Rev. Lett.*, **80**, 1786 (1998).
15. H. Hotani, *J. Mol. Biol.*, **178**, 113 (1984).
16. H. J. Deuling and W. Helfrich, *J. Physique (Paris)*, **37**, 1335 (1976).
17. K. H. deHaas, C. Blom, D. van den Ende, M. H. G. Dutis and J. Mellema, *Phys. Rev. E*, **56**, 7132 (1997).
18. H. Hotani and H. Miyamoto, *Adv. Biophys.*, **26**, 135 (1990).
19. J. A. Spudich and S. Watt, *J. Biol. Chem.*, **246**, 4866 (1971).
20. S. D. MacLean-Fletcher and T. D. Pollard, *Biochem. Biophys. Res. Commun.*, **96**, 18 (1980).
21. H. Miyata and H. Hotani, *Proc. Natl. Acad. Sci. USA*, **89**, 11547 (1992).
22. H. Miyata and K. Kinosita, Jr., *Biophys. J.*, **67**, 922 (1994).
23. R. Furukawa, R. Kundra and M. Fechheimer, *Biochemistry*, **32**, 12346 (1993).
24. L. Onsager, *Ann. N.Y. Acad. Sci.*, **52**, 627 (1949).
25. H. Isambert, P. Venier, A. C. Maggs, A. Fattoum, R. Kassab, D. Pantaloni and M.-F. Carlier, *J. Biol. Chem.*, **270**, 11437 (1995).
26. M. Bouchard, C. Pare, J.-P. Dutasta, J.-P. Chauvet, C. Giquaud and M. Auger, *Biochemistry*, **37**, 3149 (1998).
27. A. L. Weisenhorn, B. Drake, C. B. Prater, S. A. C. Gould, F. Ohensorge, M. Egger, S.-P. Heyn and H. E. Gaub, *Biophys. J.*, **58**, 1251 (1990).
28. I. Yahara, F. Harada, S. Sekita, K. Yoshihara and S. Natori, *J. Cell Biol.*, **92**, 69 (1982).
29. S. J. Smith, *Science*, **242**, 708 (1988).
30. J. A. Cooper, *J. Cell Biol.*, **105**, 1473 (1987).
31. G. M. Marriott, K. Zechel and T. M. Jovin, *Biochemistry*, **27**, 6214 (1988).
32. E. Evans and W. Rawicz, *Phys. Rev. Lett.* **64**, 2094 (1990).
33. G. Marriott, *Biochemistry*, **33**, 9092 (1994).
34. D. V. Zhelev, D. Needham and R. M. Hochmuth, *Biophys. J.*, **67**, 696 (1994).
35. M. P. Sheetz and J. Dai, *Trends Cell Biol.*, **6**, 85 (1996).
36. D. C. Shutt, D. Wessels, K. Wagenkenecht, A. Chandrasekhar, A. L. Hitt, E. J. Luna and D. R. Soll, *J. Cell Biol.*, **131**, 1495 (1995).

# Chapter 25

# Light-induced Shape Transitions of Giant Vesicles

*PETER G. PETROV AND HANS-GÜNTHER DÖBEREINER*
*Max-Planck-Institut für Kolloid- und Grenzflächenforschung, Golm, Germany*

## 1. VESICLE SHAPES

Since the late 1980s, the morphology of giant vesicles has become well understood, both theoretically [1–3] and experimentally [1,4,5], on the basis of the bending elasticity of the bilayer membrane. Briefly, the main determinants of vesicle shape are the volume-to-area ratio $v$ and the effective spontaneous curvature $\bar{c}_0$ of the vesicle. The latter quantity includes contributions from both bilayer and monolayer bending [2]. Equilibrium vesicle shapes are those that globally minimize the total elastic energy at a given vesicle volume and membrane area. At a fixed volume-to-area ratio, a large spontaneous curvature favours outward curved shapes, whereas a small or negative spontaneous curvature results in a more inward curved morphology. A detailed discussion of the relevant shape parameters and the phase diagram of vesicle shapes is given in Chapter 10.

This contribution focuses on the response of vesicle shapes to a light-induced chemical change in their aqueous environment. The next section describes the particular system under investigation and presents some prototypical experiments. Some general remarks are given on controlling membrane morphology by chemical reactions.

*Giant Vesicles* Edited by P. L. Luisi and P. Walde

## 2. PHOTOCHEMICAL MORPHOLOGY SWITCH

Single-component giant vesicles were prepared from 1-stearoyl-2-oleoyl-*sn*-glycero-3-phosphocholine (SOPC) with purity greater than 99% (Avanti Polar Lipids, USA). Potassium ferrocyanide, $K_4Fe(CN)_6 \cdot 3H_2O$ (Sigma, Germany, ACS reagent grade), was used as the photochemical compound. To explore the impact of the photochemical reaction on the vesicle shapes, two types of systems were investigated:

- Vesicles swollen in 1 mM potassium ferrocyanide solution and then incubated in 4 mM sucrose solution at a volume ratio of 1 : 4. In this case, the ferrocyanide solution is predominantly *inside* the vesicles.
- Alternatively, vesicles were swollen in 20 mM sucrose solution and then incubated in a solution containing 1 mM $K_4Fe(CN)_6 \cdot 3H_2O$ and 18 mM sucrose. For this preparation the potassium ferrocyanide is *outside* the vesicles.

Vesicles were observed by phase contrast microscopy using a Zeiss Axiovert 135 TV microscope equipped with a Plan Neofluar objective (40 × Ph2 nA = 0.75). The epi-illumination system of the microscope (100 W mercury arc lamp) was used as a light source to initiate the photochemical reaction in the sample. Images were recorded by a Hamamatsu C5985 CCD camera connected to the microscope, a videorecorder, and a workstation.

It is known that under illumination in aqueous solution $Fe(CN)_6^{4-}$ undergoes photoaquation [6,7]:

$$Fe(CN)_6^{4-} + H_2O \overset{h \cdot v}{\rightleftharpoons} Fe(CN)_5H_2O^{3-} + CN^- \tag{1}$$

This photochemical process is completely reversible upon switching off the light [6]. The cyanide ion released in the course of reaction (1) is protonated:

$$CN^- + H^+ \rightleftharpoons HCN \tag{2}$$

As a result, the solution pH increases ($pK_a = 9.31$ for HCN).

The main effect reported here, is shape transitions in giant SOPC vesicles caused by reaction (1) and/or subsequent chemical processes following it. An example of this coupling is demonstrated in Figure 25.1 where a light-induced prolate-to-stomatocyte transition is shown. In this case, the light-sensitive $Fe(CN)_6^{4-}$ ion is situated predominantly in the vesicle interior. With no illumination (Figures 25.1(a) and (b)), the prolate shape of this particular vesicle is stable. After initiating the photochemical process by illuminating the sample, the vesicle undergoes a fast shape transition within a few seconds to a stomatocyte (Figures 25.1(c) and (d)). Note that in this case the final shape is more inwardly curved, i.e. the effective spontaneous curvature $\bar{c}_0$ is decreased.

The opposite case of increasing $\bar{c}_0$ is presented in Figure 25.2. Here, the photochemical process is induced in the vesicle exterior. Consequently, we observed budding of an oblate vesicle during illumination (Figures 25.2(a)–(c); compare these

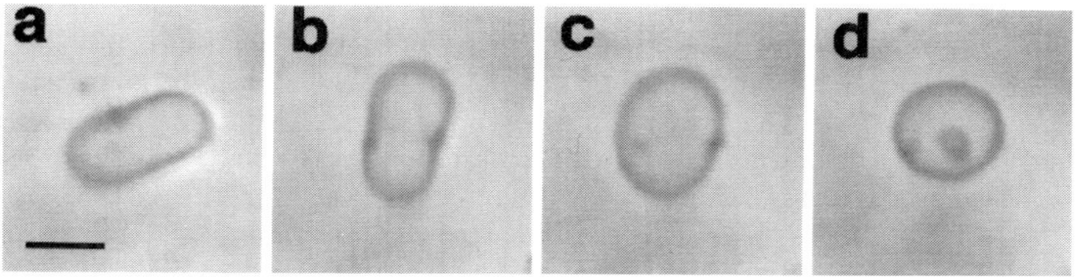


**Figure 25.1** Light-induced prolate-to-stomatocyte transition of an SOPC vesicle. (a) and (b): Snapshots of a fluctuating prolate vesicle which contains excess ferrocyanide in its interior. (c) and (d): Snapshots of the same vesicle under illumination undergoing shape transition to a stomatocyte. The light had been switched on shortly before (c) was taken. The structure visible in the interior of the vesicle in (d) corresponds to an inside bud. The bar denotes 5 μm.

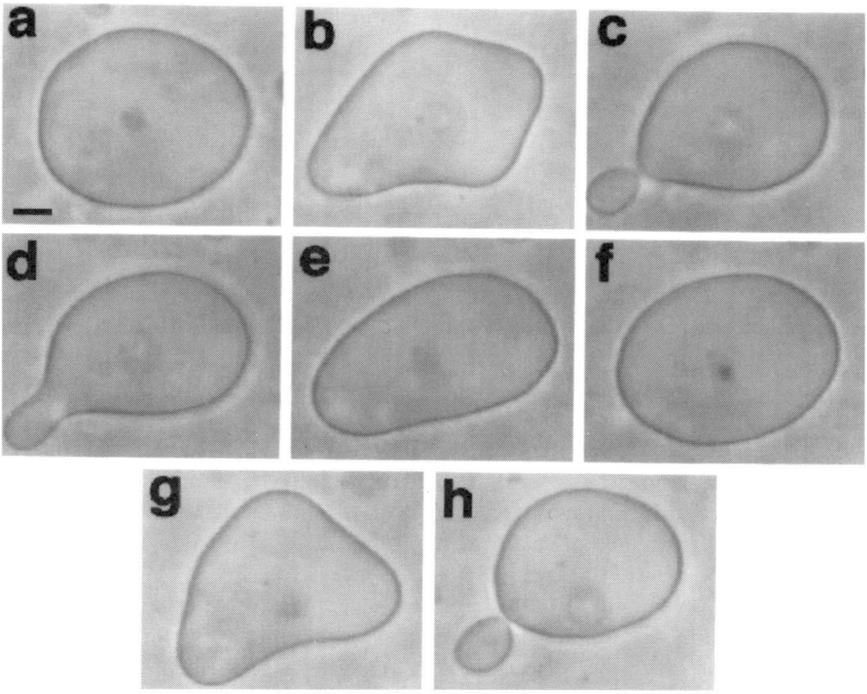


**Figure 25.2** Light-induced budding of an SOPC vesicle. In this case the potassium ferrocyanide solution is in the vesicle exterior. The snapshots show (a) an oblate vesicle viewed from the bottom; (b) and (c) after being illuminated the vesicle buds by passing through the nonequilibrium shape (b). The light had been switched off shortly before (c) was taken. (c)–(f) Upon switching off the light, the process is reversed. (g) and (h) Further illumination of the vesicle again induces the budding transition. The bar denotes 5 μm.

to the shape phase diagram in Chapter 10). The two resulting vesicles are still connected to each other by a narrow neck. Stopping the illumination drives the reverse shape transition (Figures 25.2(c)–(f)). This is evidence for a reversible interaction of the products of the photochemical process with the phospholipid membrane. This important fact is further confirmed by the repeatability of this shape sequence over many cycles (more than 20) observed by us with other vesicles.

The interpretation of these experimental results within the area difference–elasticity model of vesicle shapes [1,2] leads to the conclusion that the main effect of the photochemical reaction that causes the observed shape transitions is a change of the spontaneous curvature of the lipid bilayer. As a possible mechanism, it is suggested that adsorption occurs of one or more species produced after the initiation of the photochemical reaction (1) onto the phospholipid–solution interface. Recent theoretical work has shown that asymmetric adsorption on the vesicle surface will cause a change in bilayer curvature [8]. In addition, if the adsorbate is an ion, it will charge the interface. This also leads to a change in the spontaneous curvature, as predicted previously [9]–[11]. Further investigations are currently being pursued to clarify the mechanism underlaying the strong effect reported here.

## 3. CONTROL OF MEMBRANE CURVATURE BY CHEMICAL REACTIONS

The preceding section reports that fluid lipid vesicles show quite a strong morphological response to a small asymmetric change in the chemical environment inside and outside of the vesicle. This is true in general. Any asymmetry in molecular species and/or other (density) fields across the bilayer will result in a curved equilibrium configuration of the membrane [8,12–15]. Small differences in interfacial free energies result in a pronounced curvature effect due to the large ratio of vesicle radius, typically $10^2$–$10^5$ nm, to membrane thickness, which is 5 nm [2,5]. It is the same geometrical mechanism that drives the bending of a bimetal sheet. Asymmetry may be created by surface fields, like (mixed) amphiphile monolayer densities and molecular configuration (charge and steric) or adsorbed molecules, as well as boundary values of bulk fields at the interface, such as electrolyte concentration and solution pH [16].

The observed reversibility of the shape response, together with the possibility of alternating the direction of the $\bar{c}_0$ change can have some very important consequences. For instance, this effect makes the systems described here potential candidates for applications where tuning of the curvature of soft objects is desirable. The ability of chemically altering interfacial curvature on a nanometer to micrometer length scale is of great importance both to materials science [17] and biology [18]. Of particular interest is the coupling of morphology to oscillating chemical reactions. It should be possible to produce a periodically changing membrane curvature controlled by the (surface) density of chemical species.

## 4. ACKNOWLEDGMENTS

We thank W. Fenzl, R. Lipowsky, S. Mann, and R. Netz for helpful discussions. PGP (from the Institute of Biophysics, Sofia, Bulgaria) gratefully acknowledges financial support from the Max Planck Gesellschaft.

## 5. REFERENCES

1. U. Seifert, *Adv. Phys.*, **46**, 13 (1997).
2. L. Miao, U. Seifert, M. Wortis and H.-G. Döbereiner, *Phys. Rev. E*, **49**, 5389 (1994).
3. S. Svetina, M. Brumen and B. Žekš, *Stud. Biophys.*, **110**, 177 (1985).
4. J. Käs and E. Sackmann, *Biophys. J.*, **60**, 825 (1991).
5. H.-G. Döbereiner, E. Evans, M. Kraus, U. Seifert and M. Wortis, *Phys. Rev. E*, **55**, 4458 (1997).
6. Y. Mori, G. Rábai and I. Hanazaki, *J. Phys. Chem.*, **98**, 12968 (1994).
7. A. Kaminaga, G. Rábai, Y. Mori and I. Hanazaki, *J. Phys. Chem.*, **100**, 9389 (1996).
8. R. Lipowsky and H.-G. Döbereiner, *Europhys. Lett.*, **43**, 219 (1998).
9. M. Winterhalter and W. Helfrich, *J. Phys. Chem.*, **92**, 6865 (1988).
10. D. J. Mitchell and B. W. Ninham, *Langmuir*, **5**, 1121 (1989).
11. T. Chou, M. V. Jarić and E. D. Siggia, *Biophys. J.*, **72**, 2042 (1997).
12. B. L.-S. Mui, H.-G. Döbereiner, T. D. Madden and P. R. Cullis, *Biophys. J.*, **69**, 930 (1995).
13. C. Hiergeist and R. Lipowsky, *J. Phys. II France*, **6**, 1465 (1996).
14. H.-G. Döbereiner, A. Lehmann, W. Goedel, O. Selchow and R. Lipowsky, in *Materials Science of the Cell*, (eds B. Mulder, C. F. Schmitt and V. Vogel). Proceedings of the Materials Research Society Symposium held December 1–4, 1997.
15. H.-G. Döbereiner, O. Selchow and R. Lipowsky, *Eur. Biophys. J.*, **28**, 174 (1999).
16. P. Walde, *Curr. Opinion Coll. Interface Sci.*, **1**, 638 (1996).
17. S. Mann and G. A. Ozin, *Nature*, **382**, 313 (1996).
18. E. Sackmann, *FEBS Lett.*, **346**, 3 (1994).

# Chapter 26

# Changes in the Morphology of Giant Vesicles under Various Physico-chemical Stresses

*MARIE-ALICE GUEDEAU-BOUDEVILLE AND ANNE-LAURE BERNARD*
*Collège de France, Paris, France*

*JEAN-CLAUDE BRADLEY*
*Drexel University, PA, USA*

*ALOK SINGH*
*Naval Research Laboratory, Washington, DC, USA*

*LUDOVIC JULLIEN*
*Ecole Normale Supérieure, Paris, France*

## 1. INTRODUCTION

Bilayer lipid structures are known to transform between a variety of morphologies. These transformations are controlled by the geometry or the chemistry of the molecular structures and by the presence of external stimuli. Giant vesicles, in the form of spherical lipid bilayers, are used as simplified models of biological cells. Their *in vitro* study using optical microscopy allows their responses to be observed when exposed to stresses such as temperature changes, UV irradiation, electric or osmotic shocks, or the addition of an external agent, such as amphiphilic polymers.

*Giant Vesicles* Edited by P. L. Luisi and P. Walde

In the present article, such stress-induced changes are studied and a number of phenomenological observations are reported.

## 2. EFFECT OF TEMPERATURE QUENCH ON POLYMERIZABLE VESICLES

Tubular vesicles made of $C_{14}$ butadienic lipids (Figure 26.1) containing two conjugated double-bonds per chain, ($T_m < 20°C$) reversibly transform into chains of pearls when the temperature is decreased from 20°C to 10°C [1]. These structures are called unduloids and are known to present an intermediate state between cylinder and sphere necklace morphologies. By polymerization using irradiation of ultraviolet light at 254 nm, this transformation becomes irreversible and vesicles remain in the chain of pearls configuration indefinitely.

In contrast, spherical vesicles made of diacetylenic phospholipids $DC_{6,15}$ PC (Figure 26.2) which have two conjugated triple bonds per chain ($T_m = 58.9°C$), transform into nonvesicular structures such as tubules, helices, shards (combination of fragmented tubule, helix and vesicle) when slowly cooled below their chain-melting transition temperature. Earlier studies have shown distinct, temperature-induced morphological changes that depend on the size of the diacetylenic vesicles. Large vesicles (> 1 μm) transform into tubules when the temperature is reduced, whereas small vesicles (50–200 nm) do not [2]. Furthermore, giant vesicles (10–300 μm) display a remarkable stability of vesicular morphology down to 32°C when cooled from 70°C at a rate of 1°C $min^{-1}$. This is remarkable in that it is 26°C below their phase transition temperature ($T_m = 58.9°C$). Suddenly, at 32°C, they collapse irreversibly within a two-degree temperature range. They turn into rigid nonspherical vesicles or oddly branched structures [3]. Figure 26.3 shows a series of video images taken as a 260 μm diameter GUV begins to rupture and collapse. From 70°C to 32°C, the vesicle remained spherical, stable and without visible defects, but at 31.9°C a dark spot appears at the equator of the vesicle. Disintegration of the membrane begins at this point, spreading rapidly all around the circumference of the vesicle, resulting in a lipid aggregate of ill-defined morphology. Circular dichroism studies have shown that the molecular association of diacetylenic lipids leads to a

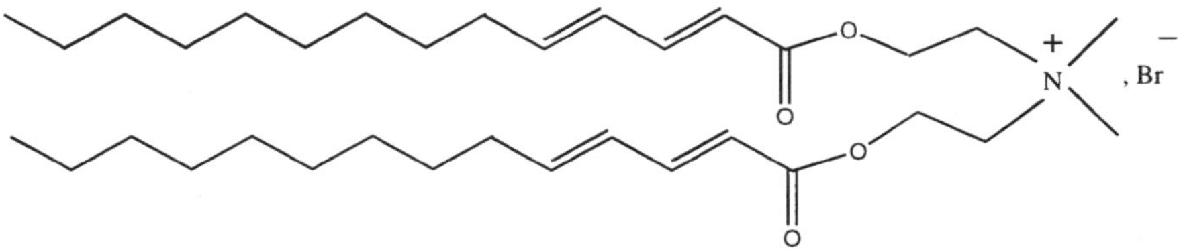


**Figure 26.1** Structure of ditetradecadienoyl dimethyl ammonium bromide (DTDIAB), $C_{14}$ polymerizable lipid

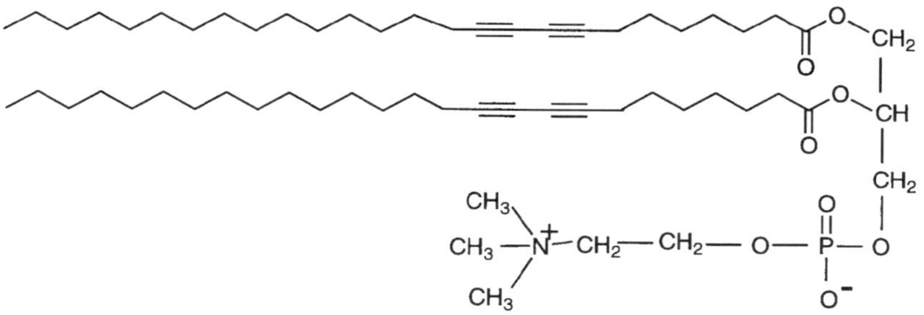

**Figure 26.2** Structure of 1,2-bis(alka-8,10-diynoyl)-*sn*-glycero-3-phosphocholine ($DC_{6,15}$ PC).

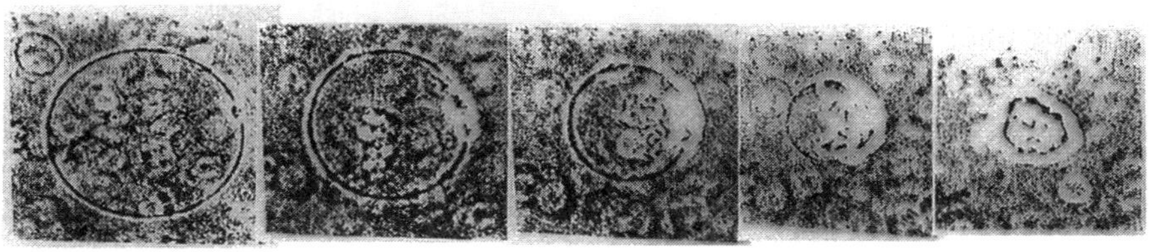

31.9°C 31.7°C 31.6°C 31.55°C 31.5°C

**Figure 26.3** Shrinkage of GUV made from $DC_{6,15}$ PC upon cooling from 31.9°C to 31.5°C. The scale bar represents 50 μm.

helical organization of structures [4]. It is believed that this oddly branched morphology corresponds to the microscopic structures in the transition towards tubules formation.

In marked contrast to the behavior seen in the previous two examples, spherical vesicles made of saturated lipids 1,2-distearoyl-***sn***-glycero-3-phospho-choline DSPC, ($T_m = 54.9$°C), show only a liquid to gel phase transition at 47.8°C when

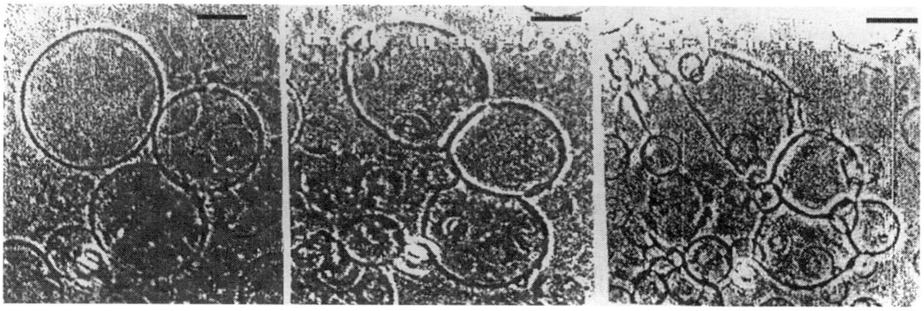

(a) (b) (c)

**Figure 26.4** Gelation of GUV made from 1,2-distearoyl-*sn*-glycero-3-phosphocholine (DSPC) upon cooling: (a) 48.8°C, (b) 46.5°C, (c) 45.3°C. The scale bar represents 50 μm.

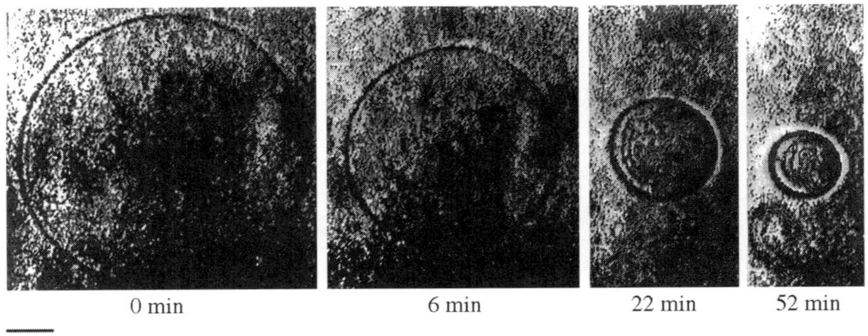


**Figure 26.5** Shrinkage of GUV made from $C_{14}$ by UV irradiation at 254 nm with time. The scale bar represents 10 μm.

cooled from 60°C at a rate of 1°C min (Figure 26.4). They turn into agglomerated vesicles with angular shapes which remain stable without any more evolution below 43.8°C.

## 3. UV POLYMERIZATION OF $C_{14}$ BUTADIENIC LIPID MEMBRANES

UV irradiation at 254 nm of $C_{14}$ vesicles leads to a decrease of vesicle radius and to an increase of membrane thickness as a function of the time of exposure [5]. The first polymerization step, corresponding to the disappearance of the conjugated double bond, observed on UV spectra after 6 min of UV irradiation, provokes a slight decrease of the vesicle radius (Figures 26.5(a) and (b)). However, a second polymerization step, due to the reticulation of the remaining nonconjugated double bonds, leads to vesicle shrinkage. During this process, the membrane thickens, as favored by the creation of defects. The vesicle becomes porous and the intravesicular water can leave the structure allowing a more extensive shrinkage of the membrane (Figures 26.5(c) and (d)).

## 4. APPLICATION OF ELECTRICAL PULSES ON $C_{14}$ POLYMERIZABLE VESICLES

When an external electric field is applied across a spherical vesicle, the membrane behaves as a capacitor, because the conductivities of the internal and external media are much larger than for the membrane. Intravesicular ions tend to accumulate at the poles of the vesicle facing the electrodes. The maximal local electric field experienced across the lipid bilayer can be several orders of magnitude larger than the applied external field and is directly proportional to the ratio of the vesicle radius and to the membrane thickness. The aim of this work was to insert transmembrane lipids locally in the membrane of $C_{14}$ unpolymerized vesicles and to immediately polymerize them to prevent the diffusion of the inserted lipid into the whole membrane [6]. Various phenomena were observed during electroporation of the unpolymerized vesicles (Figure 26.6).

In these experiments, vesicles are swollen in a 50 mM sucrose solution and diluted in a 50 mM glucose solution in order to obtain a better contrast and to observe membrane permeability to sucrose and glucose with phase contrast microscopy.

In Figures 26.6(a) and (b), electroporation leads to the formation of a large pore facing the electrodes (pulse periods are 10 ms and 1 ms, respectively). Figures 26.6(c) and (d) show the electroadhesion of two vesicles, during a 1 ms pulse. Electrofusion of these two vesicles increases the diameter of the biggest vesicle from 54 to 58 μm (Figures 26.6(e) and (f)). Another 1 ms pulse provokes the electro-expulsion of an internal vesicle (Figures 26.6(g) and (h)). After each pulse, the pore shuts off, whereas when the $C_{14}$ vesicle is UV polymerized, the pore remains open.

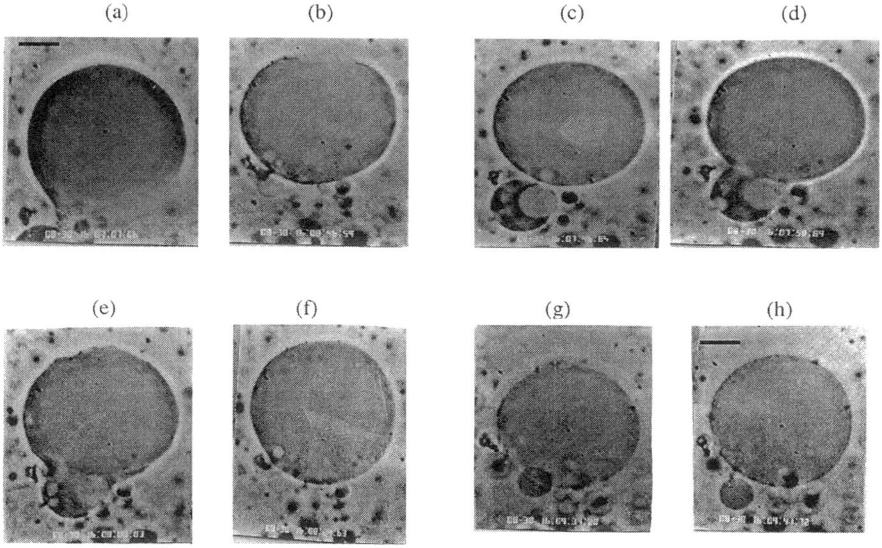


**Figure 26.6** Electroporation of GUV made from $C_{14}$: (a) and (b) formation of a large pore with pulse durations of 10 ms and 1 ms, respectively; (c) and (d) electroadhesion of two vesicles; (e) and (f) electrofusion in which diameter increases from 54 to 58 μm; (g) and (h) electroexpulsion of an internal vesicle. The scale bar represents 10 μm. In these pictures, the anode is placed at the upper side of the field of view with the cathode at the lower part, and the electric field between the electrodes is 0.5 kV $cm^{-1}$.

## 5. ACTION OF EXTERNAL OSMOTIC PRESSURE

In 1981, Helfrich [7] studied the effect of the external osmotic pressure on egg yolk phosphatidylcholine (EPC) giant vesicles by adding 15 mM of salt or glucose in the external medium of the vesicles. The vesicle radius appears to decrease linearly with time according to the law $dR/dt = -\alpha P \Delta c$, where $P$ is the membrane permeability coefficient to water, $\alpha$, is the water molar volume, and $\Delta c$, the difference of molar concentrations. The water permeability coefficient for EPC bilayers was found to be $41\ \mu m\ s^{-1}$.

In this case, a controlled osmotic stress up to 250 mM applied to the giant vesicles leads to an interesting endocytosis phenomenon that we named the 'raspberry effect'. The experimental set-up is described in Figure 26.7.

Kit(-)vesicles are made of a mixture of 70% egg phosphatidyl choline (EPC), 20% dicetylphosphate, and 10% cholesterol, (lipid mixture called negatively charged Liposome Kit from Sigma, USA) and are swollen in 50 mM sucrose. They are diluted in a 50 mM glucose solution. The cover top of the cell is an anopore filter (purchased from Alltech France), with 0.2 μm diameter pores. This filter is used as a dialysis membrane between the vesicle suspension and the concentrated glucose reservoir. The reservoir is filled with a 300 mM glucose solution which can diffuse through the anopore filter in the outer medium of the vesicle. This slow osmotic stress makes water leak out of the vesicle in order to equilibrate the osmolarity of the internal and external media. Vesicles are destabilized and rearrange their excess surface area into daughter vesicles through an endocytosis process. The daughter vesicles are present only inside the mother vesicle. Indeed, no daughter vesicle appears outside the mother vesicle when focusing at the equatorial plane of the vesicle in phase contrast microscopy studies (Figure 26.8).

The daughter vesicles are monodisperse in size, with diameters of 4–5 μm, and the mother vesicle reaches its final equilibrium radius after 15 min of dialysis (Figure 26.9). The daughter vesicle diameter is found to be independent of the imposed

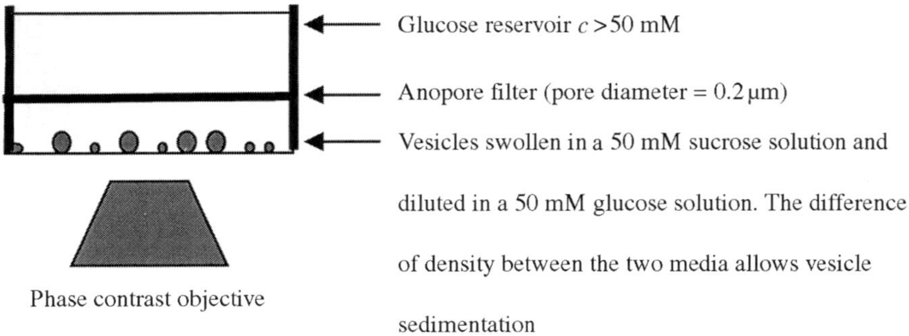


**Figure 26.7** Scheme of the experimental set-up used to obtain a controlled osmotic stress.

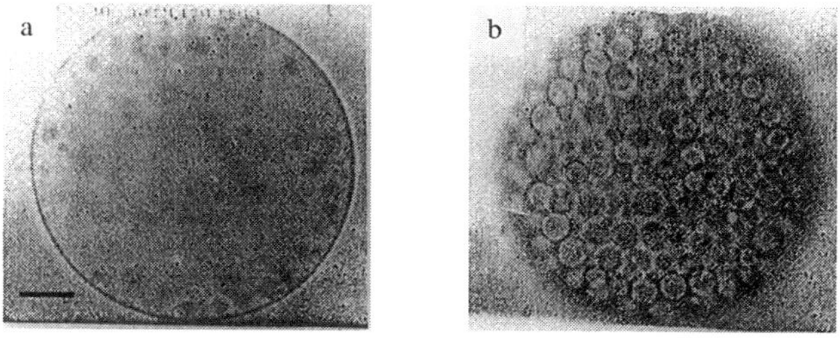


**Figure 26.8** Kit(-)vesicle in 50 mM sucrose solution after 15 mn of dialysis with 300 mM glucose solution; (a) view at the equator of the vesicle; (b) view above the equator. The scale bar represents 10 μm.

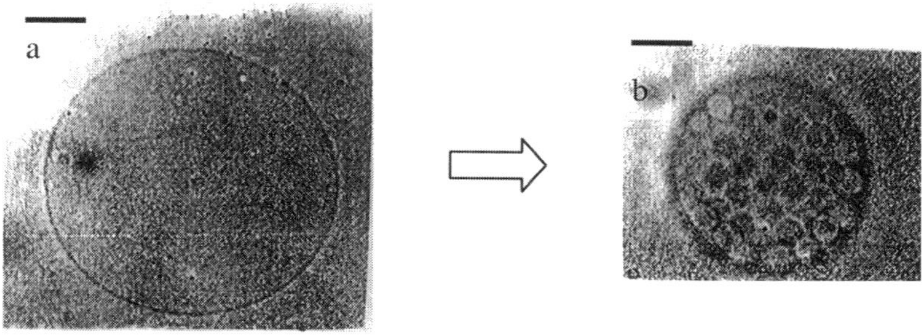


**Figure 26.9** (a)Kit(-)vesicle in 50 mM sucrose solution before dialysis (diameter = 52 μm); (b) after 15 min of dialysis with 300 mM glucose (diameter = 32 μm). The scale bar represents 10 μm.

osmotic pressure difference, but their number and their rate of formation increase with this pressure difference.

## 6. INTERACTION OF OBLATE VESICLES WITH HYDROPHILIC POLYMERS CONTAINING HYDROPHOBIC SIDE CHAINS

In order to study the effect of hydrophilic–hydrophobic polymers on the morphology of oblate vesicles, we have grafted dextran polymers of various molecular weights with different percentages of palmitoyl chains (Figure 26.10).

These palmitoyl dextrans [PD] are all soluble in water. Vesicles are obtained by slow hydration of 1-stearyl-2-oleoyl-*sn*-glycero-3-phosphocholine (SOPC, two $C_{18}$ chains, one saturated and the other one unsaturated) in pure water.

The addition of palmitoyl dextran (MM 162 000, concentration 0.04 mg ml$^{-1}$) outside the vesicles induces a rapid morphological change (Figure 26.11). The hydrophobic part of the polymer anchors inside the membrane and leads to a

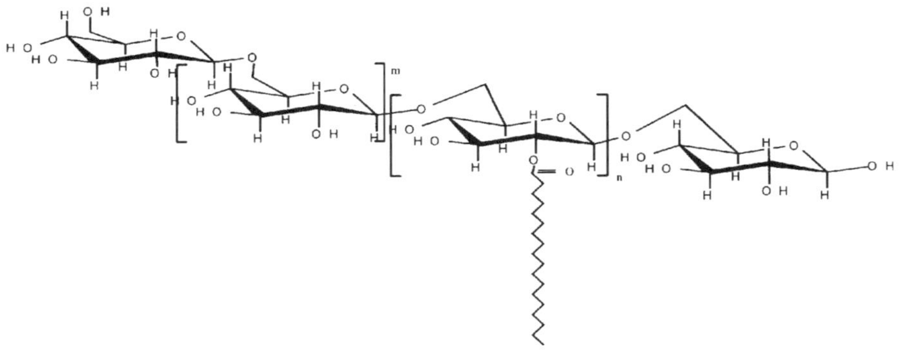

**Figure 26.10** Structure of palmitoyl dextran (PD)

budding instability, creating fingers all around the rim of the vesicle. These buds progressively turn into pearls until a complete degradation which occurs when too large amounts of alkyl chains are anchored in the membrane [8].

These different examples have shown that it is possible to induce well-controlled morphological changes of giant vesicles by different stresses such as temperature

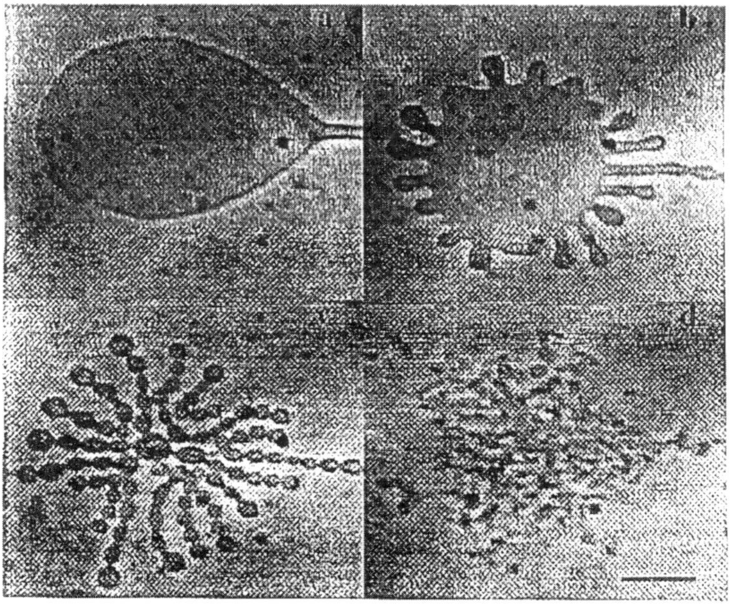


**Figure 26.11** Oblate vesicle made from SOPC in a polymer solution of large concentration, [PD] = 0.05 mg ml$^{-1}$, at different times: (a) no polymer, (b) 45 s after addition of polymer, (c) 75 s, (d) 190 s. The scale bar represents 20 μm.

quenches, UV irradiation, electric or osmotic shocks. Finally the addition of the amphiphilic polymer palmitoyl dextran in the external medium of the vesicles has been shown to provoke a spectacular series of changes due to the anchoring of the hydrophobic chain in the membrane. Planned studies of this last topic include the use of grafted polymers of different molecular weights to determine the thresholds of concentration and the number of anchors per backbone, allowing budding and pearling instabilities to be induced. This might provide greater insight into the mechanisms of these instabilities.

## 7. REFERENCES


1. M. A. Guedeau-Boudeville, *C. R. Acad. Sci. Paris*, **315**, 541–544 (1992).
2. T. G. Burke, A. S. Rudolph, R. R. Price, J. P. Sheridan, A. W. Dalziel, A. Singh and P. E. Schoen, *Chem. Phys. Lipids*, **48**, 215–230 (1988).
3. A. Singh, P. E. Shoen and M. A. Guedeau-Boudeville, *Chem. Phys. Lipids*, **94**, 53–61 (1998).
4. J. M. Schnur, *Science*, **262**, 1669–1676 (1993).
5. M. Dvolaitzky, M. A. Guedeau-Boudeville and L. Leger, *Langmuir*, **8**, 2595–2597 (1992).
6. J. C. Bradley, M. A. Guedeau-Boudeville, G. Jandeau and J. M. Lehn, *Langmuir*, **13**, 2457–2462 (1997).
7. E. Boroske, M. Elwenspoek and W. Helfrich, *Biophys. J.*, **34**, 95–109 (1981).
8. V. Frette, I. Tsafrir, M. A. Guedeau-Boudeville, L. Jullien and J. Stavans, unpublished results.

Chapter 27

# Magnification of Shape Fluctuations of Active Giant Unilamellar Vesicles

*Jean-Baptiste Manneville, Patricia Bassereau, Daniel Lévy, and Jacques Prost*

*Institut Curie, Paris, France*

## 1. INTRODUCTION

In the physics literature, membranes have mostly been studied as bilayers of surfactant molecules fluctuating around their mean equilibrium shape due to Brownian motion [1]. Lipid membranes, for instance, have been investigated both experimentally and theoretically as vesicles of spherical geometry. A wide range of problems have been considered, such as shape transitions and budding, phase separation in two-component bilayers [2], interactions of fluctuating vesicles with polymers [3], microtubules [4] or actin shells [5], surface charge and electrostatic interactions [6]. In all these studies, such *passive* fluctuating membranes are in thermodynamic equilibrium: the only noise source is thermal noise, which is uncorrelated both in time and space.

Biological membranes, however, are not in thermodynamic equilibrium because they contain active proteins inside the lipid bilayer [7]. Ion pumps, voltage- or ligand-dependent channels, for instance, can dissipate energy provided by concentration gradients, ATP hydrolysis, or an external light source. When the proteins are activated, ions are translocated from one side of the membrane to the other in the case of a pump, or ions and water flow through the membrane pore in the case of a channel. This ion translocation or ion flow exerts a localized force on the membrane.

*Giant Vesicles* Edited by P. L. Luisi and P. Walde

The noise associated with these local forces is non-thermal and, as a result, does not satisfy the fluctuation–dissipation theorems. Thus, a non-thermal noise source adds to thermal noise, making biological membranes non-equilibrium, i.e. active membranes.

From the point of view of non-equilibrium statistical physics, understanding the behavior of such active membranes is challenging. Recent theoretical results [8,9] have shown that the shape fluctuations of active membranes should differ both qualitatively and quantitatively from those of passive membranes. The presence of active proteins imbedded in the membrane is expected to induce a magnification of shape fluctuations due to a modification of the fluctuation spectrum. In search of the simplest model to test the theory [8,9], the shape fluctuations of giant lipid vesicles have been studied with the active protein, bacteriorhodopsin (BR) incorporated inside the lipid bilayer.

## 2. EXPERIMENTAL PROCEDURE

Bacteriorhodopsin (BR) [10] is a 27 kDa protein, with a light-driven proton pump activity. It is purified from the purple membrane of the bacteria *Halobacterium salinarium* using a classical procedure [11]. The BR structure and function, as well as its photocycle, are well known and characterized [12]. Briefly, a photon is absorbed maximally at 568 nm and induces a change of conformation of the protein leading to the translocation of a proton in about 5 ms from one side of the BR to the other. In the dark, or when illuminated with light with a wavelength far from the absorption maximum, the BR has no pumping activity. Electroformation [13] was used to grow giant (radius >20 μm) unilamellar phospholipid (Egg Phosphatidylcholine, Avanti Polar Lipids) vesicles, modified [14] in order to incorporate BR in the membrane. The lipids (0.5 mg ml$^{-1}$) are first solubilized in diethylether, then a concentrated solution of purple membrane (18 mg ml$^{-1}$) is added to the lipids and the lipid–protein mixture is sonicated for a few seconds at 0°C. A few microliters of solution are deposited on ITO (indium tin oxyde) glass slides and the film is dried under vacuum overnight. The ITO slides are sealed and a 50 mM sucrose solution is injected in the chamber in order to hydrate the film. The electric field (1.5 V AC) is applied across the chamber by connecting the ITO slides to copper electrodes (Figure 27.1(a)). Giant vesicles are obtained in about 2 h and transferred in a manipulation chamber filled with 50 mM glucose to enhance the optical contrast between the inside and the outside of the vesicles.

Quantifying the amount of BR incorporated inside the vesicles turned out to be rather difficult. From the ratio of lipids to BR in a given initial dried film, it was possible to infer an *a priori* amount of BR in the vesicles grown from that film, ranging from about 1000 to 60 lipids per BR, which corresponds to about 2–20% of the area of the vesicles covered by the BR. This assumes that the mixed lipids–BR film is homogeneous and that each vesicle contains the same amount of BR. Using fluorescent labeling of the BR [15], this was proved not to be the case: the intensity

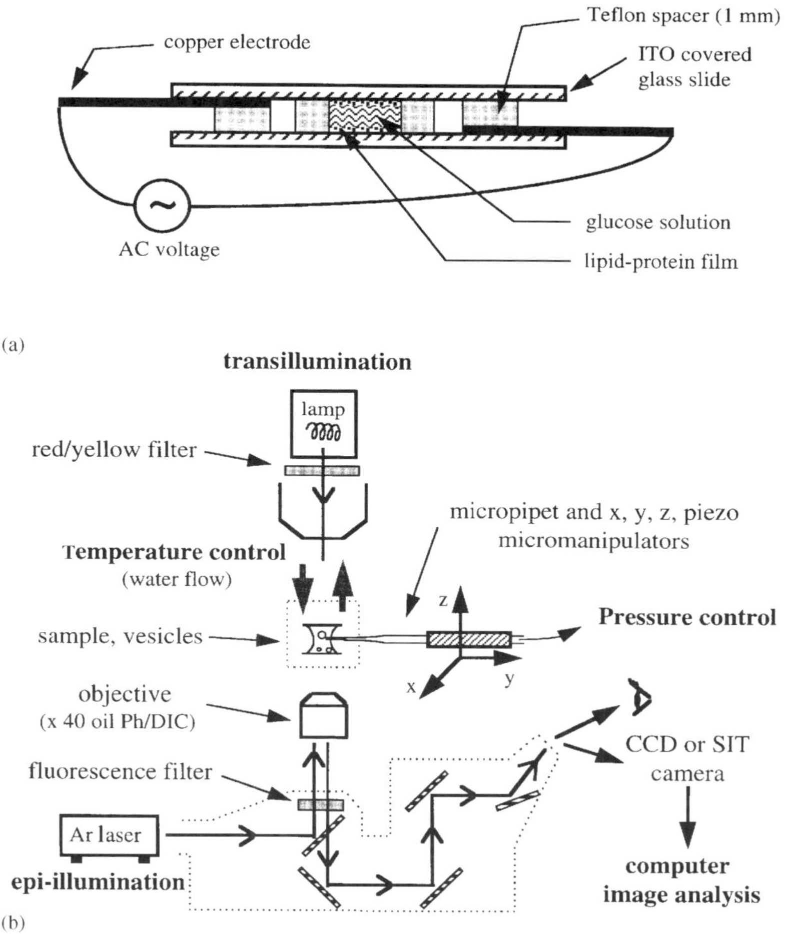


**Figure 27.1** (a) Diagram of the electroformation chamber (ITO = indium tin oxyde). (b) Diagram of the experimental set-up (CCD = charged couple device, SIT = silicium intensified target). The oil immersion objective (×40) can perform both phase contrast (Ph) or differential interference contrast (DIC) microscopy.

of fluorescence, and thus the amount of BR, differs from one vesicle to the other, even if these vesicles are grown from the same film with a given *a priori* amount of BR. Consequently, the relative intensity of fluorescence was used to quantify the amount of BR incorporated inside the vesicles.

The micropipet technique [16] was used to measure the excess surface area due to the shape fluctuations of the vesicles around their mean spherical shape. The experimental set-up is directly derived from that of Evans [17], including a temperature control at 15–16°C to prevent evaporation of the vesicle solution, and an inverted microscope (Axiovert 135 from Zeiss, Germany), with phase contrast or differential interference contrast (DIC) optics (Figure 27.1(b)). The vesicles were imaged and were made respectively passive or active by shining respectively red light (high-pass filter 650 nm) or green-yellow light (high-pass filter 500 nm) through the transillumination pathway. This provided a very simple system with which to monitor only the effects of the wavelength change. Fluorescence images were obtained with an Argon laser (488 nm) through the epi-illumination pathway and were then quantified by computer image analysis.

In the micropipet technique, a pressure difference is created between the inside and the outside of a micropipet to create tension in the membrane and to pull out the excess area due to the shape fluctuations inside the micropipet. In the low-tension regime [16,18], for a passive membrane, the slope of the logarithm of the tension $\sigma$ versus the areal strain $\alpha$ is proportional to the bending modulus $\kappa$ of the membrane,

$$\alpha \frac{8\pi\kappa}{k_B T} = \ln \frac{\sigma}{\sigma_i} \tag{1}$$

where $k_B$ is the Boltzmann constant and $\sigma_i$ is the initial tension corresponding to $\alpha = 0$. This results directly from the expression of the fluctuation spectrum given by $\langle u^2(q) \rangle = k_B T/(\kappa q^4 + \sigma q^2)$, where $u(q)$ is the displacement of the membrane compared to its mean position for a given wavelength $q$. The bending modulus $\kappa$ can be deduced from eq. (1) for a passive membrane. Qualitatively, because of the expected magnification of the fluctuations due to the activity of the proteins, for the same tension, we should be able to pull out more excess area when the membrane is active than when it is passive. The slope of the logarithm of the tension versus the areal strain should thus be smaller as compared to the passive case. If eq. (1) is used to analyse the data, the active membrane should appear more flexible, with an effective bending modulus $\kappa_{eff}$ smaller than its passive counterpart.

## 3. RESULTS

Micropipet experiments were first carried out on vesicles containing an increasing amount of BR, to investigate the influence of BR incorporation on the bending modulus of the membrane. The vesicles were illuminated with red light so that the BR was inactive. In Figure 27.2 the bending modulus $\kappa$ is plotted against the fluorescence intensity. For pure egg phosphatidylcholine (EPC) vesicles, the standard value of $\kappa = 10 \pm 1 k_B T$ was obtained [19]. Furthermore, the bending modulus does not show any measurable dependence with the BR concentration in the vesicles. EPC is a mixture of natural phosphatidylcholines with predominantly $C_{16}$ and $C_{18}$ chains and is known to match very well the length of the BR [20]. It is

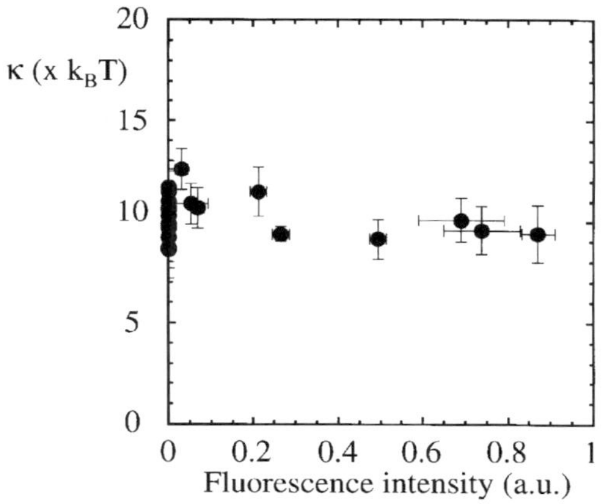


**Figure 27.2** Plot of the passive bending modulus $\kappa$ against the fluorescence intensity (as a measure of the BR concentration in the membrane).

thus expected that the presence of BR in the membrane should not modify significantly the mechanical properties of the bilayer, as long as the BR concentration remains below some crystallization limit.

For each vesicle, the micropipet experiment was performed twice: first with the membrane active (i.e. the vesicle was illuminated with green-yellow light), then it was made passive (i.e. the vesicle was illuminated with red light). For a given value of the tension, the intrusion length of the membrane inside the micropipet is larger when the vesicles are illuminated with green-yellow light than when they are illuminated with red light (Figure 27.3), indicating that it is possible to pull out more excess area from the fluctuations when the membrane is active. Consequently, the effective bending modulus $\kappa_{eff}$, defined in the previous section for an active membrane, is about twice smaller than the bending modulus $\kappa$ of the corresponding passive membrane (Figure 27.4). For pure lipid vesicles containing no BR, the active effective bending modulus equals the passive bending modulus within the error bars of the method. For the lowest content of BR reached in the experiments, the effective bending modulus drops to about 50% and ceases to depend on the concentration of BR in the vesicle, even for high protein contents.

## 4. DISCUSSION

This is thought to be the first experimental evidence of an effect of protein activity on the shape fluctuations of a lipid membrane. Micropipet experiments prove that the activity by itself induces an enhancement of the amplitude of the fluctuations, as indicated by a strong decrease in the effective bending modulus of the membrane.

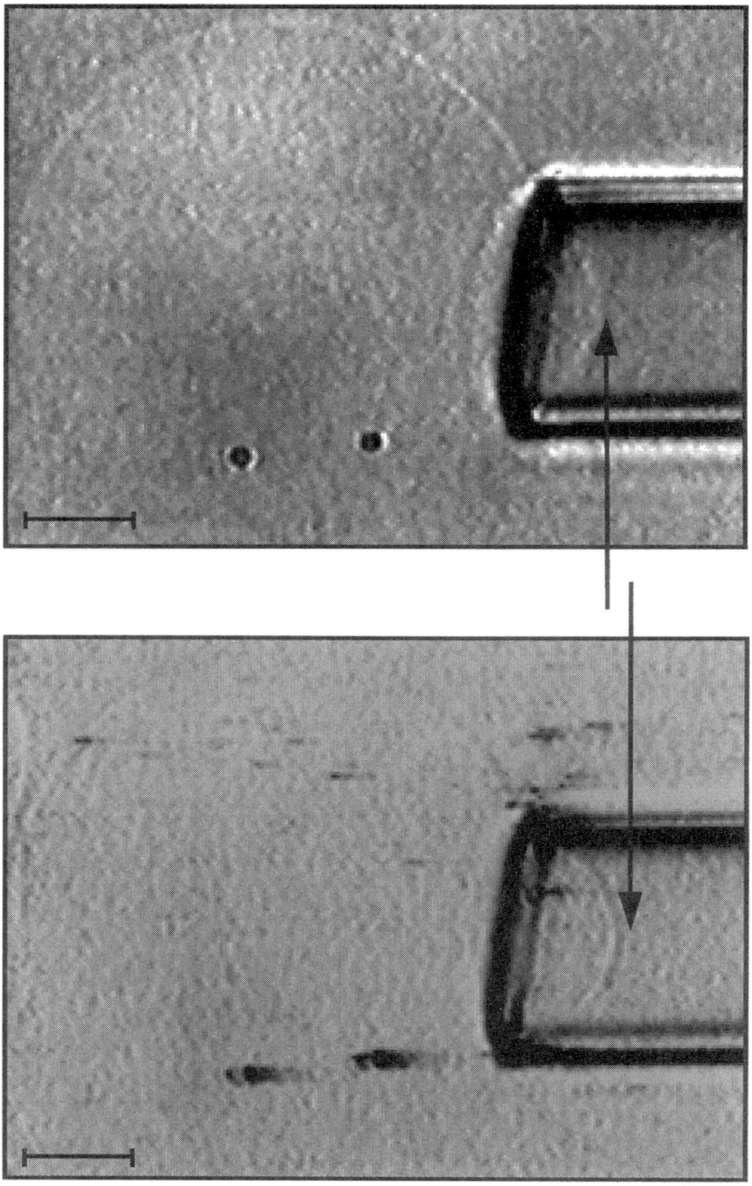

**Figure 27.3** Comparison of the intrusion length of the same vesicle inside the micropipet (arrows) with non-activating red light (upper DIC image) and activating green-yellow light (lower DIC image) at the same tension $\sigma = 4.2\,\mu\text{N}\,\text{m}^{-1}$. The bar represents $5\,\mu\text{m}$.

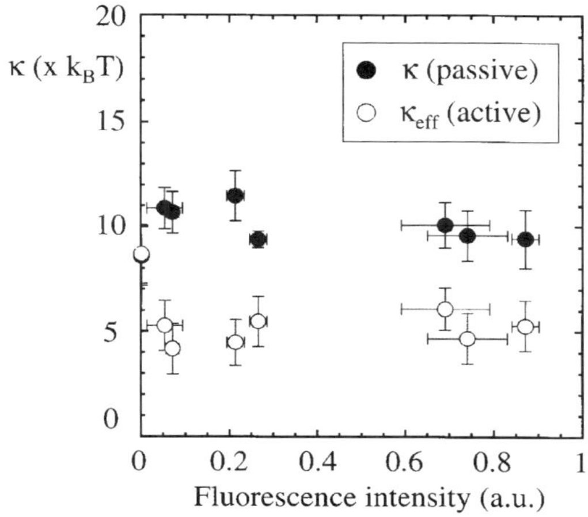


**Figure 27.4** Qualitative illustration of the effect of BR activation: plot of the passive bending modulus $\kappa$ and the corresponding active effective bending modulus $\kappa_{eff}$ as a function of the fluorescence intensity (proportional to the BR concentration in the membrane) for seven vesicles. For the pure lipid vesicle (non-fluorescent), the two circles are almost superimposed.

This effect, when switching the light from yellow-green to red, cannot be attributed to the lipid molecules themselves because no change in the bending modulus could be detected in the control experiments with pure lipid vesicles. Simple estimates also ruled out the possibility of a thermal effect due to local heating of the lipid membrane because of light absorption by the BR.

In most of the experiments described above the vesicles were first illuminated with red light, then with green-yellow light and thus showed an amplification of the fluctuations upon activation of the BR. It was checked by other experiments that illuminating the vesicles with the activating green-yellow light first, then with red light, gives the opposite effect; i.e. induces a diminution of the amplitude of fluctuations. This proves that the effect described above is a result of pumping protons through the lipid bilayer and does not arise from the change of conformation of the BR from its dark-adapted to its light-adapted form [21]. However, it was not possible to pull on a given vesicle more than twice – which would constitute a good test for the reversibility of the effect – because of adhesion of the lipid membrane to the micropipet glass, in spite of treatment with bovine serum albumin (BSA). As illustrated qualitatively by the plot of the effective bending modulus in Figure 27.4, the effect of the activity on the membrane fluctuations does not seem to depend on the BR concentration. Very recent calculations [23] show that if the coupling between curvature and diffusion of the proteins is taken into account, the dependence on protein concentration should be weaker. This calculation includes new terms that were neglected in previous models [8,9].

It should be emphasized that the reduction of the effective bending modulus when the BR is activated should only be interpreted as a qualitative illustration of the effect of the activity. The logarithmic dependence in eq. (1) has been derived for a passive membrane with a fluctuation spectrum scaling as $\langle u^2(q) \rangle \propto 1/q^4$. Equation (1) is thus a direct consequence of the equipartition theorem and should no longer be valid for an active membrane. The fluctuation spectrum of an active membrane is modified [8,9]. This new fluctuation spectrum is used to derive the dependence of the areal strain $\alpha$ with the tension $\sigma$ leading to the analog of eq. (1) for an active membrane. It is then possible to interpret the micropipet experiments in a more correct way, i.e. avoiding the concept of effective bending rigidity [22,23].

## 5. CONCLUSIONS

Reconstituted BR in giant phospholipid unilamellar vesicles provides a very simple experimental model of active membrane, because the effects of switching from the active to the passive state of BR in the membrane are easily isolated. The sensitivity of the micropipet technique allows a precise quantification of the shape fluctuations of these vesicles and gives the first experimental evidence that the fluctuations are strongly enhanced when the activity of the pump is triggered on. The observed reduction of the effective bending modulus is in good qualitative agreement with the theoretical predictions. However, the results of these micropipet experiments need to be interpreted more quantitatively so as to test the theoretical models, especially the dependence with the protein concentration [22]. For instance, a more accurate estimate of the permeation constant is needed to compute the expected orders of magnitude.

For a biological membrane, the direct implication of these findings is that the lipid bilayer should fluctuate more when proteins are turned on, although the presence of a cytoskeleton and/or a glycocalyx would certainly modify the effect. As a consequence of the amplification of the shape fluctuations upon activation, the entropic repulsion [24] between an active membrane and a wall is predicted to be of longer range [9]. This prediction could play an important role in cell–matrix or cell–cell adhesion.

## 6. REFERENCES

1. R. Lipowsky, *Nature*, **349**, 475 (1991); D. Nelson, T. Piran, and S. Weinberg (eds), *Statistical Mechanics of Membranes and Surfaces, Jerusalem Winter School*, World Scientific, Singapore, 1989; R. Lipowsky and E. Sackmann (eds), *Structure and Dynamics of Membranes* Elsevier North Holland, Amsterdam, 1995.
2. E. Farge and P. F. Devaux, *Biophys. J.*, **61**, 347 (1992); U. Seifert, *Phys. Rev. Lett.*, **70**, 1335 (1993); F. Jülicher and R. Lipowsky, *Phys. Rev. Lett.*, **70**, 2964 (1993).

3. R. Golestanian, M. Goulian and M. Kardar, *Phys. Rev. E*, **54**, 6725 (1996); C. Hiergeist, V. A. Indrani, and R. Lipowsky, *Europhys. Lett.*, **36**, 491 (1996); R. Joannic, L. Auvray, and D. D. Lasic, *Phys. Rev. Lett.*, **78**, 3402 (1997).
4. M. Elbaum, D. K. Fygenson and A. Libchaber, *Phys. Rev. Lett.*, **76**, 4078 (1996).
5. W. Häckl, M. Bärmann and E. Sackmann, *Phys. Rev. Lett.*, **80**, 1786 (1998).
6. W. Winterhalter and W. Helfrich, *J. Phys. Chem.*, **96**, 327 (1992); J. Nardi, T. Feder, R. Bruinsma and E. Sackmann, *Europhys. Lett.*, **37**, 371 (1997); T. Chou, M. V. Jarić and E. D. Siggia, *Biophys. J.*, **72**, 2042 (1997).
7. S. J. Singer and G. L. Nicholson, *Science*, **175**, 720 (1975); *Molecular Biology of the Cell*, 3rd edn (eds. B. Alberts, D. Bray, J. Lewis, M. Raff, K. Roberts, J. D. Watson), Garland Publishing, New York, 1994.
8. J. Prost and R. Bruinsma, *Europhys. Lett.*, **33**, 321 (1996).
9. J. Prost, J-B. Manneville and R. Bruinsma, *Eur. Phys. J. B*, **1**, 465 (1998).
10. R. Henderson and P. N. T. Unwin, *Nature*, **257**, 28 (1975).
11. D. Oesterhelt and W. Stoeckenius, *Methods Enzymol.*, **31**, 667 (1974).
12. R. R. Birge, *Biochim. et Biophys. Acta*, **1016**, 293 (1990); D. Oesterhelt, C. Bräuchle and N. Hampp, *Quart. Rev. Biophys.*, **24**, 425 (1991); J. K. Lanyi, *Biochim. Biophys. Acta*, **1183**, 241–261 (1993).
13. M. I. Angelova, S. Soléau, Ph. Méléard, J. F. Faucon, and P. Bothorel, *Prog. Coll. Polym. Science*, **89**, 127 (1992).
14. D. Lévy, pers. comm.
15. J. Heberle and N. A. Dencher, *Proc. Natl. Acad. Sci. USA*, **89**, 5996 (1992).
16. E. Evans and W. Rawicz, *Phys. Rev. Lett.*, **64**, 2094 (1990).
17. E. Evans and D. Needham, *J. Phys. Chem.*, **91**, 4219 (1987).
18. W. Helfrich and R-M. Servuss, *Il Nuovo Cimento*, **3D**, 137 (1984).
19. G. Niggemann, N. Kummrow and W. Helfrich, *J. Phys II France*, **5**, 413 (1995).
20. B. Piknova, E. Perochon and J-F. Tocanne, *Eur. J. Biochem.*, **218**, 385 (1993); F. Dumas, M. Sperotto, M-C. Lebrun, J-F. Tocanne and O. Mouritsen, *Biophys. J.*, **73**, 1940 (1997).
21. K. Ohno, Y. Takeuchi and M. Yoshida, *Biochim. Biophys. Acta*, **462**, 575 (1977); T. Kouyama, R. A. Bogomolni and W. Stoeckenius, *Biophys. J.*, **48**, 201 (1985).
22. J-B. Manneville, P. Bassereau, D. Lévy and J. Prost, *Phys. Rev. Lett.*, **82**, 4356 (1999).
23. S. Ramaswamy, J. Prost and J. Toner, unpublished results.
24. W. Helfrich, *Z. Naturforsch.*, **28c**, 305 (1978).

# Chapter 28

# Entrapment of Proteins in Soybean Phosphatidylcholine Vesicles

*Masanao Imai*
*Nihon University, Tokyo, Japan*

*Kazuhito Nagayama, Hikaru Tanaka, Nobukazu Osaki and Toshifusa Doi*

*Kochi National College of Technology, Kochi, Japan*

## 1. INTRODUCTION

Artificial vesicles formed by phospholipids are of interest as models of biological membranes. The inner aqueous core and the lipid bilayer of vesicles are effective media for the entrapment of biomacromolecules. Various vesicle preparation methods have been reported [1–8], and the physical properties of the vesicles formed to some degree depend on the way they are prepared [9,10].

The volume of the inner aqueous phase is a significant factor in evaluating the entrapping ability of aqueous compounds; a vesicle system with a large internal volume can encapsulate macromolecules more efficiently than systems with smaller internal volumes. The work reported here investigates the protein-trapping capability and efficiency of three different types of vesicles formed by soybean phosphatidylcholine.

*Giant Vesicles* Edited by P. L. Luisi and P. Walde

## 2. MATERIALS AND METHODS

### 2.1 Chemicals

Soybean phosphatidylcholine was kindly provided by Nihon Sieber Hegner (Lucas Meyer, Hamburg). Lipase (from *Rhizopus delemar*) was purchased from Seikagaku Kogyo (Tokyo). Hemoglobin (bovine) and cytochrome *c* (horse heart), as well as Sepharose 4B and Sephadex G-75, were supplied by Sigma (St. Louis, MO). 3,3′-Bis (*N*,*N*-di(carboxymethyl)aminomethyl)fluorescein (calcein) was obtained from Wako Pure Chemical Industries (Osaka). All other chemicals were of analytical reagent grade.

### 2.2 Preparation of vesicles

Multilamellar vesicles (MLVs) were prepared as follows. Phosphatidylcholine (197 mg) was first dissolved in 10 $cm^3$ chloroform, after which the chloroform was removed by rotary evaporation under reduced pressure. An aqueous solution containing 10 mM phosphate buffer (pH 7.5) was then added, and MLVs were formed by shaking and mixing using a vortex mixer at 323 K.

Large unilamellar vesicles (LUVs) were prepared as described by Szoka *et al.* [3]. Phosphatidylcholine (67 mg) dissolved in 10 $cm^3$ chloroform was put in a 50 $cm^3$ round-bottom flask and the organic phase was then evaporated on a rotary evaporator under reduced pressure. The lipid layers were redissolved with 5 $cm^3$ diisopropyl ether; 1.7 $cm^3$ aqueous solution (10 mM phosphate buffer, pH 7.5) was added, and the mixture was sonicated for 3 min in a bath-type sonicator (B-32H [270 W]; Branson, Danbury, CT) at 323 K. The organic phase was then removed on a rotary evaporator under reduced pressure.

Giant vesicles (GVs) were prepared according to the literature [4]. A hexane solution containing 15 mg phosphatidylcholine was put in a flat-bottom flask (60–100 mm in diameter), and the organic phase was evaporated with nitrogen. An aqueous solution (8 $cm^3$ of 10 mM phosphate buffer, pH 7.5) was then added and the flask sealed under nitrogen. The system was left for 48 h at room temperature (298 K) for vesicle formation.

### 2.3 Measurement of the trapped volume in vesicles

The trapped volume in each vesicle system was evaluated according to the method of Oku *et al.* [5]. Calcein (6.3 mg) was dissolved in 100 $cm^3$ of 10 mM Tris–HCl buffer (pH 7.3), and each vesicle system was prepared using this solution instead of phosphate buffer. Then, 40 μl of the vesicle solution was diluted with 2 $cm^3$ of 10 mM Tris–HCl buffer. The fluorescence intensity was measured before and after the addition of 20 μl of 10 mM $CoCl_2$ solution. Finally, 20 μl of 20% Triton X-100 was added to the solution and the fluorescence intensity was measured again.

### 2.4 Measurement of protein trapping efficiency

To determine the protein trapping efficiency of each vesicle system, the protein trapped in the vesicles was separated from free protein by chromatography on a sephadex G-75 column (1.5 × 30 cm). The amount of free protein was measured with a UV–visible spectrophotometer (UV-1600; Shimadzu, Kyoto) at 410 nm in the case of cytochrome *c* and at 406 nm in the case of hemoglobin. Lipase was quantified by using a spectrofluorometer (FP-750; Jasco, Tokyo) at excitation and emission wavelengths of 280 nm and 335 nm, respectively.

## 3. RESULTS AND DISCUSSION

### 3.1 Formation of vesicles

A microscopic photograph of vesicles typical of the GV system is shown in Figure 28.1, in which spherical and elliptical vesicles with a multilipid layer are seen to be connected to each other. The diameters of the vesicles formed in the GV system were in the range 10–30 μm.

Figure 28.2 shows the elution profiles of the MLV and LUV systems on a sepharose 4B column. A profile similar to that of the MLV system has been reported in the literature [6]. It can be seen that the profile of the LUV system differed from that of the MLV system. Judging from the maximal absorbance of the two systems, the vesicles of the LUV system were larger than those of the MLV system. However, the vesicle size in both the MLV and LUV systems was estimated to be below 10 μm.

### 3.2 Trapped volumes in vesicle systems

As shown in Table 1, the trapped volumes of the LUV and GV systems were greater than that of the MLV system. The GV system was effective for obtaining a large inner volume at a lower phosphatidylcholine concentration. Trapped volumes similar to that observed in our LUV system have been reported for systems using phosphatidylglycerol, phosphatidylcholine from egg yolk, and mixtures of these with cholesterol [3].

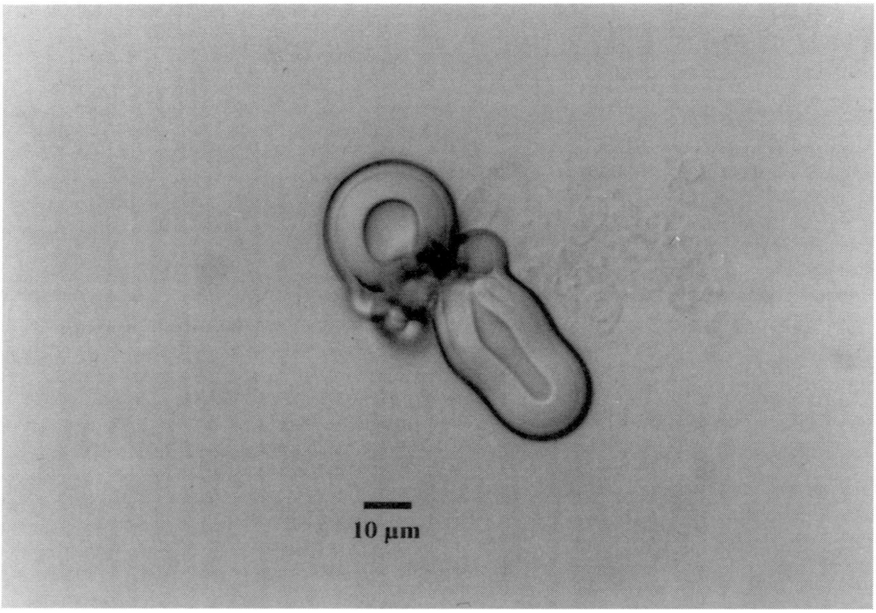


**Figure 28.1** Microscopic photograph of vesicles in the GV system.

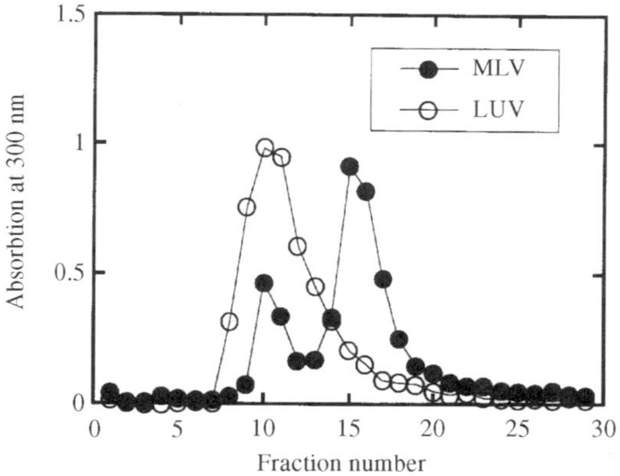


**Figure 28.2** Elution profiles of MLV and LUV systems on a Sepharose 4B column ($1.5 \times 30$ cm). The vesicle solution (7 $cm^3$) was applied onto the column and 2 $cm^3$ fractions were collected.

**Table 1** Comparison of trapped volumes in the three vesicle systems. The MLV and LUV systems were prepared at a lipid concentration of 50 mM and the GV system at 2 mM

| System | Trapped volume (μl/μmol lipid) |
|---|---|
| MLV | 2.0 |
| LUV | 7.2 |
| GV | 11.5 |

### 3.3 Protein trapping efficiency

The protein trapping efficiencies of the MLV and LUV systems were determined. Figure 28.3 shows the hemoglobin elution profiles for the LUV system and the buffer solution without vesicles. The profiles indicate that hemoglobin molecules were encapsulated in the LUV system. In addition, free hemoglobin could be separated from the vesicles because the absorbance of fraction 5, indicating protein trapped in the vesicles increased, whereas that of fraction 8, indicating free protein, decreased. Similar results were obtained with cytochrome *c* and lipase.

The protein trapping efficiency (Table 2) was determined using the protein concentration of each fraction as follows:

$$\text{trapping efficiency}(\%) = \frac{\text{amount of protein trapped in vesicles}}{\text{amount of protein in original solution}} \times 100$$

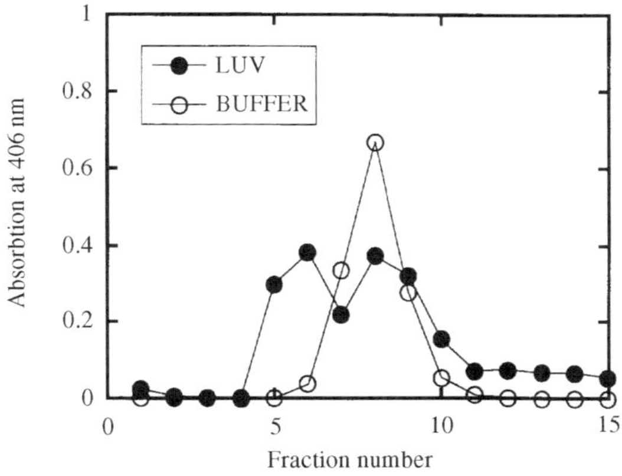


**Figure 28.3** Elution profiles of hemoglobin in the LUV system and buffer solution on a Sephadex G-75 column (1.5 × 30 cm). The vesicle solution (100 μl) containing 1 mg protein was applied onto the column and 4 cm$^3$ fractions were collected.

**Table 2** Comparison of protein trapping efficiency in the MLV and LUV systems

| Protein | Trapping efficiency (%) | |
|---|---|---|
| | MLV | LUV |
| Hemoglobin | 17 | 44 |
| Cytochrome *c* | 6 | 28 |
| Lipase | 12 | 35 |

Of the three test proteins used, hemoglobin was most effectively entrapped in the vesicle systems. In general, the trapping efficiency of proteins of the LUV system was about three-fold greater than that of the MLV system, which parallels the results of the trapped volume experiment. In terms of the ratio of LUV to MLV, the entrapment of cytochrome *c* was particularly effective in the LUV system.

## 4. CONCLUSIONS

Three different types of vesicles were formed – by the shaking method (MLVs), the reverse-phase evaporation method (LUVs), and the standing method (GVs)–using soybean phosphatidylcholine. The vesicles in the GV system were 10–30 μm in diameter, whereas in the MLV and LUV systems they were estimated to be below 10 μm. The trapped volumes of the LUV and GV systems were greater than that of the MLV system. The GV system was most effective for obtaining a large trapped volume at a low concentration of phosphatidylcholine. The trapping efficiency for hemoglobin, cytochrome *c*, and lipase was about three-fold higher in the LUV system compared to that of the MLV system.

## 5. ACKNOWLEDGMENTS

The authors are grateful to Nihon Sieber Hegner K. K. and Lucas Meyer Co. for providing phosphatidylcholine from soybean.

## 6. REFERENCES

1. A. D. Bangham, M. M. Standish and J. C. Watkins, *J. Mol. Biol.*, **13**, 238 (1965).
2. G. Adrian and L. Huang, *Biochemistry*, **18**, 5610 (1979).
3. F. Szoka and D. Papahadjopoulos, *Proc. Natl. Acad. Sci.USA*, **75**, 4194 (1978).
4. A. Darszon, C. A. Vandenberg, M. Schöeld, M. H. Ellisman, N. C. Spitzer and M. Montal, *Proc. Natl. Acad. Sci. USA*, **77**, 239 (1980).

5. N. Oku, D. A. Kendall and R. C. MacDonald, *Biochim. Biophys. Acta*, **691**, 332 (1982).
6. P. Bucher, A. Fischer, P. L. Luisi, T. Oberholzer and P. Walde, *Langmuir*, **14**, 2712 (1998).
7. M. I. Angelova and D. S. Dimitrov, *Prog. Colloid Polymer Sci.*, **76**, 59 (1988).
8. C.-H. Huang, *Biochemistry*, **8**, 344 (1969).
9. R. Wick and P. L. Luisi, *Chem. Biol.*, **3**, 277 (1996).
10. J. Suurkuusk, B. R. Lentz, Y. Barenholz, R. L. Biltonen, and T. E. Thompson, *Biochemistry*, **15**, 1393 (1976).

# Chapter 29

# Study on Stress-mediated Behavior and Preparation of Giant Vesicles

*TOSHINORI SHIMANOUCHI, HIROSHI UMAKOSHI AND RYOICHI KUBOI*

*Osaka University, Japan*

## 1. INTRODUCTION

Vesicles, which are self-assembled and closed phospholipid bilayers, have been widely studied as a typical model for the lipid matrix of the biological membrane and also as drug delivery systems [1,2], artificial cells [3], and for many other purposes. Among their various functions, the transport of proteins across the lipid membrane is an important issue.

To date, most research on translocation, including the lipid–protein interaction, has been on translocation by way of channels constituted by membrane proteins [4,5], on the signal peptide model [6], and on interaction with chaperones across the cytoplasmic membrane [7,8]. In all cases, the translocation of proteins is thought to be triggered by the hydrophobic interaction of the binding sites among targeted proteins, lipid membranes, and chaperones. The protein translocation may therefore be controlled by controlling the hydrophobic interaction. It is expected that the mechanism of protein translocation may be clarified by using the quantitative information on hydrophobic properties of such biomolecules.

From this viewpoint, the heat-induced translocation of cytoplasmic $\beta$-galactosidase across the inner membrane of *Escherichia coli* has been investigated [9–12]. The cell membrane [13,14] under stress condition has also been studied, where the possible roles of GroEL and other chaperones have also been discussed [12]. In order

*Giant Vesicles* Edited by P. L. Luisi and P. Walde

to clarify the function of the lipid membrane on the protein translocation, the heat-induced translocation of protein across phospholipid bilayer membrane of large unilamellar vesicles (LUV), in diameter 200 nm, has been investigated [15]. The phospholipid membrane has been also reported to assist the refolding of denatured protein of the partly unfolded states [16]. Based on previous results, the conformational change of proteins is confirmed to be required for translocation, refolding, and hydrophobic lipid–protein interaction. A giant unilamellar vesicle (GUV), that can be observed on the stage of a microscope, is thought to be a useful tool for the study of the above interactions because it can visualize the interaction between vesicle membrane and biomolecules, such as proteins, peptides, and other synthetic polymers. Therefore, through the study on the mechanism of the formation of the giant vesicles, the detailed mechanism of the translocation and refolding of proteins across and on vesicle membranes may be clarified. The electroformation method of Angelova and Dimitrov is suitable to monitor the formation process of GUVs on electrodes on the photomicroscopy stage [17], see also Chapters 3 and 4.

The major objective of this work was to investigate the effect of biomolecules such as proteins and enzymes under stress conditions on GUV formation. First, the preparative behavior of GUV was observed in the presence of proteins, polymers and salts. Second, the surface properties of vesicles and proteins at various conditions were evaluated by using the aqueous two-phase partitioning method [18]. Then the effect of stressed proteins against GUV formation was discussed in relation to the hydrophobic surface properties of vesicles and proteins.

## 2. MATERIALS AND METHODS

### 2.1 Materials

The phospholipid used was 1-palmitoyl-2-oleoyl-*sn*-glycero-3-phosphocholine (POPC) (Avanti Polar Lipids, Birmingham, U.K.). Poly(ethylene glycol) (PEG1540, 4000, and 6000, MW = 1 k, 3 k, and 6 kDa) and dextran (Dex)100–200 k (MW = 100–200 kDa) were obtained from Wako Pure Chemicals Ltd. (Osaka, Japan). $\alpha$-glucosidase from human liver, $\beta$-galactosidase from *Escherichia coli* and carbonic anhydrase from bovine (CAB) were purchased from Sigma (MW = 80 k, 464 k and 30 kDa). All other reagents were of analytical grade.

### 2.2 Preparation of giant vesicles by electric field

The lipids were dissolved in diethyl ether : chloroform 9 : 1 (v/v) solvent. Two drops of this solution (1 µl each) were deposited on two parallel platinum electrodes (diameter 0.2 mm, separation distance 3 mm). The solvent was then evaporated under nitrogen. An electric field (1.5 V, 10 Hz) was applied and distilled water or aqueous solution added. The observations were performed under a phase contrast inverted microscope [17,19]. In some cases, a fluorescent probe, octadecylrhodamine B (R-18), was added to improve visualizing the thin-walled vesicles.

### 2.3 Membrane fluidity

The membrane fluidity of liposome was determined as the fluorescence polarization $P$ using the fluorescence of DPH (1,6-diphenyl-1,3,5-hexatriene) [20] with a spectrofluorometer (FP2010, JASCO). Membrane fluidity was defined as the reciprocal of polarization, $1/P$, of the hydrophobic fluorescence probe DPH.

### 2.4 Measurement of local hydrophobicity

Local hydrophobicities (LH), of various proteins were evaluated by using the aqueous two-phase partitioning method. The basic composition of the systems was PEG4000 (9 wt %):Dex100 k–200 k (9 wt %). The difference of partition coefficients of various proteins in the phase systems with and without 1 mM Triton X-405 gives their local hydrophobicities [18,21].

## 3. RESULTS AND DISCUSSION

### 3.1 Osmolarity and salt effect

The formation of GUV under various stress conditions such as osmolarity and salt was investigated in order to verify the effect on the formation of GUV under the neutral pH condition (pH 7.5). Micrographs of GUV formed on electrodes are shown in Figure 29.1. The diameters of these GUVs were measured by using a software HPD-CP extended from HAMAMATSU PHOTONICS (Japan). Figure 29.2 shows the variation of the size distribution of vesicles in the presence of PEG and $Na_2SO_4$ with their molecular weights or concentrations. At all conditions (with PEG and salts), the GUV formation was found to be inhibited. These results were in good agreement with the previous reports by Angelova and Dimitrov [17]. Moreover, the variation of the mean GUV diameter was also investigated in the presence of various additives at pH 7.5. Figure 29.3 shows the effect of additive concentration on the final diameter of the giant vesicles. In the case of PEG, dextran and salts, such an effect was detected. The formation of GUV was inhibited by increasing the concentration of polymers and salts. However, no inhibitory effect was detected in the case of biomolecules such as $\alpha$-glucosidase, $\beta$-galactosidase, and CAB.

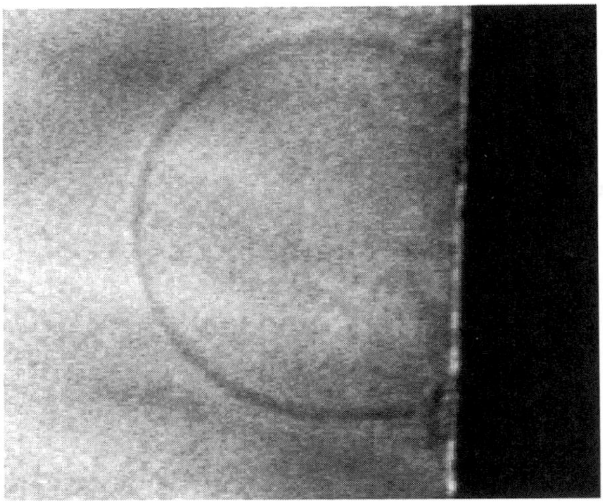

**Figure 29.1** Micrographs of giant vesicles formed on electrodes. Giant vesicles were prepared in aqueous solution of CAB by applying an electric field of 1.5 V at 10 Hz.

## 3.2 pH stress against the formation of GUV

After the preparation of GUV was performed using protein solutions with respective pH values, the diameters of generated GUVs were measured. The size distributions of GUVs at various pH in the presence of CAB are shown in Figure 29.4. No noteworthy shift of peaks was recognized at pH 2 or 7.5. The distribution peak at pH 4 shifted to the larger diameter range. In the case of α-glucosidase and β-galactosidase, similar results were obtained (data not shown). Mean diameters of GUVs, $D_{mean}$, reached a maximum at pH 4 in the presence of CAB. A slight increase in the mean diameter $D_{mean}$ was also observed with decreasing the pH values in the

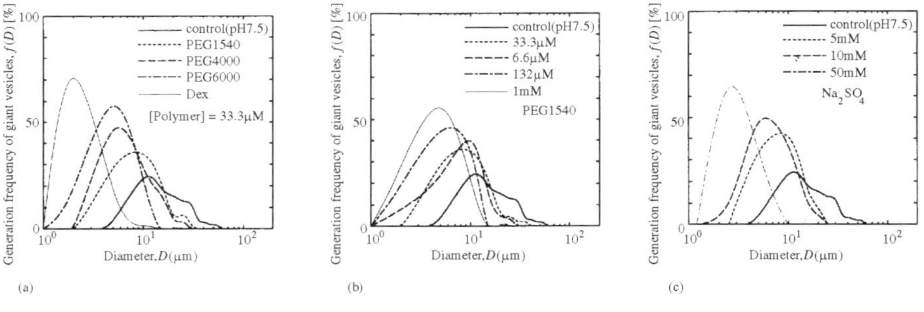


**Figure 29.2** Size distribution of GUVs in the presence of (a) various polymers, (b) PEG1540 and (c) $Na_2SO_4$. An electric field (1.5 V, 10 Hz) was applied to the chamber. In the respective experiment, diameters of more than 500 vesicles were measured.

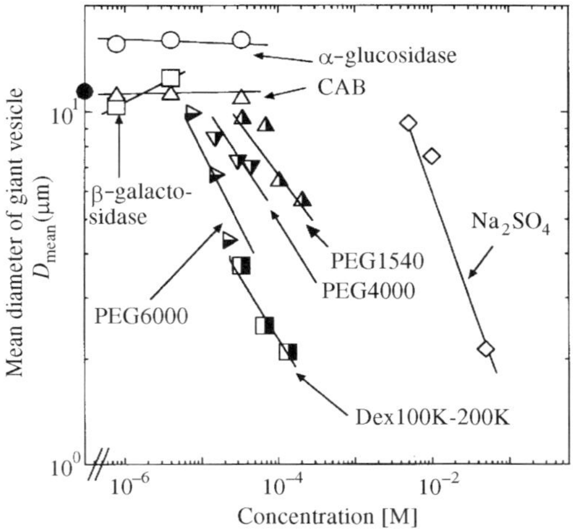


**Figure 29.3** Concentration dependency of the mean diameter of GUVs in the presence of proteins, synthetic polymers, and salts. $Na_2SO_4$ was employed as salt stress. An electric field (1.5 V, 10 Hz) was applied to the chamber.

presence of α-glucosidase. In the case of β-galactosidase, no significant change of $D_{mean}$ was observed, in contrast to that of CAB.

### 3.3 Surface properties of various proteins under stress

An important factor for the enhancement of GUV formation is likely to be stress-induced change in the surface properties, followed by their conformational changes.

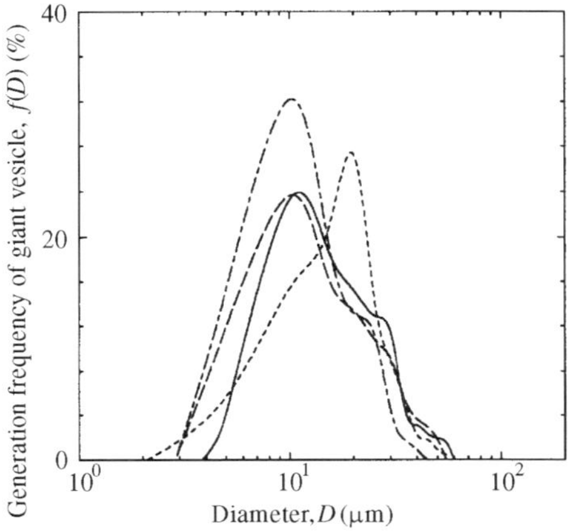


**Figure 29.4** Size distribution of GUVs under various pH stress conditions. The pH range selected was from 2 to 7.5. An electric field (1.5 V, 10 Hz) was applied to the chamber. Experiments performed were: (——) control, (---) CAB at pH 2, (– – –) pH4, (——) pH 7.5.

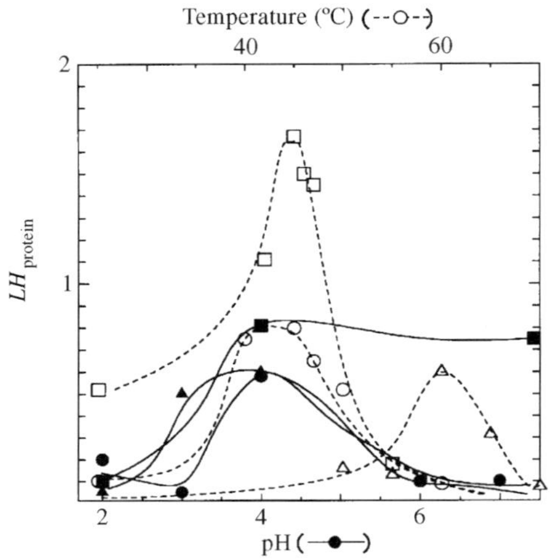


**Figure 29.5** The pH and temperature dependency of local hydrophobicity (LH) for various proteins. The pH range selected was 2–7.5. LH values were measured by the aqueous two-phase partitioning method [18,21]. Symbols are (○,●) α-glucosidase, (□,■) β-galactosidase and (△,▲) CAB. Open and closed symbols indicates heat and pH stresses, respectively.

The behaviors of their conformational changes were investigated after various proteins were added to the solutions in the pH range 2–7.5. The variation of surface properties (local hydrophobicity, LH) caused by such conformational change were analyzed previously using the aqueous two-phase partitioning method [17,21]. LH values of protein were plotted against pH and temperature as shown in Figure 29.5. The LH values of CAB (closed symbol) reached a maximum value at pH 4. The behaviors of protein LH values have already been reported to give corresponding their conformational states [15]. In the case of heat stress, the LH value (open symbol) of the intermediate (molten globule) state is much higher than that of other native and unfolded states. The CAB, having the higher LH value around pH 4, is therefore thought to be a molten globule state. The protein of such a state has been confirmed to interact easily with the hydrophobic binding sites of the vesicle membrane surface. These results are in good agreement with those obtained by Shin *et al.* [22]. As the strong hydrophobic interaction between vesicle membranes and proteins at the molten globule state is induced, such interactions are thought to be related to the GUV formation behavior in the presence of proteins.

### 3.4 Surface properties of vesicle membranes

In order to verify the surface property of vesicle membrane, the membrane fluidity ($1/P$) was investigated. As shown in Figure 29.6, the $1/P$ values were plotted against

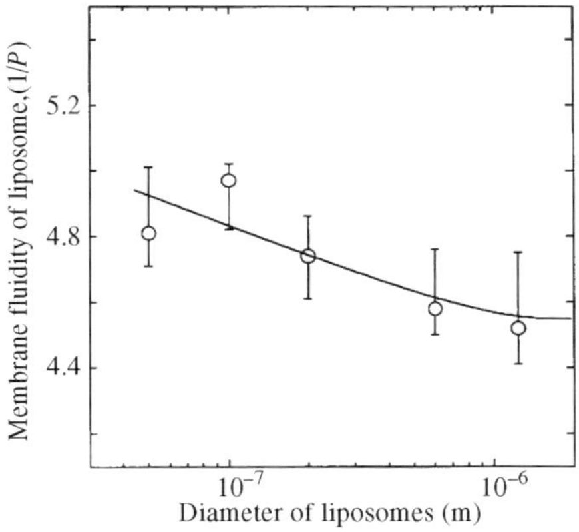


**Figure 29.6** Diameter dependency of membrane fluidity of vesicles. Membrane fluidity was defined as the reciprocal of polarization of hydrophobic fluorescence probe DPH (diphenyl hexatiene). The lipid : DPH molar ratio was maintained at 250 (([DPH]$_{final}$ = 0.4 mM). Exciting at 360 nm, the emission at 430 nm was measured as the fluorescence intensity.

the diameter of GUVs. In the diameter range of GUVs (above $10^{-6}$ m), the membrane fluidity may be regarded as constant. The temperature dependency of membrane fluidity previously reported shows a linear increase with increasing heating temperature, and that heat stress modifies the physical state of vesicle membrane, especially the membrane fluidity, and that the increase in membrane fluidity corresponds well with the increase in surface hydrophobicity of the membrane surface [15].

### 3.5 Effect of vesicle–protein interaction on GUV formation under stresses

The variation of GUV formation behavior in the presence of proteins is thought to be a kind of stress response of the vesicle membrane due to vesicle–protein interaction. This is on the basis of the hydrophobic interaction, and the strength can be estimated by the product of LH values of vesicle membrane and those of proteins. As LH is the dynamic property, the LH of the vesicle membrane corresponds well with the membrane fluidity ($1/P$). The corresponding LH values of GUVs of specific diameter can, therefore, be obtained by extrapolating the $1/P$ values. The values of mean diameter, $D_{mean}$, were plotted against the product of both LH values under pH stress (Figure 29.7), together with the maximum translocated amount of various proteins, $TR_{max}$, under heat-stress conditions. As reported previously, translocation phenomena are also a kind of stress response of the vesicle membrane induced by the heat stress [15], and both $D_{mean}$ and $TR_{max}$ are related to the product of their LH values and these results were comparable with those of the heat-induced

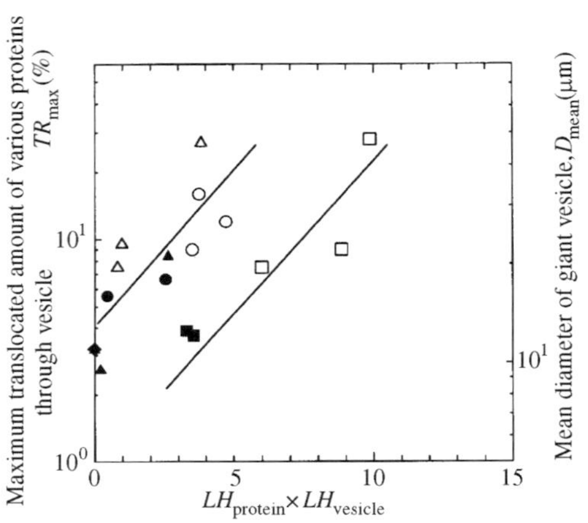


**Figure 29.7** Effect of vesicle–protein interaction against the GUV formation and heat-induced translocation across phospholipid bilayer. The membrane fluidity was used as the vesicle LH value as it was linearly correlated with the LH [16]: (○,●) α-glucosidase, (□,■) β-galactosidase, and (△,▲) CAB. Open and closed symbols indicates heat and pH stresses respectively.

translocation of proteins across the phospholipid bilayer membrane, suggesting that enhancement of GUV formation is due to the stress response based on the hydrophobic interaction between vesicles and proteins.

## 4. CONCLUDING REMARKS

The formation of GUVs in the presence of the various additives under various stress conditions has been studied quantitatively. The presence of CAB in the molten globule state at pH 4 was observed to favor the formation of the larger size of giant unilamellar vesicles. The hydrophobic interaction between vesicles membranes and proteins was found to be a key factor for the formation.

## 5. REFERENCES

1. C. Diagnostic and O. Potential, *Drugs*, **45**, 15 (1993).
2. R. Ho, B. T. Rouse and L. Huang, *Biochemistry*, **25**, 5500 (1986).
3. T. Oberholzer, T. M. Albrizo and P. L. Luisi, *Chem. Biol.*, **2**, 677 (1995).
4. R. N. McElhaney, *Biochim. Biophys. Acta*, **864**, 361 (1986).
5. R. George, R. N. A. H. Lewis and R. N. McElhaney, *Biochem. Cell Biol.*, **68**, 161 (1990).
6. G. Blobel and B. Dobberstein, *J. Cell. Biol.*, **67**, 835 (1975).
7. D. N. Collier, V. A. Bankaitis, J. B. Weiss and P. J. Bassford, *Cell*, **53**, 273 (1988).

8. E. Crooke, L. Brundage, M. Rice and W. Wickner, *EMBO. J.*, **7**, 1831 (1988).
9. M. B. Yatvin, *Biochim. Biophys. Acta.*, **901**, 147 (1987).
10. M. B. Yatvin, K. M. Smith and F. L. Siegel, *J. Biol. Chem.*, **261**, 8070 (1986).
11. M. Kitagawa, Y. Matumura and T. Tsuchido, unpublished results.
12. H. Umakoshi, R. Kuboi, I. Komasawa and T. Tsuchido, *Biotechnol. Prog.*, **14**, 218 (1998).
13. H. Umakoshi, R. Kuboi and I. Komasawa, *Solv. Extr. Res. Dev.*, **3**, 74 (1996).
14. H. Umakoshi, R. Kuboi and I. Komasawa, *J. Ferment. Bioeng.*, **84**, 572 (1997).
15. H. Umakoshi, M. Yoshimoto, T. Shimanouchi, R. Kuboi and I. Komasawa, *Biotechnol. Prog.*, **14**, 218 (1998).
16. R. Kuboi, M. Yoshimoto, P. Walde and P. L. Luisi, *Biotechnol. Prog.*, **13**, 828 (1997).
17. M. I. Angelova and D. S. Dimitrov, *Progr. Colloid. Polym. Sci.*, **76**, 59 (1988).
18. R. Kuboi, K. Yano, H. Tanaka and I. Komasawa, *J. Chem. Eng. Jpn.*, **26**, 286 (1993).
19. R. Wick, M. I. Angelova, P. Walde and P. L. Luisi, *Chem. Biol.*, **3**, 105 (1996).
20. A. G. MacDonald, K. W. J. Wahle, A. R. Cossines and M. K. Behan, *Biochim. Biophys. Acta*, **938**, 231 (1988).
21. R. Kuboi, K. Yamahara and H. Ota, *J. Chem. Eng. Jpn.*, **30**, 1119 (1997).
22. I. Shin, D. Kreimer, I. Silman and L. Weiner, *Proc. Natl. Acad. Sci. USA*, **94**, 2848 (1997).

Chapter 30

# Molecular Organization on Giant Unilamellar Vesicles

*Susanne Gangl, Susanne Stark, Peter Mayrhofer, Dominik Rünzler, Caroline Thurner, Arno Lukas, Bernd Mayer and Gottfried Köhler*
*University of Vienna, Austria*

## 1. INTRODUCTION

Molecular organization at membraneous interfaces is a key element in biochemical organization and provides potential for application in biomolecular materials design [1]. The conceptual basis for the formation of such higher-order chemical aggregates with defined functionality is provided by emergent design theory [2,3], denoting the minimal object complexity of chemical elements which is sufficient to develop functionality in a cooperative and dynamic manner.

Giant unilamellar vesicles (GUVs) [4,5] are an excellent substrate with which to study molecular organization at membraneous interfaces and to provide realistic and easy-to-handle cell-like compartments. This chapter presents the docking and insertion of various chromophors to study the properties of the membranes themselves, and the attachment of helical peptides mediating GUV clustering.

## 2. EXPERIMENTAL SYSTEM

### 2.1 Chemicals

1-Palmitoyl-2-oleoyl-*sn*-glycero-3-phosphocholine (POPC) and egg yolk phosphatidylcholine (egg-PC) were from Sigma (USA). Anilinonaphthalene-1-sulfonic acid

*Giant Vesicles* Edited by P. L. Luisi and P. Walde

ammonium salt (ANS) was from Fluka (Switzerland), 7-diethylamino-4-trifluoromethylcoumarin (coumarin 6) was from Radiant dyes Chemie (Germany), 7-diethylamino-3-(2-benzothiazolyl)-coumarin (coumarin 35), and *meso*-tetraphenylporphyrine (TPP) were from Sigma (USA), *N*-(5-dimethylaminonaphthalene-1-sulfonyl)-*sn*-glycero-3-phosphoethanolamine, triethylammonium salt from egg phosphatidylethanolamine (dansyl egg PE) and lissamine rhodamine B 1,2-dihexadecanoyl-*sn*-glycero-3-phosphoethanolamine, triethylammonium salt (rhodamine DHPE) were from Molecular Probes (USA).

### 2.2 Experimental set-up

Vesicle formation and incorporation of probes and peptides were studied with an Axiovert S100TV inverted light microscope from Zeiss (Germany), using phase contrast and fluorescence mode and ×20 or ×40 long working distance Achroplan lenses. Images were recorded with an intensified CCD camera ITE/CCD-576 from Princeton Instruments (USA), pictures were recorded using a PC with WinView software and processed by X-View and Adobe Photoshop Pro. The AC electric field applied for GUV formation was produced by the function generator 3311A, controlled by an oscilloscope (HF 54601 A), both from Hewlett Packard (USA).

### 2.3 GUV formation

Three microliters of a 260 μM phospholipid solution (0.2 mg of lipid per ml in diethyl ether : methanol = 9 : 1, v/v) were deposited on both platinum wires of the investigation chamber. The lipid film was dried by first blowing a stream of nitrogen over the wires for about 1 min, followed by storage under reduced pressure for at least 1 h.

The platinum wires of the preparation cell [5] were connected on the stage of the light microscope with the function generator, 1 ml of an aqueous solution was added, and the electric field was switched on (2 V at 10 Hz for POPC, and 3 V at 15 Hz for egg yolk phosphatidylcholine).

## 3. RESULTS

### 3.1 Chromophore–GUV interaction

The following fluorescent molecules were used to study GUV membrane properties: ANS, coumarin 6, coumarin 35, and the labeled lipids rhodamine DHPE and dansyl egg PE (Figures 30.1 and 30.2). Such probes are excellent tools with which to analyse membrane dynamics as well as to monitor the interaction of membranes and proteins.

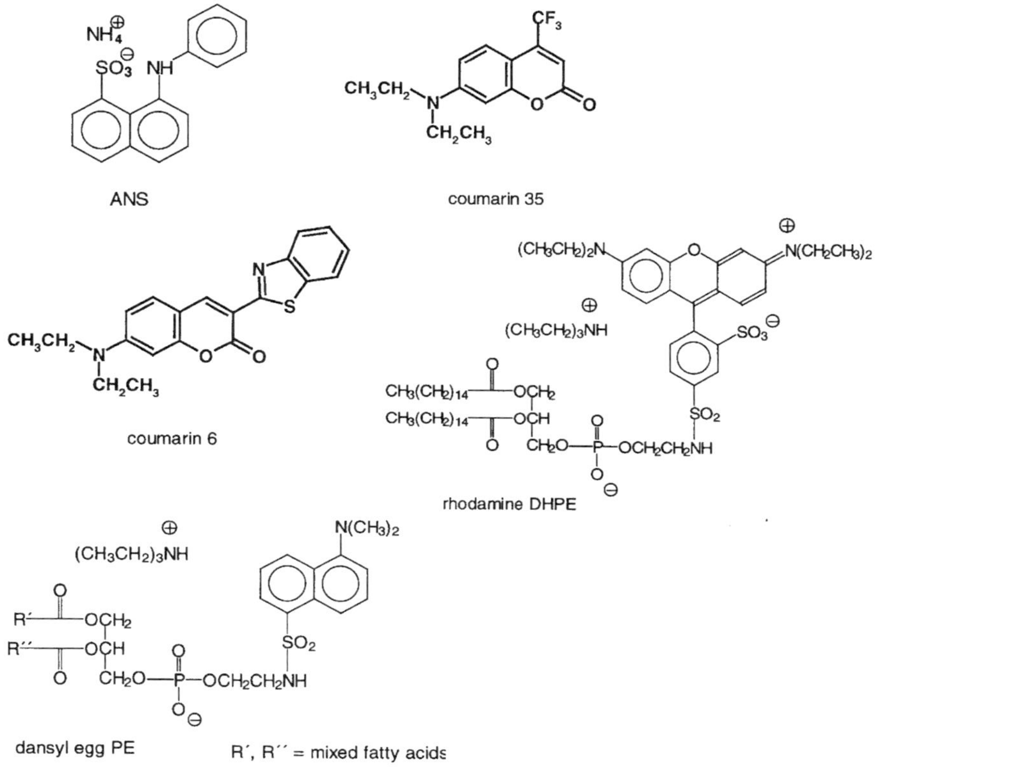


**Figure 30.1** Chemical structures of the fluorescent dyes

The incorporation of chromophores was achieved by either codissolving the probe and the organic solution of the lipids before the lipid film is deposited on the platinum wires, or by adding 10–80 μl of a concentrated solution of the probe in water or ethanol to the preformed GUVs.

The use of fluorescent probe molecules results in information on the molecular environment in the vicinity of the chromophore, interpreting spectral shifts, changes in fluorescence lifetimes and fluorescence polarization.

ANS is essentially not fluorescent in a polar solvent such as water, but the fluorescence quantum yield increases considerably in an apolar environment, such as within membranes (Figure 30.2(A)). Therefore ANS is a perfect label for such artificial membranes [6].

Both coumarins (Figures 30.2(B) and (C)) are located near the lipid–water interface with spectral shifts and fluorescence lifetimes similar to their solutions in ethanol or acetone [7]. The fluorescence properties of these polar probes are strongly

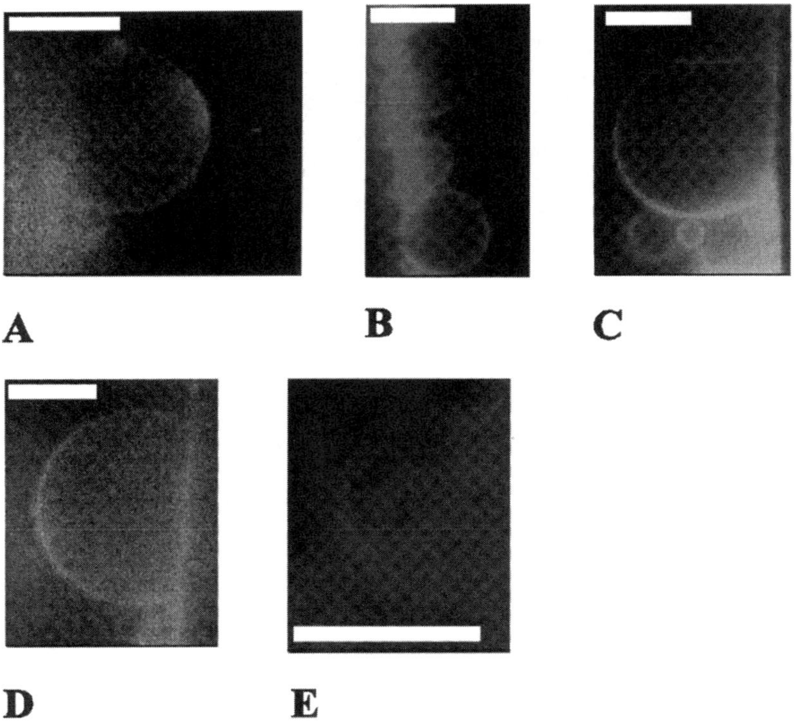


**Figure 30.2** Fluorescence images of POPC GUVs (bar = 50 μm) labeled with: (A) ANS ($6 \times 10^{-5}$ M), (B) coumarin 6 ($1 \times 10^{-6}$ M), (C) coumarin 35 ($6 \times 10^{-6}$ M). Fluorescence images of POPC GUVs and the following colipids: (D) dansyl-egg PE (mol ratio POPC : labeled lipid = 40 : 1), (E) rhodamine DHPE (mol ratio POPC : labeled = lipid 270 : 1).

dependent on the local environment and they thus provide specific information on the dynamics in the polar head group region of the bilayer.

Dansyl-labeled lipids (Figure 30.2(D)) are useful probes for investigating the assembly of proteins on and within membranes, either by monitoring the spectral shift of dansyl fluorescence or via fluorescence resonance energy transfer measurements using protein-inherent tryptophan as donor [9]. Rhodamine DHPE (Figure 2(E)) does not easily transfer between separated lipid bilayers. Therefore this label is extensively used for membrane fusion assays [8].

### 3.2 Peptide–GUV interaction

Alanine-based peptides of the form [AKAAAAKAAAAKAAAAW] were used to study (i) peptide fold into an α-helical structure, (ii) attachment to membranes, and

(iii) membrane insertion and oligomerization. Such peptides represent a good model for the formation of transmembrane protein aggregates, such as channel elements. The peptides given above form stable α-helices in a polar environment as shown by CD spectroscopy [10] and extended conformational analysis [11].

Figure 30.3 shows the docking of the model peptide to POPC GUVs. Figure 30.3(A) gives the fluorescence image of the dansyl-labeled model peptide attached at the membrane–water interface. Although the peptide is, due to charged lysine residues at pH 7, highly soluble in water, a sufficient electrostatic interaction between lipid head groups and peptide mediates stable adhesion. The docking of the peptide is monitored by the location of fluorescent signals at the GUV boundary as well as by the increase of dansyl fluorescence due to the decreased polarity at the membrane–water interface compared to bulk water.

Longer incubation of the GUV–peptide mixture results in the higher-order membrane cluster products shown in Figure 30.3(B) and (C). One possible explanation of this effect is complementary charge saturation of charged lysines and the lipid phosphate groups. The helical structure of the peptide results in a stacked order of the lysines (Figure 30.3(D)) which would mediate membrane–membrane adhesion.

Aggregation of the vesicles can also be interpreted in the light of results on lamellar-to-inverted hexagonal phase transitions in phosphatidylethanolamine, as obtained recently [12]. The initial step in the process of membrane fusion must be similar to the first step in the $L_\alpha$–$H_{II}$ phase transition: small connections between opposed membranes are formed in this process, and subsequently convert into fusion pores. The structure of the first intermediates in membrane fusion has been predicted to be that of the 'stalk', which is thought to transform into a hemifusion intermediate (also called a transmonolayer contact (TMC)). These stalks and hemifusion intermediates have also been proposed as intermediates in protein-mediated fusion [13].

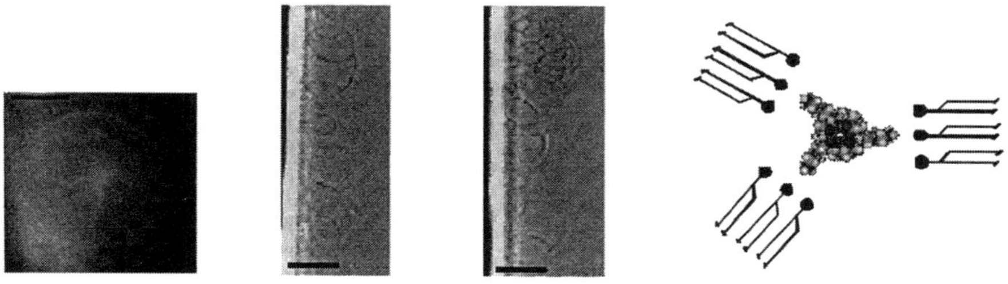

A B C D

**Figure 30.3** (A) Fluorescence image of a POPC GUV and the membrane-attached dansyl-labeled model peptide (molar ratio POPC : peptide = 30 : 1, bar = 50 μm). (B) and (C) Phase contrast image of peptide-mediated GUV–GUV clustering (incubation time > 45 min, bar = 50 μm). (D) Diagram of peptide (helical wheel representation) – membrane interaction triggering GUV–GUV attachment.

The peptides might lower the rupture tension of the TMC diaphragm. This probably occurs through a peptide effect on the cohesion of the lipid membrane (for example through a local change of bilayer thickness around an adsorbed peptide).

## 4. CONCLUDING REMARKS

Due to their size and curvature, giant unilamellar vesicle (GUV) membranes exhibit dynamical properties comparable to real cells. GUV model membranes are therefore most suitable for the study of structural as well as dynamical properties of general membrane–guest systems.

Peptide–membrane interactions, as discussed in this paper, represent a major functional class of biochemical systems. Spectroscopic as well as microscopic techniques on GUV membranes, applied via fluorescent probes as reporter molecules, provide a wide spectrum of analysis to better understand these complex multicomponent processes.

## 5. ACKNOWLEDGEMENT

S. G. and B. M. are indebted to the Austrian Academy of Sciences for financial support.

## 6. REFERENCES

1. H. C. Hoch and L. W. Jelinski, *Integrating Materials Science, Engineering and Biology*, (ed.) H. C. Craighead, Cambridge University Press, 1996.
2. B. Mayer and S. Rasmussen, in *Artificial Life VI*, (eds) C. Adami, R. Belew, H. Kitano and C. Taylor, MIT Press, 1998, p. 123.
3. N. A. Baas, in *Artificial Life III*, Proceedings of the SFI Studies in the Science of Complexity, (ed.) Ch. G. Langton, Vol. XVII, 1993, p. 515, Addison-Wesley.
4. D. S. Dimitrov and M. I. Angelova, *Progr. Colloid. Pol. Sci.*, **73**, 48 (1987).
5. R. Wick, M. I. Angelova, P. Walde and P. L. Luisi, *Chem. Biol.*, **3**, 105 (1996).
6. J. Slavic, *Biochim. Biophys. Acta*, **694**, 1 (1982).
7. K. Rechthaler and G. Köhler; *Chem. Phys.*, **189**, 99 (1994).
8. P. Vanderwerf and E. F. Ullman, *Biochim. Biophys. Acta*, **596**, 302 (1980).
9. P. Wu and L. Brand, *Anal. Biochem.*, **218**, 1 (1994).
10. S. Marqusee, V. H. Robbins and R. L. Baldwin, *Proc. Natl. Acad. Sci. USA*, **86**, 5286 (1989).
11. Ch. Th. Klein, B. Mayer, G. Köhler and P. Wolschann, *J. Mol. Struct. (Theochem)*, **370**, 33 (1996).
12. D. P. Siegel and R. M. Epand, *Biophys. J.*, **73**, 3089 (1997).
13. G. B. Melikyan and L. V. Chernomordik, *Trends Microbiol.*, **5**, 349 (1997).

# Chapter 31

# Membrane Properties of Archæal Phospholipids: Effect of Macrocyclization

*OLIVIER DANNENMULLER, YOICHI NAKATANI AND GUY OURISSON*
*Université Louis Pasteur, Strasbourg, France*

*KENJI ARAKAWA, TADASHI EGUCHI AND KATSUMI KAKINUMA*
*Tokyo Institute of Technology, Japan*

*SYLVIE BLANC AND ANNE-MARIE ALBRECHT*
*Ecole Européenne de Chimie, Polymères et Matériaux, Strasbourg, France*

## 1. INTRODUCTION

Archæa, the third major kingdom of living organisms, have attracted much interest as their cell structures and biocomponents are markedly different from those found in Bacteria and Eucarya. Archæal membrane lipids in particular have been studied extensively because of their characteristic chemical structures: the hydrophobic part is composed of $C_{20-25}$ polyprenyl chains linked by ether functions to a variety of head groups, based on glycerol or other polyols and phosphoric acid derivatives. They sometimes comprise macrocyclic structures obtained by $\omega,\omega'$-dimerization of $C_{20}$ chains into a $C_{40}$ one, the termini of which are linked either to the same, or to both head groups, thus giving either 36- or 72-membered rings [1–3].

*Giant Vesicles* Edited by P. L. Luisi and P. Walde

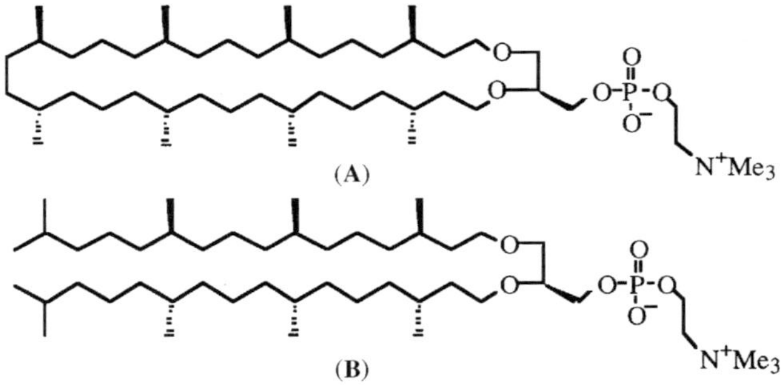


**Figure 31.1** Structures of archæal-type phosphatidylcholines

Of interest are the roles of the characteristic macrocyclic ring structure in the archæal lipids, in particular the 36-membered ring, the presence of which in a new deep-sea hydrothermal vent methanogen, *Methanococcus jannaschii*, was reported by Comita *et al.* [3]. A $\sigma$-bond formation between the termini of two chains could affect the conformation (*trans* : *gauche* ratios), rotational motions, and lateral diffusion of the chains. Consequently, there would be a change in three-dimensional structure, lipid packing, membrane permeability and stability.

Phosphatidylcholine (**A**), bearing the natural archæal $C_{36}$ macrocyclic glycerol skeleton, has been synthesized recently (Figure 31.1) [4,5]. A comparison is given here of its physicochemical properties with those of its non-cyclized analogue (**B**), carrying the common basic skeleton of membrane lipids in *Halobacteria* [2]. The structures obtained in water include vesicles large enough to be observed by optical microscopy. Various methods were employed to study the membrane properties of the vesicles of these phosphatidylcholines: Nomarsky and confocal optical microscopies, a stopped-flow light-scattering method, and NMR (not presented here).

## 2. RESULTS AND DISCUSSION

### 2.1 Light microscopies

Using video-enhanced optical (Nomarski type) [6] and confocal [7] microscopies, different types of self-organized structures obtained from a water suspension of archæal phospholipids were observed at 298 K. Vesicles formed at the edges of the solid. For instance, uni- and multilamellar vesicles (diameter = 0.2–20 μm), oligovesicular vesicles, tubules, and membrane aggregates were seen. Nile Red as a neutral fluorescent hydrophobic membrane probe (1 mol %) was especially useful in observing various types of morphological changes such as transformation of a tubular structure into a vesicle, fusion, and fission of vesicles. Unilamellar vesicles and tubules were observed in addition to the multilamellar structures shown in Figure 31.2. Membranes appear to be thick because of the diffraction-limited resolution of the optical system. The transformation of a tubular structure to an oligo-vesicular vesicle was observed: a tubular vesicle at 0 s; bending at 17 s; fusion at 25 s; oligovesicular vesicle formation at 32 s. Simple thermal convection in the sample solution, even between microscope slide and coverglass, is apparently enough to bring about such a transformation.

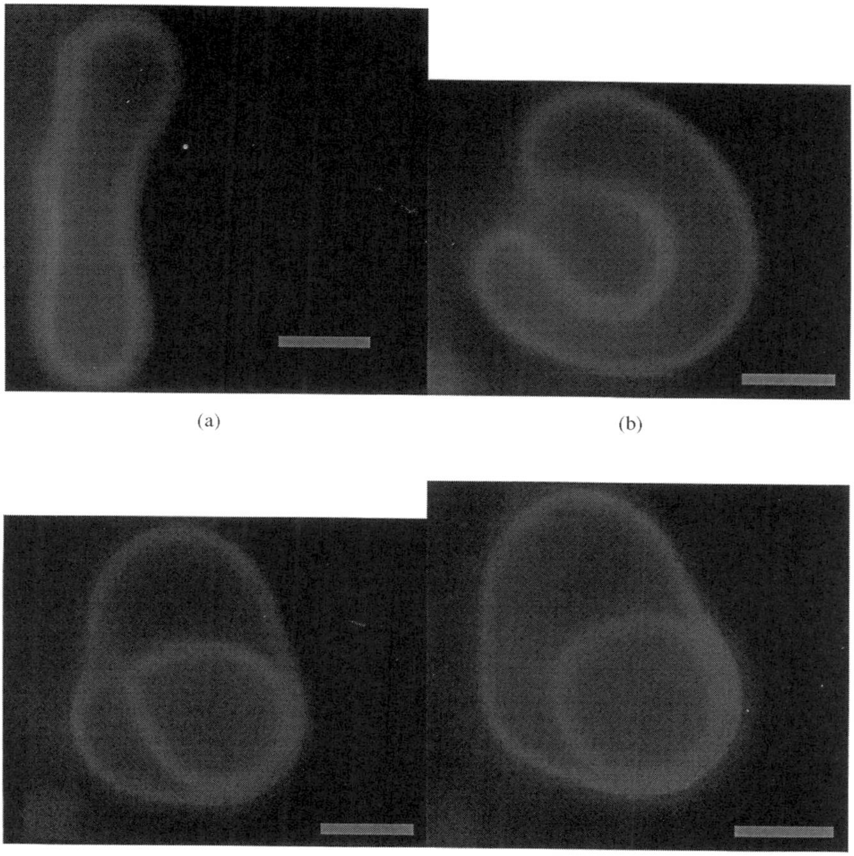

(a) (b) (c) (d)

**Figure 31.2** Observation by differential interference contrast (Nomarski) fluorescence optical microscopy of the transformation from a tubular structure to an oligovesicular structure of phospholipid (**A**) suspended in a phosphoric buffer ($T = 298$ K, $pH = 7.0$, $\lambda_{ex} = 485$ nm, $\lambda_{em} = 525$ nm, Nile Red 1 mol %), scale bar 10 μm: (a) 0 s, (b) 17 s, (C) 25 s, (d) 32 s.

The three-dimensional structure of an ellipsoidal vesicle was investigated by fluorescence (Nile Red) confocal microscopy. The ellipsoidal wall structure with an inner water phase was confirmed by the cross-sections of *xy*, *yz*, and *zx* planes (Figure 31.3(a)). The cross-section images along the *xy* plane at different *z* values are shown in Figure 31.3(b). The images confirm a quasi-spherical morphology of the vesicle. Top and side views of the vesicle obtained from the cross-sectional analysis are represented in Figures 31.3(c) and (d).

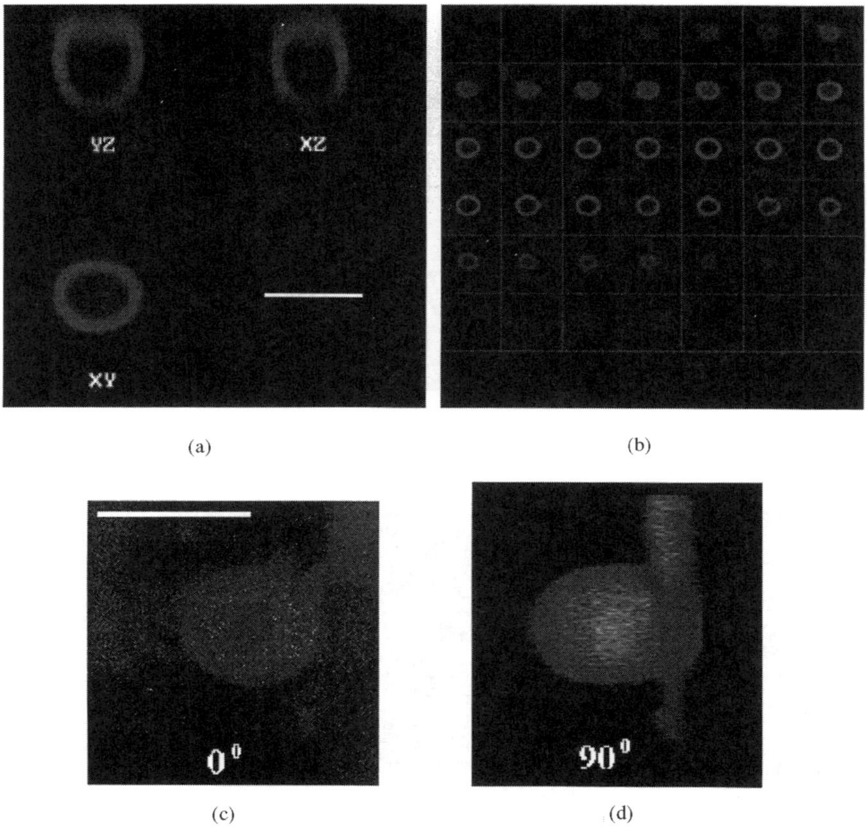


**Figure 31.3** Fluorescence confocal microscopy images of an ellipsoidal vesicle (5 μm small axis, 8 μm large axis) of phospholipid (**B**) (T = 298 K, pH = 7.0, $\lambda_{ex} = 551$ nm, $\lambda_{em} = 636$ nm, Nile Red 1 mol %). Also seen is the nonincorporated hydrophobic fluorescent probe which spreads on the glass slide. (a) Vesicular confocal section in three perpendicular planes, scale bar 5 μm. (b) IR serial vesicular optical sections at 0.2 μm incremental depth in the *xy* plane. (c) and (d) Face (0°) and profile (≈90° rotation) views of the three-dimensional reconstruction, scale bar 5 μm.

**Table 1** Water permeability ($t_{1/2}$) for vesicles of a given phosphatidylcholine at $(33.0 \pm 0.1)$°C

| Lipids | Swelling experiments | | | Shrinking experiments | | |
|---|---|---|---|---|---|---|
| | Vesicle mean diameter (SD) (nm) | Rate constant ($s^{-1}$) | $t_{1/2}$ (ms) | Vesicle mean diameter (SD) (nm) | Rate constant ($s^{-1}$) | $t_{1/2}$ (ms) |
| **(A)** | 179 (narrow) | $3.0 \pm 0.1$ | $228.0 \pm 9.0$ | 181 (39) | $2.0 \pm 0.1$ | $351.9 \pm 3.6$ |
| **(B)** | 193 (52) | $10.2 \pm 0.3$ | $68.1 \pm 1.7$ | 178 (47) | $7.0 \pm 0.1$ | $99.0 \pm 1.1$ |

### 2.2 Water permeability of phosphatidylcholines

A previous study [8] has shown that the stopped-flow light-scattering method is well adapted to the evaluation of permeability of the lipid bilayer to water. The theoretical model shows that the swelling of vesicles resulting from an osmotic shock should follow first-order kinetics. Its characteristic half-time, $t_{1/2}$, is the half-time of permeation of water through the membrane [8,9].

Table 1 shows that barrier properties of vesicles to water are strikingly improved for phospholipid (**A**) in both swelling and shrinking experiments. This improvement, due to the presence of the linkage at the termini of the chains, may be interpreted by the fact that this linkage prevents the two chains from engaging in independent motion and ensures a more close-packed assembly in the liquid-crystal state. These results are in a good accordance with those obtained with straight-chain 36-membered analogues both by Menger's group [10] and ourselves [11].

In conclusion, these results show that macrocyclization remarkably improves the barrier properties to water and the membrane stability, which may be related to the adaptation of *Methanococcus jannaschii* to the extreme conditions of the deep-sea hydrothermal vents.

## 3. ACKNOWLEDGMENTS

This work was supported in part by the JST-ULP Supermolecules Joint Research Project. We thank the Japanese Ministry of Education (Monbusho) for a sabbatical grant to T. E. and the Université Louis Pasteur for a guest position to T. E. For confocal microscopy experiments, we are indebted to Dr. Sylvette Chasserot-Golaz.

## 4. REFERENCES

1. M. De Rosa and A. Gambacorta, *Prog. Lipid Res.*, **27**, 153 (1988).
2. M. Kates, *Membrane Lipids of Archæa*, Elsevier, Oxford, 1993, p 261.

3. P. B. Comita, R. B. Gagosian, H. Pang and C. E. Costello, *J. Biol. Chem.*, **259**, 15234 (1984).
4. T. Eguchi, T. Terachi and K. Kakinuma, *J. Chem. Soc. Chem. Comm.*, 137 (1994).
5. T. Eguchi, K. Arakawa, T. Terachi and K. Kakinuma, *J. Org. Chem.*, **62**, 1924 (1997).
6. S. Inoué, *Video Microscopy*, 4th edn, Plenum, New York, 1989.
7. G. Pozzi, V. Birault, B. Werner, O. Dannenmuller, Y. Nakatani, G. Ourisson and S. Terakawa, *Angew. Chem., Int. Ed. Engl.*, **35**, 177 (1996).
8. A. Milon, T. Lazrak, A. M. Albrecht, G. Wolff, G. Weill, G. Ourisson and Y. Nakatani, *Biochim. Biophys. Acta*, **859**, 1 (1986).
9. R. Lawaczeck, *Biophys. J.*, **45**, 491 (1984).
10. F. M. Menger, X. Y. Chen, S. Brocchini, H. P. Hopkins and D. Hamilton, *J. Am. Chem. Soc.*, **115**, 6600 (1993).
11. K. Taguchi, K. Arakawa, T. Eguchi, K. Kakinuma, Y. Nakatani and G. Ourisson, *New J. Chem.*, 63 (1998).

# Cumulative Author Index

This index comprises the names of contributors to Volumes 1–6 of **Perspectives in Supramolecular Chemistry**.

# Cumulative Title Index

This index comprises the titles and authors of all chapters appearing in Volumes 1–6 of **Perspectives in Supramolecular Chemistry**

# Index